LABORATORY MANUAL

STEPHANIE DILLON
Florida State University

SANDRA CHIMON-PESZEK
DePaul University

GENERAL CHEMISTRY ATOMS FIRST

Second Edition

JOHN E. McMURRY
ROBERT C. FAY

PEARSON

Boston Columbus Indianapolis New York San Francisco Upper Saddle River
Amsterdam Cape Town Dubai London Madrid Milan Munich Paris Montréal Toronto
Delhi Mexico City São Paulo Sydney Hong Kong Seoul Singapore Taipei Tokyo

Editor in Chief: Adam Jaworski
Senior Acquisitions Editor: Terry Haugen
Senior Marketing Manager: Jonathan Cottrell
Project Editor: Jessica Moro
Assistant Editor: Erin Kneuer
Managing Editor, Chemistry and Geosciences: Gina M. Cheselka
Senior Project Manager: Beth Sweeten
Operations Specialist: Jeffrey Sargent
Supplement Cover Designer: Seventeenth Street Studios
Cover Image Credit: *Molecular Chair,* Design by Antonio Pio Saracino—www.antoniopiosaracino.com

Credits and acknowledgments borrowed from other sources and reproduced, with permission, in this textbook appear on the appropriate page within the text.

1 2 3 4 5 6 7 8 9 10—**VHO**—17 16 15 14 13

www.pearsonhighered.com

ISBN-10: 0-321-81337-5; ISBN-13: 978-0-321-81337-4

Table of Contents

To the Student:

Welcome to general chemistry laboratory. This is a brand new laboratory manual designed to accompany *General Chemistry: Atoms First 2e* authored by McMurry and Fay. What is so exciting about an atoms first approach, especially for a laboratory, is our ability to discuss molecules and their properties in depth at a very early stage in your chemistry education. This means that we can describe not only what happens in a reaction but also why and how the structure and physical properties of each atom in a molecule influences the outcome of your experiments.

As always with any lab manual I write I insist that the labs:

1) Are investigative in nature (you will always be given a puzzle to solve or unknown to identify.)

2) Teach a new technique as often as possible.

3) Accompany the chapters in the text in a manner that enhances your understanding of the concepts being taught in your lecture course.

To this end, we have made sure there is at least one lab experiment for each chapter in the textbook.

Each laboratory experiment in this manual provides the student with four sections of information: Introduction, Background, Procedure and Report. Additionally, there is a Pre-Laboratory and Post-Laboratory exercise at the end of each experiment that often requires collection of information or calculation of data necessary for the completion of the laboratory experiment. (NOTE: you need to read the entire laboratory experiment ***before*** attempting the pre-lab) The pre-laboratory exercises are also designed to give you practice with the calculations you will need to complete to produce your lab report. The post-laboratory exercises are what we refer to as "thinking" questions that ask you to reflect on the experiment you have just completed and extrapolate your findings to the real world. All the sections of the laboratory manual are focused on helping you get the most out of your laboratory experience.

As with any new material there is always a possibility for error or typos that we have not caught in editing. If you find any mistakes please let me know (sdillon@chem.fsu.edu) as it is very important to me that you receive a quality product that is as error free as possible. I hope that you will find the experiments in this manual both enjoyable as well as educational. Good luck with your semester and happy experimenting!

Stephanie R. Dillon
Coordinator of General Chemistry Laboratories
Florida State University

Laboratory Safety

Safety is of paramount importance in any laboratory. The laboratory safety rules here should be considered as a starting point for safe laboratory practice. Additional safety notes are included in each of the experiments in this manual. In addition, your instructor will give specific guidelines on the safe conduct of each experiment at the beginning of the laboratory period. Non-attendance during this safety briefing can result in students being barred from participation in the laboratory experiment.

The following comments on the general rules as listed may help to clarify any questions which arise. Please inquire of your instructor if there are further questions.

1. **Working without supervision is forbidden.**

   A student should never stay in the laboratory alone or work without an instructor present.

2. **Performing unauthorized experiments or any experiment at unauthorized times is forbidden.**

   Unanticipated results to an unauthorized or altered experiment can be quite hazardous. Materials should never be taken from the laboratory. Horseplay and pranks are never acceptable in the laboratory.

3. **Approved safety eye wear must be worn at all times.**

   Consult your instructor about the policy for wearing contact lenses in your laboratory.

   Certain solvents in use during some experiments can have adverse effects on soft contacts and students should take appropriate precautions following those labs.

4. **Loose hair and clothing must be restrained.**

   Long hair and loose clothing that can get caught on glassware or otherwise be a hazard in lab should be pulled back.

5. **Appropriate attire must be worn in lab, including shoes which cover the entire foot and clothing which covers the legs and torso.**

   Clothing items such as halter and midriff tops, as well as shorts, skirts and capri pants, are inappropriate for laboratory work, as are sandals and flip-flops. Clothing worn should be capable of providing a barrier between your skin and the chemicals you are working with. Coverage of the lower legs and feet are necessary in case of dropped glassware.

6. **Eating, drinking, smoking, chewing tobacco and applying cosmetics are forbidden in the laboratory.**

   These activities should take place only outside the laboratory doors.

7. **All accidents and breakage must be reported to an instructor.**

   Any accident requires attention to clean up. Alert your neighbors immediately if there is a spill as to its nature and extent before you leave the area.

8. **Pipetting by mouth suction is forbidden.**

   Pipet bulbs and/or dial-up pipetters will be provided when necessary.

9. **When needed, gloves must be worn.**

   Gloves should be of a material and thickness appropriate for the reagents being used. However, gloves provide only a temporary layer of protection against chemicals on your skin and may be permeable to some chemical reagents without visible deterioration. If your gloves come in contact with a chemical reagent, remove them, wash your hands, and get a new pair immediately. Hands must be washed just before leaving the laboratory.

10. **All laboratory workers must know the location and proper use of all laboratory safety equipment, including eyewash, safety shower, fire extinguishers, and telephone.**

    Your instructor will show you the location of and proper use of laboratory safety equipment.

11. **All laboratory workers must know how to safely evacuate the laboratory in the event of an emergency.**

    You should note all possible exits from your laboratory. Emergency classroom evacuation requires that the class reassemble away from the building to call roll and check for complete evacuation. Check with your instructor to confirm where the class should reassemble.

ADDITIONAL SAFETY NOTES

The following items were not covered in the general safety rules detailed above, but should be noted in connection with general chemistry laboratories.

1. Items such as book bags, backpacks, purses and coats should be put out of the way off the floor and lab bench.

2. Cell phones should be turned off before entering the laboratory.

3. Read labels and dispose of any waste in an appropriate waste container. Ask questions if you are not sure what to do.

4. Clean up in the laboratory is essential. Make sure you wash and put away all equipment before you leave the laboratory.

6. Inform your instructor at the beginning of the term if you have any special medical conditions that may need attention during the laboratory, such as known allergies or seizures.

Laboratory Notebooks and Reports

Accurate records and reporting of experimental results and conclusions are an indispensable part of any scientific work. A laboratory notebook should be written and organized so that it can be understood by anyone conversant with the subject. All data taken in the laboratory should be recorded in the notebook. NEVER write something on a loose sheet or scrap of paper to be recorded later. On the other hand, laboratory reports allow time for thoughtful analysis of the experiment performed and the data collected. Hence, the two are similar to taking notes prior to writing a term paper; the first is an immediate record, the second is what you present as your original work. Examples of the two follow at the end of this discussion.

LAB NOTEBOOKS

1. A bound notebook with numbered pages is the standard style of notebook used for research. Carbonless copy notebooks are often preferable as they allow your instructor to keep a copy of your data. Your instructor will tell you which type of laboratory notebook is required for your class. Make sure you read the rules on how to keep a notebook before writing in it for the first time. Some general rules are listed below:

   Notebook Organization:

   The **Title Page,** included with the **Table of Contents,** is the front page of the notebook. Make sure you fill in the following information as soon as possible so that your notebook can be returned to you if misplaced:

   Name
   Local Address
   Local Phone
   E-mail Address
   Lab course section, time and place of meeting
   Instructor or Teaching Assistant's name

   The **Table of Contents** should have the date, title and beginning page number of each experiment.

   Each **Experiment** should be numbered, and have the title, date, and your name on each page, as well as your class and section number.

   If it is not pre-numbered, write the page number on every page of the notebook before using the notebook.

2. Before coming to lab, make sure to read the experiment to be done. Make notes in your notebook and work out a tentative procedure and data tables to be filled in.

3. Never remove the original numbered pages.

4. Take all data in ink.

5. Relative "neatness" is important. If an error is made in recording data, draw a single line through the mistake and correct it nearby. If a substantial portion of a page is to be discarded, cross the material out with an X.

6. All entries should be appropriately labeled with units. If there are calculations which need to be worked out in class, do them in the notebook.

7. Sign and date each page. Be sure that the name of your lab partner(s) is recorded if applicable.

8. Have your notebook initialed by the instructor at the end of each lab period and turn in a copy of your data before you leave, if required.

9. Neatness and care in taking notes is quite important– notes should be intelligible, but an *immediate record.*

LABORATORY REPORTS

One of the important phases of carrying out any experimental project is the reporting of the results to interested individuals. Rightly or wrongly, in this course as well as in future professional activities, performance will be judged frequently on the basis of written reports. It is therefore very important that students learn to prepare and submit well-organized reports. The following format may help organize the information desired into useful reports:

1. ***Title.***
   Always submit a report with the formal title of the experiment. Include the names of lab partners if there are any, as well as the date, your instructor's name and the lab section number.

2. ***Purpose or Abstract.***
   Give a brief but complete statement describing the major goals of the experiment. This should not simply be a regurgitation of what is written in the lab manual but your own thoughts on what will constitute a successful conclusion to the experiment being performed.

3. ***Procedure.***
   In the formal report of an experiment *you* have designed, this section describes how you went about doing the experiment. Since in most cases for this lab, the procedure is already described, students need only cite the reference to the laboratory manual. Remember to note in this section any changes or deviations made to the procedure referenced. (Complete sentences are best.)

4. ***Data.***
   Present summarized experimental data carefully and neatly. This need not be as complete a data table as in the lab notebook, but should contain all the pertinent data used in calculations. Headings should be enough that they stand alone without reference to the manual. It may be convenient to include calculated values and results in this section for clarity and ease of reference.

5. ***Calculations.***
   Show all equations and calculations (neatly organized and documented) used to arrive at final results, and show the final results clearly. For repetitious calculations, you may show the general formula and one example. This section should also contain any graphs and further tables. Units must be used correctly in all calculations.

6. ***Conclusions.***
   Prepare a brief conclusion in which the results of the experiment are discussed. Was the purpose stated at the beginning achieved? If not, why? Include in this section a discussion of major possible errors and their projected impact on the experiment. Also in this section, answer any questions or exercises included as part of the experiment.

NOTE TO STUDENTS: Constructive written criticism of the experiments and/or their reports is welcome and will be given serious consideration.

Some Common Laboratory Glassware & Tools

Glassware or Tools	Common Use	Glassware or Tools	Common Use
 **Beaker**	Beakers are used to contain volumes of liquids. They are not volumetric and they're accurate only to +/- 5% of their graduation. They should not be used to measure.	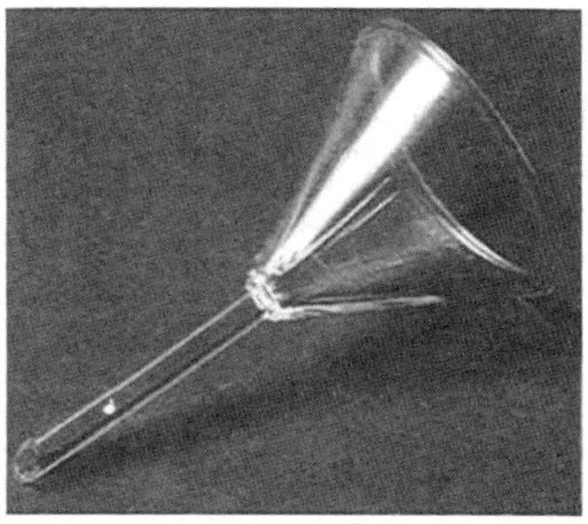 Funnel	Funnels are used in chemistry for the same purpose that they are used in everyday life—to prevent spillage.
 **Buret & Buret Stand**	A buret is used to deliver solution in precisely-measured, variable volumes. Burets are used primarily for titration, to deliver one reactant until the precise end point of the reaction is reached. Burets are volumetric and therefore are used to make accurate measurements.	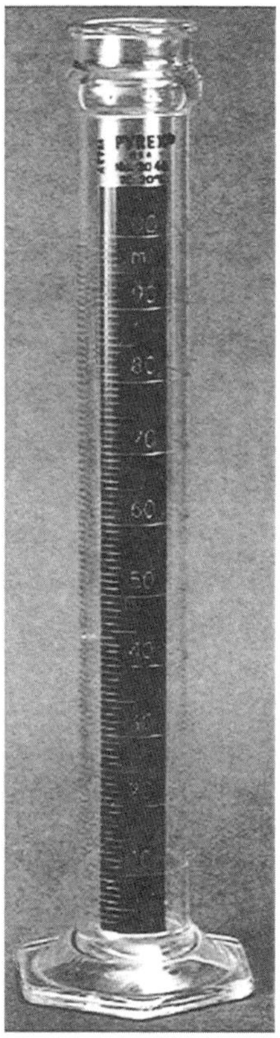 **Graduated Cylinder**	Graduated cylinders are volumetric measuring devices designed to measure and deliver accurate volumes of liquids. When measuring liquids in a graduated cylinder, you always measure from the center of the curvature in the liquid called the meniscus. Some liquids, like mercury, have a convex meniscus. Most liquids, however, will have a concave meniscus and the volume in the graduated cylinder should be read from the bottom center of this meniscus. (See below) 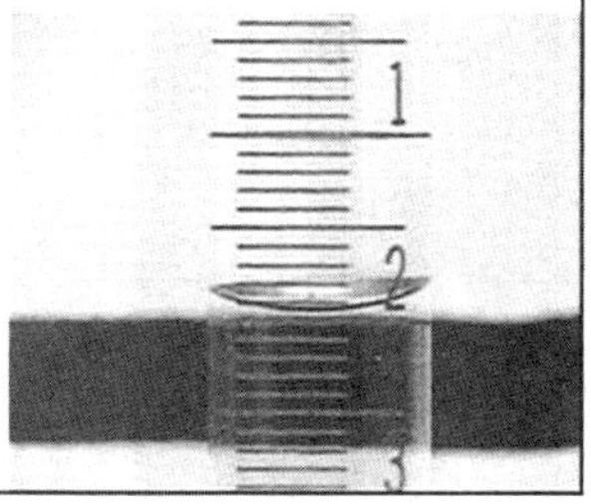

Some Common Laboratory Glassware & Tools

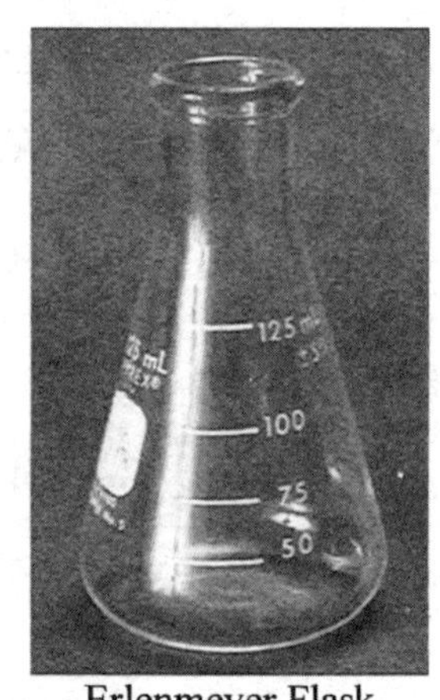 Erlenmeyer Flask	Erlenmeyer flasks are used for mixing, transporting, reacting, and filtration but not for accurate measurements. The volumes stamped on the sides are approximate and only accurate to within about 5%. Erlenmeyer flasks should not be used for measuring volumes.	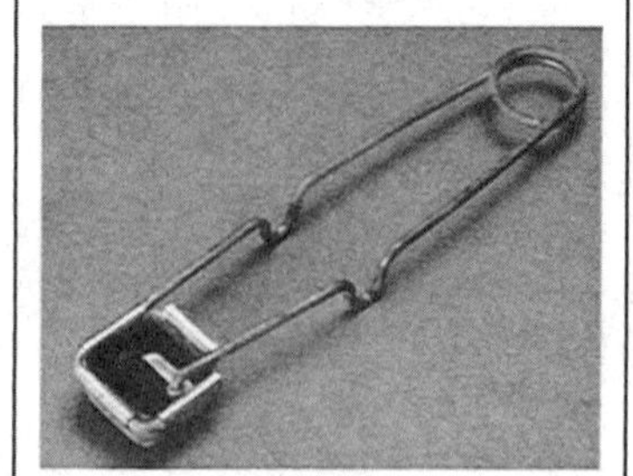 **Striker**	Strikers are used to start Bunsen burners. The striker itself is constructed with a rough surface positioned opposite to a piece of flint. When the arm of the striker containing the flint is pushed back and forth over the rough surface, sparks are produced. These sparks when created in the presence of a flammable gas such as propane will start a fire, or for our purposes a Bunsen burner.
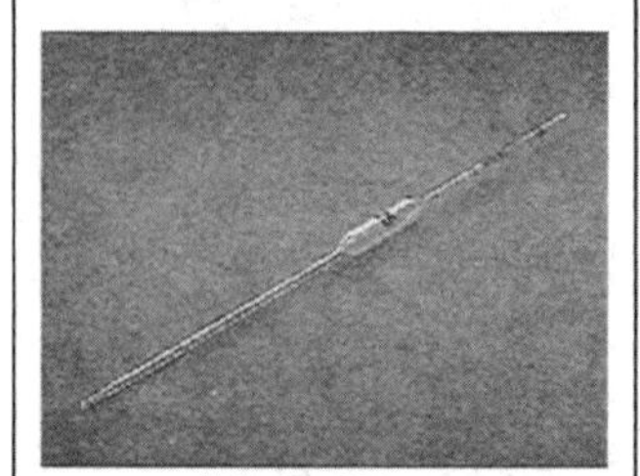 **Mohr Pipet**	A Mohr or transfer pipet is used to measure small amounts of solution very accurately. Common pipet volumes range from 0.1 mL to 10 mL.	 **Test Tubes & Test Tube Rack**	A test tube is a non-volumetric cylindrical container used to observe small volumes of reactants.
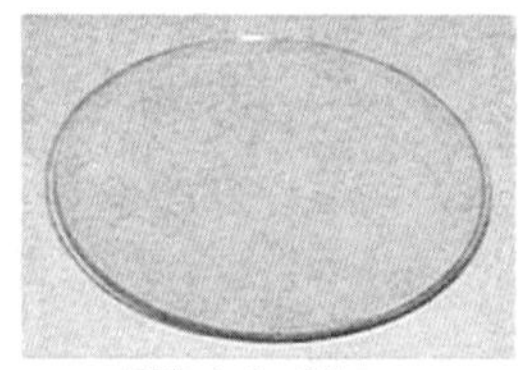 **Watch Glass**	A watch glass is a round, shallow, concave piece of glass that can be used to hold liquid or solids for evaporation, or it can be used as a lid for a beaker to prevent dirt from getting into a solution. Because it is not an airtight seal, gases can still be exchanged.	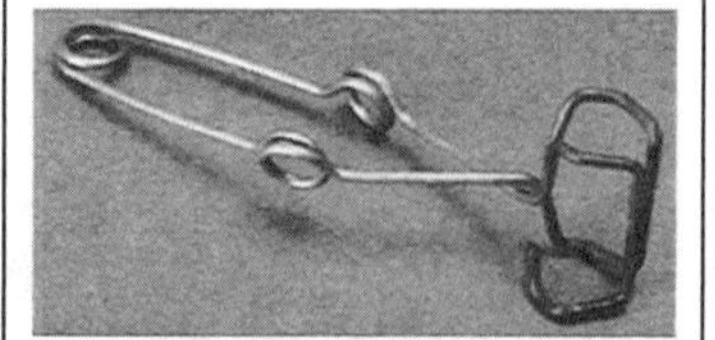 **Test Tube Clamp**	The clamp to the left is a test tube clamp. Clamps are used to allow a person to be "hands-free" during the progress of a reaction. Clamps such as the one shown are used to handle test tubes that are hot or to prevent slippage.
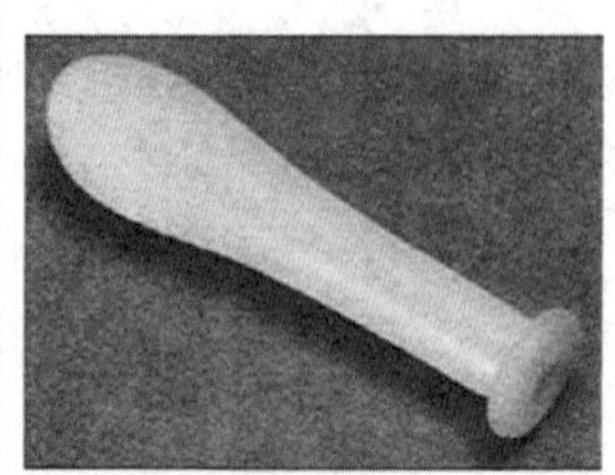 **Pipet Bulb**	Rubber bulbs are used in conjunction with pipets to transfer liquids from place to place.	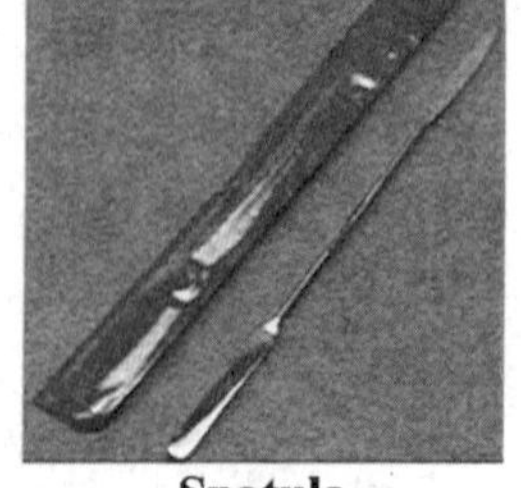 **Spatula**	Spatulas are used to transport and distribute dry chemical compounds. Spatulas are used most often when weighing out chemicals on a balance because they allow you to transfer very small quantities of the chemical at a time.

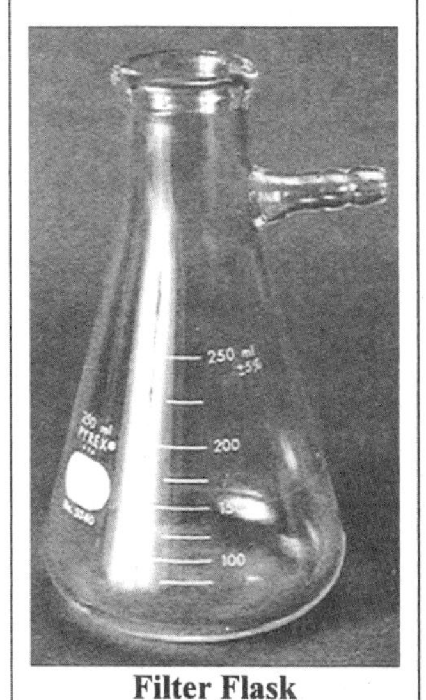 **Filter Flask**	A filter or side-arm flask is an Erlenmeyer flask made of slightly thicker glass to withstand high pressure. The side-arm is conical and used to attach a vacuum hose. A Buchner filter is placed into the neck of the flask to allow pressurized filtration of solutions.	 **Volumetric Flask**	Volumetric glassware or volumetric flasks are containers that have been calibrated to a specific volume. These flasks, unlike Erlenmeyer flasks or beakers, are marked with lines that are calibrated to a specific volume of liquid. A volumetric flask should be used whenever an accurate concentration of solution is required.
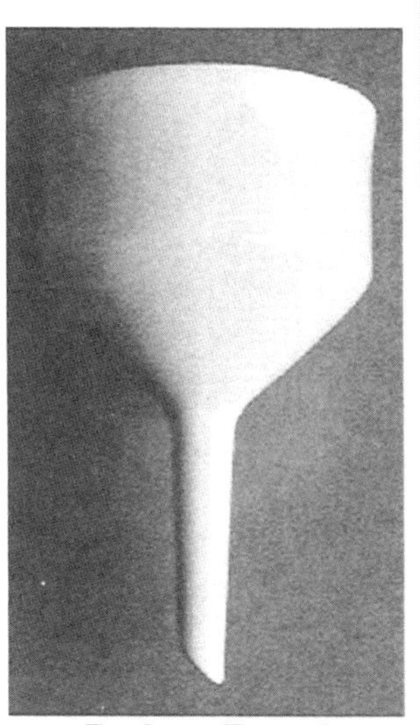 **Buchner Funnel**	A Buchner funnel is a suction or vacuum filtration funnel that is used in conjunction with a filter flask to separate particulates from solution. Filter papers of varying pore sizes are used to specify the size of the particulate retained.	 **Desiccator**	The desiccator is used to store dried samples in a dry atmosphere.
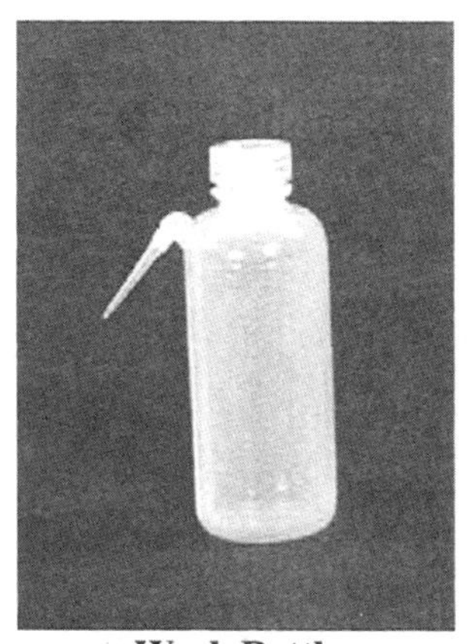 **Wash Bottle**	A wash bottle is generally filled with distilled or de-ionized (DI) water and used to make solutions and rinse glassware. The DI water should be kept fresh as exposure to air will eventually introduce CO_2 and increase the acidity. The wash bottle can be used for any liquid so always label the bottle appropriately.	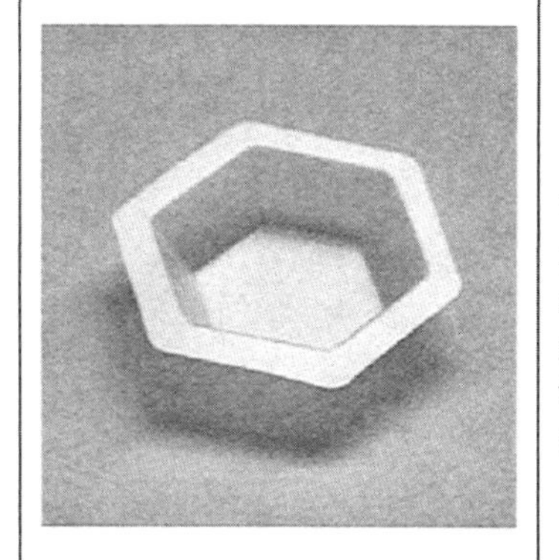 **Weigh Boats**	Weigh boats are disposable plastic containers used to prevent reagents from contacting the balance pan when they are being weighed.

Experiment 0

Measurement and Expression of Experimental Data

Purpose

As stated in your textbook the use of the "0" in the numbering of this laboratory is an indicator that this material may be considered a review although many of you may find some of the content new. The textbook covers many topics that need to be fully grasped in order to get the most out of the chapters that follow. In much the same way, this first laboratory is a review of the mathematic, statistical and graphing techniques that you will be expected to know and use during the progression of the laboratory experiments that are in this manual. The expression and analysis of experimental data is the focus of any science related field. The handling of this data, its expression, and the interpretation of its meaning is the text of most scientific journals. We, as scientists, devise experiments that will result in the collection of data that will give us the answer to a specific problem or lend support to a hypothesis. Most often, this data is numerical in form and must be dealt with by a combination of mathematical, graphical and statistical methods to allow for a reasonable and more importantly, reproducible result.

The purpose of Part A of this lab exercise is to demonstrate the proper way to manipulate and report experimental data. It will include a basic review of some mathematical applications of data as well as the basics of graphing. Part B will cover a review of the statistical methods used to evaluate experimental data and will explore more sophisticated graphing methods as well as the use of spreadsheet programs like Excel®. Both sections should serve as a reference for most of the experimental portion of this general chemistry laboratory and any that follow.

Part A

Background

In any modern scientific study of the world around us, measurements are included as a part of our observations. Because of the need to communicate observations to others, a set of agreed upon standards has evolved. The current system, established in 1960, is called the International System of Units (abbreviated SI) which has 7 basic units from which all others can be derived. They are as follows: meter (m), kilogram (kg), second (s), Kelvin (K), mole (mol), ampere (amp), and candela (cd). Some of these are quite familiar, others less so. The advantage of this system, as of the metric system which preceded it, is that through the use of various prefixes, powers of 10, and exponential notation, the units can be adjusted to fit almost any sensitivity of measurement, from the mass of the earth ($\sim 5.98 * 10^{24}$ kg) to the mass of a helium atom ($6.644 * 10^{-27}$ kg).

For work in the laboratory we will continue to use some of the older and more familiar units such as liters for volume, atmospheres, torr or mm Hg for pressure, and Celsius for temperature. By and large, these are easily converted to SI units when necessary. We must also be familiar with other units which have been used in the past, including the English system of measures. Although SI and metric units are easily convertible using powers of 10, conversions between English and SI or metric units are often inexact, depending upon the accuracy of the conversion factor between the two systems. Several useful tables are found in the Appendices of this lab manual, including tables of conversion factors, physical constants and metric prefixes.

Measurement

When measurements are taken, there is always some degree of uncertainty. The size and type of uncertainty (and therefore possible errors) depend upon the care with which a measurement is made. These errors typically take two forms: errors arising from the imperfections of the instrument used to make the measurement (mis-calibration) and errors arising from the skill or technique of the experimenter. In the first case, the errors are systematic (all in the same direction). In the second case, the errors will tend to be random (sometimes in one direction, sometimes in another). The two types of errors affect measurements differently. Systematic errors affect accuracy (closeness to the "true" value). Random errors affect precision (how well a set of data agree). In taking measurements in the laboratory, we are concerned with both accuracy and precision of measurements. One way to minimize the impact of any error is to take multiple measurements of the same value and use a statistical treatment of the set of values obtained, such as average and/or standard deviation to determine the value. While it is seldom possible to determine accuracy (the "true" value is not often known), the size of the standard deviation does give some indication of the precision of a series of measurements.

Significant Figures

In the chemistry laboratory, a number of different tools and instruments are used to make measurements. Most of these instruments are very finely machined or calibrated to produce accurate and precise values. For instance, which do you think would produce a better value for the length of a room, a measurement in notebook lengths, or a measurement produced with the use of a small ruler? No matter how well calibrated a given instrument may be, there is still some degree of uncertainty in the measurement made. Look at the smallest delineating mark on a metric ruler; the numbered marks are centimeters and the smallest mark is a millimeter (0.001 m). If an object is measured, it may fall between the 14.1 cm mark and the 14.2 cm mark. An estimate of the length is then made in the next decimal place (one place past the marks), and the object's length is determined to be 14.13 cm. In this example, the uncertainty is ± 0.01 cm, because we have estimated the final decimal place. Although the last digit is not certain, it is still reasonably reliable and should be reported. Any digits that are considered reliable are important for calculations involving a measurement, and therefore, these digits are known as **significant digits** or **significant figures.** In using any measuring device, an estimate of one decimal place past the marks should be made, thus the uncertainty in any measurement is in the last digit. On some instruments such as the barometer, there is a built-in estimating device called a Vernier scale, which allows the final decimal to be estimated more quickly. In reading various digital displays, the last digit is also assumed to be an estimate, unless the instrument states a different uncertainty. Often, very sensitive devices will state explicitly what the error in the measurement is calibrated to be.

It stands to reason that any time a series of measured values is combined mathematically, the value which is least precisely known determines the precision of the whole set. The following set of rules will help to clarify how to determine the number of significant figures in a measurement and how any combination of measured values should be rounded to show an appropriate number of significant figures in the end result.

Rules for determining significant figures

1. All nonzero digits are significant regardless of their position relative to a decimal point.

2. Any zeros surrounded by nonzero digits are significant.

   Examples: 1.101 has four significant figures
   750.25 has five significant figures
   303 has three significant figures

3. Zeros to the left of all nonzero digits are not considered significant because they serve only as decimal place holders.

   Examples: 0.08206 has four significant figures (start with the 8)
   0.0003 has only one significant figure

4. Zeros to the right of the last nonzero digit may or may not be considered significant.
   a. If there is a decimal point to the left of any of this type of zero, then they are to be considered significant.

      Examples: 0.03750 has four significant figures (start with the 3)
      69.0 has three significant figures
      10.000 has five significant figures

   b. If the decimal point is to the right of this type of zero, then it is necessary to know what the number represents in order to determine the amount of significant figures.

      Example: The number 100 could have one or three significant figures depending on its context. If it represented an approximation of the volume of water in a glass (about 100 mL), then it would only be considered to have one significant figure. However, if the water had been measured in a graduated cylinder, then all three digits would be significant.

      Often, this kind of guessing can be avoided if the measurement is recorded in scientific notation. For instance, in the above example, the number could be written as $1.00 * 10^2$ mL, indicating that there are 3 significant digits rather than one in the number.

Rounding Numbers

The number of significant figures in a measurement becomes very important when doing calculations with collected data. For example, the density of an object should not be reported to four significant figures if its weight or volume can only be recorded to three significant figures. Similarly, the density should not be reported to only two significant figures, because useful information would be discarded. **Generally, the calculated result is only as reliable as the least precisely measured value used in the calculation.** (In the Examples: below, all numbers are rounded to the hundredth place for simplicity. The least number of significant figures present in the calculation would determine the actual number to be reported).

Rules for rounding

1. If the first digit being dropped is lower than 5, then the previous digit is not changed.

   Examples: 1.234 rounds to 1.23
   0.091 rounds to 0.09

2. If the first digit being dropped is higher than 5, then the previous digit is increased by 1.

   Examples: 2.3154 rounds to 2.32
   78.987 rounds to 78.99

3. If the digit to be rounded is exactly 5 (or 5 followed by zeroes), then the number should be rounded to be even.

   Examples: 3.72500 rounds to 3.72
   0.975 rounds to 0.98

Rounding during addition or subtraction

Rounding in addition and subtraction is determined by the least precisely measured value, not just the count of significant figures. It is the placement of the decimal relative to the significant figures which determines how many significant figures the answer should have.

Examples: 45.5609 + 0.975 + 34.9 + 56.43 = 137.8659, but the answer should be rounded to 137.9, because **34.9** is the least precisely known value, therefore, the answer should be recorded to only one decimal.

8.674 - 3.09 = 5.584, but the answer should be rounded to 5.58, because **3.09** is significant to the hundredths place and the answer should be the same.

Note the addition or subtraction occurs first and then rounding follows as the final step.

Rounding during multiplication and division

As stated earlier, a calculated result can only be as reliable as its least precisely measured value. In multiplication and division, this means that the result of a calculation should have the same number of significant figures as the number with the fewest significant figures used. If this is not done, the precision of the calculated result is overestimated.

Examples: 108.14 * 0.0015 = 0.16221, is rounded to **0.16** because 0.0015 only has two significant figures.

457.2 / 625 = 0.73152, is rounded to 0.732 because **625** contains three significant figures.

The role of conversion factors in rounding

Numbers in conversion factors can be either exact or inexact. Exact numbers are used when conversions are made from one set of units in the metric system to another set also in the metric system. The prefixes in the metric system are similar to definitions. There are exactly 100 centimeters in 1 meter and because of this, both **100** and **1** are exact numbers. Exact numbers are considered to have an infinite number of significant figures. Therefore, when doing this type of conversion, the conversion factors do not limit the number of significant figures the answer can have. Conversions from the metric system to the English system (and vice versa) involve inexact numbers. There are approximately 30.5 cm in 1 foot. The **1** foot is the quantity being defined, so it is an exact number. However, **30.5** is the rounded value, which makes it an inexact number. Inexact numbers do affect the number of significant figures that can be present in the answer.

Examples: **Exact:** 355 mL * (1 L/1000 mL) = 0.355 L

Inexact: 27.9654 cm * (1 ft/30.5 cm) = 0.916893607 ft., but this is rounded to 0.917 ft. because of the inexact number in the conversion.

Scientific Notation

One way to express very large or very small quantities is to use exponential or "scientific" notation. Using the power of 10 is a reasonable way to specify the precision of a measurement. Very small numbers have a negative exponent, and very large numbers have a positive exponent. The rules on rounding remain the same. Remember in addition and subtraction, to adjust the values to have the same exponent before performing the operation.

Examples: $1760 = 1.760 * 10^3$, showing 4 significant figures
$0.000000530 = 5.30 * 10^{-7}$, showing 3 significant figures

Addition/Subtraction

$2.761 * 10^2 + 1.32 * 10^1 =$
$2.761 * 10^2 + 0.132 * 10^2 = 2.893 * 10^2$

$1.1941 * 10^{-2} - 8.62 * 10^{-3} =$
$11.941 * 10^{-3} - 8.62 * 10^{-3} =$
$3.321 * 10^{-3} = 3.32 * 10^{-3}$

Multiplication/Division

$(2.761 * 10^2) * (1.32 * 10^1) =$
$3.64452 * 10^3 = 3.64 * 10^3$

$(1.1941 * 10^{-2}) / (8.62 * 10^{-3}) = 1.38366 * 10^1 = 1.38 * 10^1$

Significant figures using logarithms

Logarithms have two parts: 1) The characteristic is the portion of a logarithm to the left of the decimal point, and reflects the exponent. 2) The mantissa is the portion to the right of the decimal and reflects the value of the measured quantity.

Example: log 559 = 2.747, **2** is the characteristic and **747** is the mantissa

The rule for determining the number of significant figures in a logarithm is that the mantissa should have the same number of significant figures as the measured quantity.

Examples: log 7 = 0.8
log 7.0 = 0.85
log 7.00 = 0.845

Likewise, the same rule applies when the operation is reversed (antilog).

Examples: antilog 0.60 = 4.0
antilog 0.602 = 4.00
antilog 0.6021 = 4.000

Graphing

Using graphs to illustrate a set of data and make predictions is a typical experimental approach. In many experiments, one parameter is varied systematically while observing what changes there are in a second parameter, such as observing the volume of a gas when pressure is increased or decreased. If a linear relationship can be found between two parameters, then predictions are possible. Thus the graphs can be used both to visually inspect a set of data for linear behavior and predict other information. In order to make the best use of graphical data, a few points on the best way to make them are in order.

1. Use a whole sheet of graph paper for each graph. If the goal is to read information from a graph, then shouldn't the graph be as large as possible?

2. Decide which of the two parameters on the graph is independent (usually this is the one which is varied at regular intervals) and which is dependent (the one which is being observed.) The independent variable is plotted on the x axis, the dependent variable on the y axis.

3. The best graph of a set of data does not always have the origin, (0, 0). Look at the highest and lowest data points of the variables to be plotted. Choose convenient numbers slightly higher and lower than those data points and make them the extremes on the graph. Again, the goal is to spread the data out on the graph as much as possible.

4. Label axes clearly. This should include regularly spaced marks and the units of measurement for each axis. A graph will not be of much use if the axis is not linear.

5. Put a title on the graph that will explain something, not just "graph of y vs. x". For example, "Volume of argon gas as a function of pressure at 25 °C."

6. Use a ruler or curve to draw the best line through the plotted points. One way of showing the points is putting a circle around each one. Alternatively, draw error bars, showing the precision of the measurement taken.

7. If a graph is not linear, don't try to draw a straight line through the points, draw a curve. To get a straight line, try a mathematical manipulation of one or both of the variables, such as inversion (1/y) or logarithm (log y) and do another plot. It may take two or three attempts to find something that works to yield a straight line.

8. Once a straight line is obtained, the slope of the line can be found using two conveniently read points on the line, which should not be previously plotted data points. Remember from algebra that slope, m, is $\Delta y/\Delta x$.

9. Using the equation for a straight line, $y = mx + b$, also from algebra, the y intercept can be found by substituting any (x, y) point and the slope, m, and solving for b.

Report Contents and Questions

As the first experiment of the term, the exercises presented below are to be used to refresh your memory. None of the experiments are actually done in the laboratory. Therefore, since no data has been taken in the laboratory, none should be recorded in your notebook. Instead, this report should be prepared on notebook paper to be turned in at the next lab class meeting.

The lab report should consist of a title page, purpose and the exercises below. Please remember all the rules for constructing good scientific graphs. Show all of your work for full credit.

Exercises

1. For each of the following numbers: a) Indicate the number of significant figures in the number; b) Round the number to 2 significant figures; and c) Place the number into scientific notation:
   a. 0.0030405
   b. 1,237,888
   c. 0.10101
   d. 15.0005
   e. 0.000009995

2. Complete the following calculations using the correct number of significant figures and order of operations:
   a. Convert 432 K to ^{o}F.
   b. Convert 1.67 %T to Absorbance (A) using $A = -\log\left(\frac{\%T}{100}\right)$.
   c. Solve for P_2 using $\ln\left(\frac{P_1}{P_2}\right) = \frac{\Delta H_{Vap}}{R}\left(\frac{1}{T_2} - \frac{1}{T_1}\right)$, where P_1 = 1.5 atm, T_1 = 273K, T_2 = 400K, ΔH_{Vap} = 40.7 kJ/mol, and R = 8.3145 J/mol·K
   d. Calculate the total mass in grams for:

      4.445×10^{-3} g + 15.3 mg + 9.77561×10^{-2} mg =

   e. Calculate the total pressure in kPa for:

      2.34 atm + 788 Torr – 650 mmHg =

3. Calculate the average and standard deviation for the following set of calibration data:
   66.788, 66.898, 66.999, 65.998, 65.778, 67.434, 67.311, 66.009

4. The table of data below was collected using a helium lamp to calibrate a spectroscope used in this general chemistry lab. Use the data to create a calibration graph and then use the graph to calibrate the experimental data collected for other light sources:

Color	Wavelength (nm)	Spectroscope Reading (AU)
Violet	388.9	3.78
Blue	468.6	4.75
Green	501.6	4.93
Yellow	587.6	5.82
Orange	667.8	6.55
Red	706.5	6.83

a. Sodium line spectroscope reading = 5.80 , corrected wavelength = _________
b. Hydrogen line line spectroscope reading = 6.43, corrected wavelength = _________
c. Mercury line spectroscope reading = 5.69 , corrected wavelength = _________
and line spectroscope reading = 4.65, corrected wavelength = _________

5. Create a graph of the Absorbance (Optical Density) versus Molar Concentration and answer the questions about the resulting line.

Concentration (M)	Optical Density
0.19	0.15
0.34	0.34
0.49	0.52
0.64	0.68
0.79	0.81
0.94	0.92
1.09	1.02
1.24	1.09
1.39	1.14
1.54	1.17

NOTE: The Absorbance, or Optical Density, of a solution for a particular wavelength of light is given by $\log(I_o/I)$, where Io is the incident light on a sample and I is the transmitted light. Beer's Law states that the Optical Density of a substance in solution is directly proportional to its concentration (i.e. has a linear relationship). Not all substances follow Beer's Law, but a calibration graph of Optical Density versus concentration can still be used to measure concentrations of unknown solutions.

a. Does the solution in your graph follow Beer's Law?
b. What is the concentration of a solution that has an Optical Density of 0.98?
c. What is the concentration of a solution that has an Optical Density of 0.71?

6. Robert Boyle used an open-ended barometer (similar to that in the figure) for the experiments which led to the discovery of Boyle's Law, describing mathematically the effect of pressure on the volume of a gas. He determined the nature of the law by graphing his data. The law states that PV= constant, where P is the total pressure, which is the sum of $P_{atm} + P_{Hg}$. The pressure can be measured in Torr, which corresponds to the height of the Hg column in mm.

The following table contains some data in an experiment similar to that of Boyle's, though the open-end manometer is considerably larger. The volume of the gas in the closed end of the tube is measured in mL, and the pressure recorded is P_{Hg}, the height of the column, not the total pressure.

Pressure (torr)	Volume (mL)
58	16.5
162	13.9
239	12.4
316	11.1
385	10.1
468	9
536	8.3
620	7.4
733	6.5

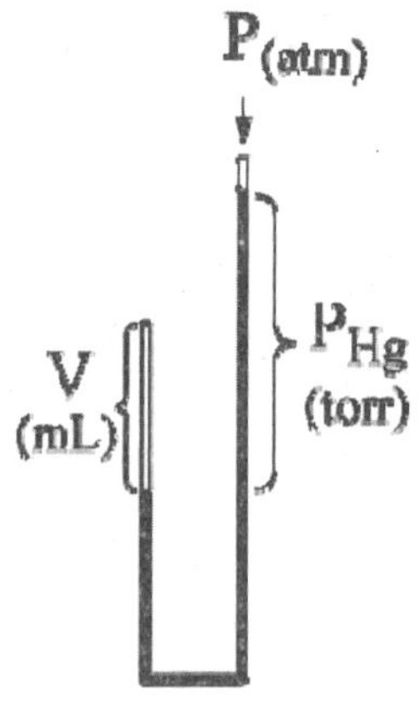

You are to analyze this data by graphing it, discovering that a direct graph **(V vs P)** gives a curved line. Scientists often try manipulating data in a way to give a straight line, and when the data is re-graphed as **(P vs 1/V)**, you will see that a straight line is produced, revealing the nature of Boyle's Law (that pressure and volume are reciprocally related). Further analysis of this straight line will also reveal the atmospheric pressure at which the measurement is made.

$$\mathbf{P_{total} * V = constant}$$
$$\mathbf{P_{total} = constant(1/V)}$$
$$\mathbf{P_{atm} + P_{Hg} = constant(1/V)}$$
$$\mathbf{P_{Hg} = constant(1/V) - P_{atm}}$$

Since the graph is of the form **y = mx + b**, where the slope, **m**, gives the value of the constant, and the **P** intercept is the negative of the atmospheric pressure.

a. What is the slope of the line (i.e. the constant in the Boyle's Law equation)? Give units in your answer.
b. What is $\mathbf{P_{atm}}$ (i.e. the negative of the **P** intercept)? Give units in your answer.

7. Using the same data as above, calculate the logarithm of the volume. Graph the Pressure versus the $\log_{10}$ Volume and answer the questions about the resulting line.

    a. If this graph is a straight line then it is of the form $P = m\text{Log}_{10}V + b$, in which m is the slope and b is the y-intercept. Is the graph a straight line or a curved line (nonlinear variation)?
    b. If possible, calculate the equation of the line for the graph. Show your work.
    c. Using the resulting equation from question 2, what would be the volume of a system that had a pressure of 3.9 atm?

Experiment 0

Part B

Graphing and Statistical Analysis

Purpose

Scientific results are accumulated in a laboratory setting but are only the backbone to the understanding of science. The ability to express and analyze laboratory results is how scientific fields continue to progress generation after generation. It is one thing for a person to go into the lab and run an experiment but those experiments mean nothing to anyone unless that data is processed, explained and defended. This data handling is the body of a scientific text and must give us answers positive or negative to a specific problem or lend support to a hypothesis.

Most often, this data is numerical in form and must be dealt with by a combination of mathematical, graphical and statistical methods to allow for reasonable and more importantly, reproducible results. The handing of data has become easier with the development of computer programs such as Excel and Spreadsheets. Spreadsheets are as useful for handling numbers as word processors are for handling the words. Spreadsheets are like giant data tables that can do any type of calculation on that data. Repetitive calculations are handled automatically, requiring less time and energy compared to those done by hand. For example, if one piece of data is changed, all calculated results involving that number are immediately and effortlessly recalculated. Excel also allows for statistical analyses and a variety of graphical representations.

The ability to use a program such as Excel will not happen overnight. In fact Excel has so many endless possibilities that it may be impossible to learn them in a lifetime. This lab though will teach you the basics of Excel and familiarize you with the program. The goal of this lab is to make your life easier throughout the semester and your lifetime. While this lab is going to take time and what may seem like hours and hours to familiarize yourself with the program IT IS going to make your life easier in analyzing data in the future. This lab exercise is designed to give you practical experience in interpreting the results of raw data from experiments similar to those done throughout the semester and in real life.

One often hears the phrase, "A picture is worth a thousand words," this is true in science also. Numerical data can be expressed in tables and paragraphs, but a majority of the time a picture would be more helpful. In science pictures are better known as graphs. Scientist use graphs as comparison tools but most importantly as a method of determining unknowns. In most cases, known mathematical expressions are rearranged in the form of a straight line, $y = mx + b$, graphed and then the slope and y-intercept are determined.

Graphical representation is commonly used in the form of a calibration graph. To use the calibration curve technique, several standards containing an exactly known concentration of the analyte are needed. The absorbance of each of analyte concentrations are then measured using an absorption technique. The concentration and the absorbance of the standards are then graphed, with concentration on the x-axis and absorbance on the y-axis. A best-fit line or trend line is added to include these points. This is known as the calibration curve. The equation of this line is obtained and used to determine the concentration of an unknown sample. This technique is used a great deal in testing water for various metals, nonmetals and ions. For example if one wanted to test the concentration of nitrate in various water sources the calibration curve technique could be used. The absorbance of several standards, with different known concentrations of nitrate would be found using a Genesys-20 and used to make a calibration graph. The absorbance values of the water samples from various sources would also be measured. The absorbance values would be used along with the equation of the line to solve for the concentration of nitrate in the water. In this case, the equation $y = mx + b$ is used to solve for x the nitrate concentration of the water sample; where y is the absorbance value of the water sample, m is the slope of the calibration curve and b is the y-intercept of the calibration curve.

Another common graphical representation is the extraction of data from the slope of a graph. The slope of the line can provide various information depending on the equation and variables being used for the graph. For example we know that Gibb's Free Energy is defined as $\Delta G = -\Delta S(T) + \Delta H$, where ΔS is the entropy, T is temperature in Kelvin and ΔH is enthalpy. This equation is already in the form of a straight line $y = mx + b$, where ΔG is y, -ΔS is m, T is x and ΔH is b. Graphically, ΔG is the y-axis, T is the x-axis, the slope of the line is equal to $-\Delta S$ and the y-intercept is equal to ΔH. This graphical representation allows for the determination of ΔS and ΔH, which are difficult to determine experimentally.

Statistical analysis is important in expressing the reliability of the data being presented. Statistics can be considered an exact treatment of uncertainty. Data values are usually presented with a statistical error. The statistical error is a value written next to the data value stating how much variation in both the positive and negative direction is being experienced. For example one could write the average concentration as 0.56 + 0.01 M, meaning that the average concentration was found to be 0.56M with a standard error of 0.01M. If one includes the error it can be said that the average concentration of the sample is between 0.55 and 0.57. Statistical errors can be caused by multiple sources; common sources are variations between trials, and limited accuracy of instrumentation.

Data presented with large statistical errors (standard deviations) are generally not well received by the scientific community. The scientific community wants data present with small statistical errors, which corresponds to good precision and accuracy. There are two areas of statistics of great importance to a chemist: the analysis of error in repeated measurements and the analysis of distributions and central tendencies. The first deals with trying to get a feel for the "true" value of a measurement, the second deals with trends in large collections of values or measurements.

The "true" value is when you measure the same thing over and over, and you often end up with slightly different results, despite the fact you know the values should be the same. For example it could be measuring the exact miles per gallon your car gets in town or maybe it's the weight of your pet dog. If you measure your dog's weight to be 105.25 lbs, and repeat the measurement five more times, you would expect the dogs weight to be the same, but it's critical to realize that every measurement has an error or uncertainty associated with it. In weighing your dog the error of uncertainty lies in the accuracy of the scale.

On the other hand if we count the pennies in a jar and fine there to be 125 pennies. The 125 pennies counted is not a measurement, it is an exact count. But take one of those pennies and measure its density ten times and you will likely get 10 "slightly" different results. These 10 numbers are not random. The density values will all be very similar and, in fact, may be identical to one or two decimal places. They are obviously clustering around a value, the "true" value of the density. There is no way for one to know the "true" value so therefore we say that we are a certain (90, 95, or 99) percent confident that the "true" value is within a certain range of the values, the mean + the standard deviation.

In addition to analyzing the results of the data graphical and statistical it is important to determine if the results are publishable in a scientific journal such as the Journal of the American Chemical Society. The decision of publication is based on the creditability and reproducibility of the data. For example an experiment with multiple trials is a must to insure the results are not just a fluke along with accurate data analysis showing the reliability of the results. Both graphical and statistical analyses are necessary in determining if the results can be published.

Background

Data Handling

In several laboratory experiments this semester you will be collecting large quantities of numerical data. In order to tabulate and express this data, you will need to become familiar with a "spreadsheet" program. There are several programs of this sort available, but the one we recommend is Excel. As a Microsoft program Excel® is available to students on most PCs and is accessible on most University computers. For this reason, this lab recommends the use of and will provide tutorials in Excel®. A tutorial on simple data handling in Excel® is available in Appendix B.

Graphing - The Calibration Graph

Beyond simple data handling, the first technique that will be introduced is how to successfully and concisely express a large quantity of data graphically. No one likes looking at a table full of hundreds of numbers and trying to figure out a particular trend or pattern. What we do like looking at are pictures! And in science that means graphs. If the same data is presented graphically, the results are much easier to observe as well as explain.

Types of graphs are as varied as the data they represent, but one of the most common types of graphs used for data analysis, at least in this lab, is the 'calibration' graph. Briefly this is a graphic representation of standard reference data used to calibrate some variable in an unknown sample. For example, this technique will be used this semester when you have to determine the concentration of phosphate in local water samples using spectrometric methods.

Looking at the data given in the table below, we see the concentration (mol/L) as well as the corresponding absorbance (AU) for a set of standard phosphate concentrations in water. If our overall experimental goal is to determine the concentration of phosphate in a water sample obtained from Lake Bradford, we need to use a method to analyze our standard data by which we can extrapolate a trend and then use that trend to determine the concentration of phosphate in our unknown sample. From what is already given, this is a rather elementary task, and will be far more laborious when you actually perform the procedure yourself. However, at this point all we have to do is generate a scatter-plot, like the one shown below, of the dependent variable (absorbance) versus the independent variable (phosphate concentration).

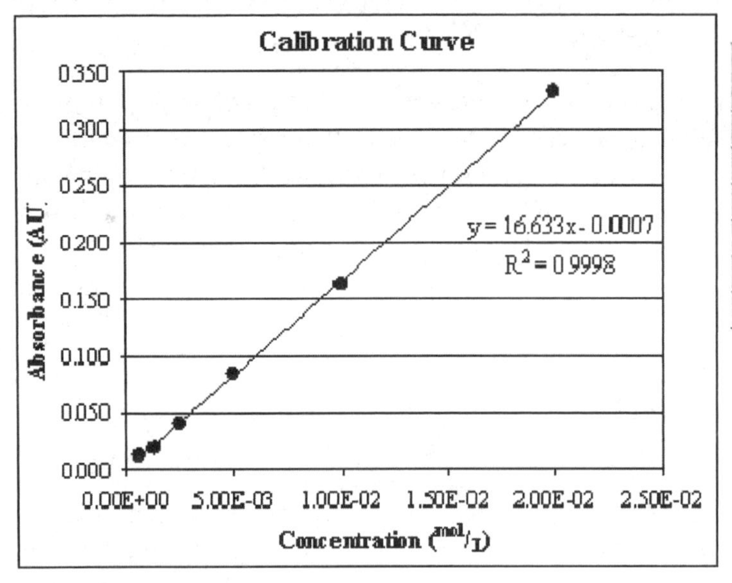

Conc. (mol/L)	Abs. (AU)
2.00 x 10-2	0.333
1.00 x 10-2	0.163
5.00 x 10-3	0.084
2.50 x 10-3	0.041
1.25 x 10-3	0.019
6.25 x 10-4	0.011

Once the scatter-plot has been generated, a "best-fit line" can be added to see if our data can be described as linear, logarithmic, polynomial, or exponential. This process is called linear regression. Remember these different possibilities, if you don't, it will come back to haunt you in later experiments. For the example above, the plotting of data results in a linear graph, and can therefore be fit with a linear equation of the $y = mx + b$ form. The actual linear fit equation is shown in the upper right hand corner of the graph, as well as a corresponding R2 value. The R2 value is a statistical measure of how well the "best-fit line" fit the data. Specifically, the closer the R2 value is to 1, the better the fit. Now that we have an equation that describes the trend in the relationship between phosphate concentration and spectrophotometric absorbance, we can use it to determine the concentration of phosphate in our unknown. After obtaining the absorbance of our sample experimentally, we can quickly calculate the corresponding phosphate concentration.

Example : Determining the Phosphate Concentration in an Unknown Sample of Water

According to the "best-fit line" equation, $y = 16.663x - 0.0007$,

$$\textbf{Absorbance} = \textbf{16.663}[PO_4^{3-}] - \textbf{0.0007}$$

We determine this equation simply by substituting in the true values of the independent (x) variable and the dependent (y) variable.

In our example we have proposed that our unknown sample has an absorbance value of **0.126**. If we plug this value into our equation and solve for **$[PO_4^{3-}]$**, we will have determined the concentration of **$[PO_4^{3-}]$** in our unknown:

$$0.126 = 16.663[PO_4^{3-}] - 0.0007$$

$$0.1267/16.663 = [PO_4^{3-}]$$

$$7.6 \times 10^{-3}\ M = [PO_4^{3-}]$$

The other part of the question asks if this answer is reasonable. There are two reasons that we can say "*yes*". First the "best-fit line" to the data has a R^2 value of 0.9998 which is very close to 1.0 showing that our line equation is well representative of the trend in the data. Secondly, the value we obtained from our calculation makes good common sense when we look back at the original data. That is to say that the absorbance value we were investigating falls between 0.840 and 0.163 of our reference data indicating that our concentration should fall somewhere between the concentrations of 5.00×10^{-3} M and 1.00×10^{-2} M, which it does.

This is therefore a good result. The process by which we just determined this to be a good calculation sets an important principle: ***You should always check to make sure that your calculated answers make sense whenever possible.*** Scientists who do not verify their results are not well respected in the scientific community and soon become ridiculed and forgotten. (And students tend to get poor grades).

Graphing - Using the Slope

Another type of graphical analysis you will be faced with in general chemistry is the extraction of data from the slope of a line. In particular, when we address the concept of kinetics, we can use this form of analysis to ascertain the individual orders of a rate equation. Reactions are categorized as zero-order, first-order, second-order, or mixed-order (higher-order) reactions. These reaction orders are important because they tell us which reactant is most important in the overall rate by which a reaction progresses. We obviously don't expect you to understand all of these concepts at this point, there will be plenty of time for that later, but by simply determining the slope of the line resulting from a plot of the rate of a reaction versus the initial concentrations of the reactants we can determine the individual reaction orders for each chemical in a reaction.

For example, let's say we want to determine the individual reaction orders for the reaction between nitrogen monoxide (NO) and oxygen (O_2) from the data presented below.

	$[NO]^o$ (mol/L)	$[O_2]^o$ (mol/L)	Instantaneous Rate (mol/L h^{-1})
Trial #1	0.020	0.010	0.028
Trial #2	0.020	0.040	0.114
Trial #3	0.020	0.020	0.057
Trial #4	0.040	0.020	0.227
Trial #5	0.010	0.020	0.014

What do we need to do? Well, first we need a balanced chemical reaction:

$$2NO(g) + O_2(g) \leftrightarrows 2NO_2(g)$$

Second, we need a rate equation:

$$\text{Rate} = k[NO]^x[O_2]$$

Third, we need this equation rearranged such that it is in a linear form:

$$\text{LogRate} = x\text{Log}[NO] + \text{Log}k[O_2]$$

and

$$\text{LogRate} = x\text{Log}[O_2] + \text{Log}k[NO]$$

Please notice that the above equation is of the form y = mx + b, where x is the slope and reaction order in each case. You can look forward (if you would like) to the lab on kinetics to see a more complete breakdown of how these two equations were developed, but for our purposes here we just need to know that they are linear equations.

At this point, two graphs must be generated; both with the form of log (Rate) versus log (Reactant). It is imperative that in the each graph one of the reactant's concentrations must be constant. Looking at the original data this means that we will plot Trials #3, #4, and #5 to generate our first graph ($[O_2]$ is constant), while Trials #1, #2, and #3 will be used in the second graph ([NO] is constant). Linear regression (e.g. adding a "best-fit" line) will allow us to determine the slope for each equation. Further, *x* in both equations not only represents the slope of the line and but also corresponds to the individual reaction order for NO and O_2 respectively.

More data can still be gathered from our graphs. Observing the equations carefully, we notice the common factor of *k*, which, in short, is the rate constant. We can extract a value of *k* from each graph, average it, and report an overall rate equation.

$$\textbf{Rate} = (7.4 \times 10^3\ \text{mol/L}^{-2}\ \text{s}^{-1})\ [NO]^2[O_2]^1.$$

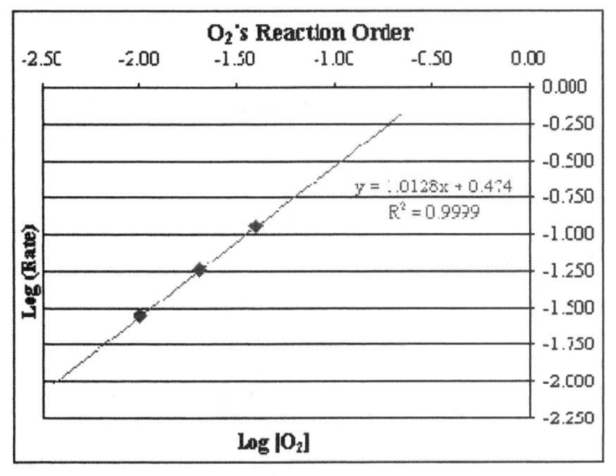

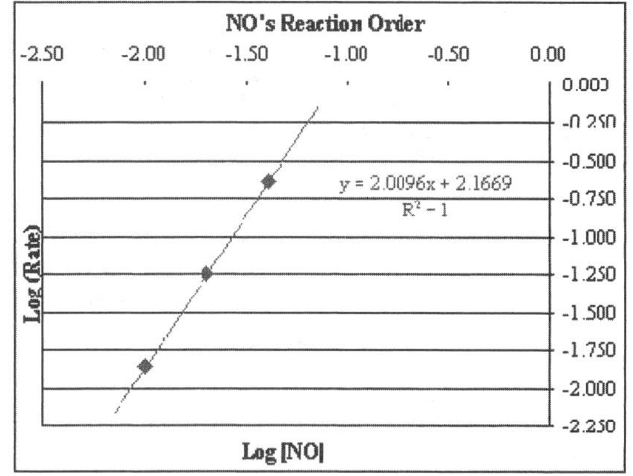

Example Graphs of Log [O_2] versus Log (Rate) and Log [NO] versus Log (Rate) used to determine reaction orders

Statistics

Analyzing our data graphically is only half of the battle, as this type of analysis rarely gives information on how 'accurate' or 'precise' our data may be. For this aspect, we turn to statistical analysis. Statistics are a way for us to easily express the reliability of the data and analysis we present. With that being said, the rest of this portion of the experiment will directed towards introducing several statistical methods that must be mastered in order express our data in a manner that can easily be critiqued by us and the scientific community.

Overall, there are two areas of Statistics of great importance to a chemist: the analysis of error in repeated measurements and the analysis of distributions and central tendencies. Basically, the first application deals with trying to ascertain the 'true' value of a measurement, while the latter deals with observing patterns or trends in large collections of values or measurements.

Error Analysis

When you measure the same thing over and over again you often end up with slightly different results no matter how hard you try. This is solely due to the fact that every measurement has some error or uncertainty associated with it. Even advanced instrumentation such as radar guns have errors associated with them. To help understand how to handle this type of data, here are 5 measurements taken to determine the mass of carbohydrate in 50.0-g of a particular protein:

12.62g, 11.91g, 13.07g, 12.73g, and 12.59g.

If we were asked to report this scientifically, we would never just list the 5 values, we would give a 'true' value and its associated error.

The Mean

The first step to correctly report our finding is to figure out the mean of our data. The mean is the average of our data set:

$$Mean(\overline{x}) = \frac{\sum_i x_i}{N}$$

For our example, the **mean** is found by adding each individual data (***x_i***) and dividing by the size of the sample (***N***) as shown:

$$x = \frac{12.02 + 11.91 + 13.07 + 12.73 + 12.59}{5} = 12.58g$$

Standard Deviation

The next statistic we need to calculate is the standard deviation. Specifically, the standard deviation measures how closely data are clustered about the mean value, and is technically defined as:

$$StDev(\sigma) = \frac{\sqrt{\Sigma(x - x_i)^2}}{N - 1}$$

In general, the smaller the standard deviation, the closer the mean will be to the 'true' value. In particular, the numerator in this equation calculates the residual for each piece of data. In other words, the numerator calculates how much an individual measurement differs from the mean. As for the squares and the square root, they take care of the fact that some of our data points are larger than the mean while others are smaller than the mean.

NOTE: Some statistical approaches require the variance of our data. This is just simply the square of our standard deviation.

Again using our example data:

$$StDev(\sigma) = \frac{\sqrt{\Sigma(12.58 - 12.62)^2 + (12.58 + 11.91)^2 + (12.58 - 13.07)^2 + (12.58 - 12.73)^2 + (12.58 - 12.59)^2}}{4} = 0.422_3$$

Confidence and the Student's t

Finally, we can calculate the confidence (μ) with which we present our data. This statistical value incorporates several other values including the mean, standard deviation in the mean, and the Student's t. This value is calculated by taking the mean and adding the corresponding confidence interval. The generalized form of the equation is shown below, where t is the value of the **Student's t** at a given number of degrees of freedom and confidence:

$$\mu = x \pm \frac{t \cdot \sigma}{\sqrt{N}}$$

For our example:

$$\mu = 12.58 \pm \frac{t \cdot 0.422_3}{\sqrt{5}}$$

The value of t here is vital, as scientific data is generally expressed at 95% and 99% confidence. Using the table shown below, and the fact that we have 4 degrees of freedom (N – 1); we see that our Student's t value will be either 2.78 or 4.60.

$$\mu_{95\%} = 12.58 \pm 0.053 grams$$

or

$$\mu_{99\%} = 12.58 \pm 0.087 grams$$

Confidence table taken from *Analytical Chemistry, An Introduction, 7th Edition: Table 7-2 p. 152.* Q Table taken from *Quantitative Chemical Analysis, 5th Edition: Table 4-5 p. 82.*

Degrees of Freedom	90%	95%	99%
1	6.31	12.70	63.7
2	2.92	4.30	9.92
3	2.35	3.18	5.84
4	2.13	2.78	4.60
5	2.02	2.57	4.03
6	1.94	2.45	3.71
7	1.90	2.36	3.50
8	1.86	2.31	3.36
9	1.83	2.26	3.25
10	1.81	2.23	3.17
11	1.80	2.20	3.11
12	1.78	2.18	3.06
13	1.77	2.16	3.01
14	1.76	2.14	2.98
Infinite	1.64	1.96	2.58

Q Table (90% Confidence)	*Q Table* (95% Confidence)	*Q Table* (99% Confidence)	Number of Obs. (*N*)
0.941	0.970	0.994	3
0.765	0.829	0.926	4
0.642	0.710	0.821	5
0.560	0.625	0.740	6
0.507	0.568	0.680	7
0.468	0.526	0.634	8
0.437	0.493	0.598	9
0.412	0.466	0.568	10

What we have calculated is essentially error bars. We can report our data with 95% certainty that the amount of carbohydrate in this particular protein is within +/- 0.53 grams of 12.58 grams, while we know with 99% certainty that the amount of carbohydrate is within +/- 0.87 grams of 12.58 grams. In other words, we can conclude that this particular protein will contain ~12.00 to ~13.00 grams of protein at 95% confidence, and anywhere from ~11.40 to ~13.40 grams at 99% confidence.

Outliers

Utilizing our same example, what if we took one more experimental measurement and found the carbohydrate content to be 17.64 grams. Immediately you notice that this value is well outside our confidence interval just reported, but being a good scientist you include it in your calculations. Shockingly, you notice that its presence dramatically affects both your mean and standard deviation. Luckily, you are about to learn an approach where you can statistically prove this data point to be an outlier, a piece of data that is 'far away' from the rest of the data.

The Q-Test

The tool that will help you with this situation is the Q-test, a method dedicated entirely to determining if one particular data point can be rejected from the others. Shown below is the equation for Q Calculated which can be compared to a value from the Q Table. If the value of Q Calculated is found to be greater than Q Table, then the data can be discarded.

$$Q_{calculated} = \frac{Gap}{Range}$$

For our example:

$$Q_{calculated} = \frac{16.64 - 13.07}{16.64 - 11.91} \cdot 0.755$$

In our example, *Q Calculated* > *Q Table*, we can therefore reject this data!

Report Contents and Questions

As the first experiment of the term, the exercises presented below are to be used to refresh your memory. None of this experiment is actually done in the laboratory. Therefore, since no data has been taken in the laboratory, none should be recorded in your notebook. Instead, this report should be prepared on notebook paper to be turned in at the next lab class meeting.

The report should consist of a title page, purpose and the exercises below. Please remember all the rules for constructing good scientific graphs. Show all of your work for full credit.

Exercises

1. Consider the following experiment in which you wish to determine the molarity of the concentrated HCl in your new bottle of reagent. You carefully measure out 100.0 mL of the concentrated acid and dilute it to 1.000 L (in a one-liter volumetric flask). You then use a 5.00 mL volumetric pipette to remove 9 aliquots for titration with 0.500 M sodium hydroxide solution. The following are the volumes of base needed for each sample.

14.75 mL 14.70 mL 14.81 mL 14.73 mL 14.73 mL

14.76 mL 14.48 mL 14.55 mL 14.69 mL

Calculate Q_{exp} for both the maximum and minimum values in this data.

(a) Q_{exp} for 14.81 mL =
(b) Q_{exp} for 14.48 mL =
(c) Can you reject 14.81 mL with 90% confidence?
(d) Can you reject 14.81 mL with 95% confidence?
(e) Can you reject 14.81 mL with 99% confidence?
(f) Can you reject 14.48 mL with 90% confidence?
(g) Can you reject 14.48 mL with 95% confidence?
(h) Can you reject 14.48 mL with 99% confidence?

Average your measurements (throwing out the outlier(s) if justified with 90% confidence), and calculate the molarity of the original concentrated HCl solution.

2. You are asked to calibrate a 10-mL volumetric pipette by weighing to the nearest 0.1 mg the mass of water delivered by the pipette. You weigh five samples of water delivered by the pipette and convert the mass of each to volume by dividing by the density of water at 25°C (0.997048 g/mL). Following are your measurements:

9.9755 mL 10.0029 mL 9.9899 mL 10.0302 mL 9.9963 mL

Calculate the following statistical measures for this data:

Mean (x) = ____________________ Standard Deviation (s) = ____________________

Variance (s^2) = ____________________ Standard Error of the Mean (s_{mean}) = ____________

90% Confidence Interval = 9.9899 ± ________________
(The "true value" of the volume will lie within this interval 90% of the time).

95% Confidence Interval = 9.9899 ± ________________

(The "true value" of the volume will lie within this interval 95% of the time).

99% Confidence Interval = 9.9899 ± _______________
(The "true value" of the volume will lie within this interval 99% of the time).

3. The table below shows some data in which the peak height of triplicate samples of a substance is listed as a function of the quantity of material injected. You would like to know first how good the precision of the method is by determining the standard deviations and confidence intervals of each set of triplicate samples. Then you would like to know whether the means of the triplicate samples show a linear relationship with sample size, and the equation for this relationship.

Gas Chromatograph			**Calibration**
Sample Mass μg	**Peak Height (cm)**		
	Trial 1	**Trial 2**	**Trial 3**
15	1.4	1.2	1.3
25	2.6	3.1	3.7
35	5	5.4	4.8
45	7.8	7.6	7.3
55	10.1	9.8	10.3
65	12.5	11.9	12.6

a. Using Excel or another similar spreadsheet program, graph the average peak height versus Sample mass. Apply a best fit line to the resulting graph.

b. Again using the spreadsheet program, calculate *sample mean, variance, standard deviation, standard error of the mean,* and *confidence intervals* for the data above.

c. Answer the following questions:
   i. For the 45 μg sample, what is the sample standard deviation and 90% confidence interval of the three peak height measurements?

   ii. Excel gives the best straight line fit to the data as an equation y=mx+b. What are the slope and intercepts of this line? (Be sure to include units)

4. During this semester, you will encounter several sections that discuss kinetics. Kinetics is the study of the rate at which a reaction proceeds from reactants to products. These rates can only be determined experimentally by direct observation of the appearance or disappearance

of products or reactants, with respect to time. Catalysts are compounds that increase the rate at which a reaction proceeds and inhibitors are compounds that slow down or stop reactions.

Enzymes are biological catalysts. In discussing the properties of an enzyme, certain values, or parameters are determined experimentally under steady state conditions. The values are determined through kinetics studies and include:

V_{Max} : The so-called maximal rate of the catalyzed reaction. The enzyme's active site is saturated.

K_M : The Michaelis constant is the substrate concentration at which the reaction rate is one half its maximum value. Also known as the turnover number.

These values are determined experimentally by recording the progress of an enzyme-catalyzed reaction using fixed amounts of enzyme and a series of different substrate concentrations.

A typical data set looks like Table 1 where V_o is the initial reaction velocity and $[S]_o$ is the substrate concentration.

V_{Max} and K_M can be determined from linear regression analysis of a plot of $1/V_o$ vs. $1/[S]_o$, a so called Lineweaver-Burk plot.

The Lineweaver-Burke Equation: $\frac{1}{V_0} = \frac{K_M}{V_{Max}}\left(\frac{1}{S_0}\right) + \frac{1}{V_{Max}}$ has the familiar form y = mx + b, where $m = \frac{K_M}{V_{Max}}$ and $b = \frac{1}{V_{Max}}$.

a. For the data below, create a Lineweaver-Burk plot of the data and calculate V_{Max} and K_M.

S_o (μmol/L)	V_o (μmol/L·min)
0.07	0.155
0.075	0.185
0.09	0.285
0.15	0.555
0.25	0.795
0.55	1.135
1.05	1.375

5. In biological systems, the enzyme catalyzed reaction rates are often affected by substances that inhibit or interfere with the enzymes interaction with the substrate(s). There are three basic types of inhibition: noncompetitive, uncompetitive, and competitive inhibition.

The type of inhibition can be determined from graphical analysis of experimental data with Lineweaver-Burk plots. Inhibition experiments involve a series of experiments with fixed amount of inhibitor added to varying amounts of substrate. Lineweaver-Burk plots are constructed showing multiple lines for the various inhibitor concentrations.

The inhibition types and their graphical characteristics are summarized below and graphically in example graph 1.

Reference Table 1: Definition and Graphical Characteristics of Enzyme Inhibition

Inhibition type	Definition	Graphical characteristics
Noncompetitive	The inhibitor binds the enzyme at a different site than the substrate causing a conformational change. The conformational change affects the rate of catalysis, but the overall turnoff remains constant.	**The slopes of Lineweaver-Burk plots are different, as are the y-intercepts. Yet, the x-intercepts remain constant.**
Competitive	The inhibitor competes for the enzymes binding site with the substrate. The proportion of substrate molecules bound by the inhibitor reduces the rate of catalysis.	**The slopes of Lineweaver-Burk plots are different, yet the y-intercepts are the same.**
Uncompetitive	The inhibitor binds the active site after the substrate. Binding of the inhibitor can stimulate the binding, but produces a non-productive complex.	**The slopes of Lineweaver-Burk plots are the same and both intercepts are different.**

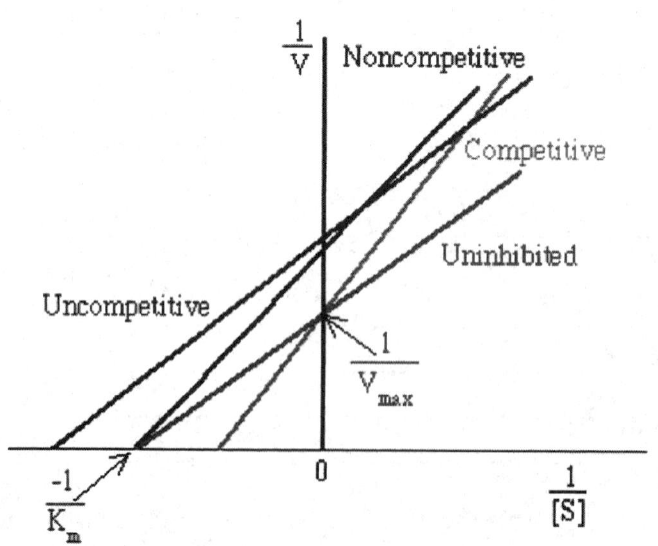

Example Graph 1: Representation of the possible types of enzyme inhibition

You have created two inhibitors (A and B) and done experiments to compare their effectiveness. For each set of data collected:

a. Create a Lineweaver-Burk plot of the data.
b. Calculate V_{Max} and K_M of the uninhibited and inhibited line.
c. Determine the type of Inhibition
d. Which is the better inhibitor? Why?

Typical Inhibitor Data: Inhibitor A

S_o (μmol/L)	Without Inhibitor V_o (μmol/L·min)	With Inhibitor V_o (μmol/L·min)
0.000005	9.1	2.8
0.000007	13.2	5.1
0.000012	21.2	10
0.000022	26.8	15.1
0.000032	32.5	21.3
0.000062	37.3	28.9
0.000092	39.2	32.5
0.000122	39.8	32.78

Typical Inhibitor Data: Inhibitor B

S_o (μmol/L)	Without Inhibitor V_o (μmol/L·min)	With Inhibitor V_o (μmol/L·min)
0.000005	9.1	0.8
0.000007	13.2	1.6
0.000012	21.2	3.2
0.000022	26.8	4
0.000032	32.5	5.5
0.000062	37.3	6.4
0.000092	39.2	6.8
0.000122	39.8	7.2

Experiment 1:
Conservation of Matter

Introduction

There is a scientific law called the Law of Conservation of Mass, discovered by Antoine Lavoisier in 1785. In its most compact form, it states:

Matter is neither created nor destroyed.

In 1842, Julius Robert Mayer discovered the Law of Conservation of Energy. In its most compact form, it is now called the First Law of Thermodynamics:

Energy is neither created nor destroyed.

In the early 20th century, Albert Einstein announced his discovery of the equation $E= mc^2$ and, as a consequence, the two laws above were merged into the Law of Conservation of Mass-Energy:

The total amount of mass and energy in the universe is constant.

What does this mean to us? Well, these laws allow us to balance chemical equations, calculate product amounts and determine whether reactions will be spontaneous.

Our entire system of stoichiometry is based on the veracity of these laws. The purpose of this lab experiment is to verify the first of these laws, the Law of Conservation of Mass.

If you were to design an experiment to confirm this law, you would want to observe two things: 1) A reaction is taking place; and 2) The total mass of all reactants is equal (within experimental error) to the total mass of all the products. In order to *observe* a reaction taking place, there must be a color change, the emission of a gas, or some other chemical change that can be visually monitored. There are reactions that we observe quite often, such as wood burning, that could be used but that are difficult to quantify because their products escape as soon as they are produced. When wood burns it is converted into water and carbon dioxide which escape as gases as soon as they are formed. More importantly, the other by-product of this reaction is extreme heat, which makes trapping the gases difficult and beyond the technology available in most introductory chemistry laboratories.

Since it produces both a color change and a gas, the reaction of copper(II) sulfate and zinc metal in aqueous HCl is useful for our purposes. The reaction can be monitored by observing the loss of blue color in the solution, by the production of hydrogen gas, and by the formation of solid copper. By quantifying the reactants and products of this reaction, we will be able to confirm that the total mass remained unchanged (within experimental error) while visually confirming that a reaction did in fact occur.

As always in lab, there are other purposes being served along with the learning of a new concept. We will also be revisiting the use of the analytical balance and producing our first chemical solution. You will notice the phrase "within experimental error" being used a couple of times above. This is because with any experiment there is a certain amount of reactant and product lost when they are transferred from flask to flask or spilled, splashed or dropped as part of the human error in the experiment. These "errors" must be taken into account when reporting the results of any experiment. Statistics are often used to indicate the relative importance of the error. For example, losing 100 grams of product would seem tremendous unless it was compared to an expected product mass of 2.5×10^6 g. Then this error seems very small indeed. We will use this experiment to practice our knowledge and use of statistics (Appendix 4) to report the error in the mass of products created from the mass of reactants used.

Background

One of the most important aspects of chemistry is the balanced chemical equation. Recall from the purpose that the law of conservation of mass states that matter can be neither created nor destroyed. Therefore, because matter is composed of atoms that are unchanged in a chemical reaction, the number and types of atoms in a chemical equation must be the same both before and after the reaction.

The Reaction

The experiment you are going to conduct involves the addition of zinc metal (Zn^0) to an acidic solution of copper(II) sulfate ($CuSO_4$) which is composed of copper ions (Cu^{+2}) and sulfate ions (SO_4^{2-}). Thus, the reaction consists of the reactants, species on the left of the arrow: Zn^0 (*s*), Cu^{2+} (*aq*), and HCl (*aq*) in the form of H^+(aq) and Cl^-(*aq*), while the products of the reaction, species found to the right of the arrow, are Zn^{2+} (*aq*), Cu^0 (*s*), and H_2 (*g*).

> You might notice that we don't mention the sulfate or chloride ions within the reactants or products. This is because in this reaction they are 'spectator' ions. They are called spectators because they don't participate in the reaction we are observing and remain at a constant concentration throughout the reaction. Just as in a math equation, any molecules or ions that remain the same on both sides of the reaction are said to 'cancel' each other and can be removed from the net reaction equation. This doesn't mean those ions are gone, just that we can ignore them since they do not influence the reaction.

If we know the initial masses of both the solution and the metal pieces before the reaction, we can track how the masses have changed after the reaction, but we should observe that the total mass doesn't change. However, before we can proceed, we need to look at the chemical equations involved so that we can fully understand just what is taking place.

The reaction you will be running can be broken down into the two 'mini' net ionic reactions shown below:

$$Zn^0(s) + Cu^{2+}(aq) \rightarrow Zn^{2+}(aq) + Cu^0(s)$$
$$Zn^0(s) + 2H^+(aq) \rightarrow Zn^{2+}(aq) + H_2(g)$$

Both of these reactions are referred to as oxidation-reduction reactions. We will return to this subject later, but for now it is simply important to note that the aqueous copper is being converted to solid metal copper. During the reaction you will observe this process by the fading of the blue color characteristic of the Cu^{2+} ions, and the accumulation of the pink copper metal onto the zinc metal pellets.

Net Reaction

The two 'mini' reactions above are then combined, kind of like an addition problem, to form one equation:

$$Zn^0(s) + Cu^{2+}(aq) + 2H^+(aq) \rightarrow Zn^{2+}(aq) + Cu^0(s) + H_2(g)$$

In this equation, you can see that in addition to solid copper (Cu^0) being produced and solid zinc (Zn^0) being ionized to aqueous Zn^{2+}, hydrogen gas (H_2) is released.

Molarity

You should already have learned about the unit called the mole in your course. Molarity (M) is simply a unit of concentration that indicates the numbers of moles of a substance, called the solute, to the volume in liters of solution it is dissolved in. Thus,

$$M = \frac{\text{mol of solute}}{\text{L of Solution}} = \frac{mol}{L}$$

Solution concentrations are very often given in units of molarity and we will use molarity to create the proper concentration of the $CuSO_4$ solution used in this experiment.

Solution Preparation

The first task in this experiment requires you to make 50.0 mL of a 0.50 M solution of $CuSO_4$ in 3.0 M HCl. Because this is one of the first solutions you will be preparing this semester, what will follow is a tutorial-like discussion of the proper method for preparing solutions.

As an example, let's pretend that we have to make 50.0 mL of a 0.75 M solution of potassium permanganate ($KMnO_4$) in 3.0 M HCl. The first step in this process is to obtain about 50 mL of the 3.0 M HCl in your 100-mL graduated cylinder. With the HCl obtained, we have to calculate how many grams of $KMnO_4$ we will need to make a 0.75 M solution. In order to do this, we first have to find the number of moles in our solution and then multiply by the formula mass of $KMnO_4$.

The formula mass for a compound is calculated by adding together all of the atomic masses found in the formula. For $KMnO_4$ this is the mass of 1 potassium atom + the mass of 1 manganese atom + the mass of 4 oxygen atoms:

$$\text{Formula Mass } KMnO_4 = 39.0983\tfrac{g}{mol} + 54.938049\tfrac{g}{mol} + 4(15.9994\tfrac{g}{mol}) = 158.04\tfrac{g}{mol}$$

Molarity is defined as the number of moles of solute divided by the liters of solution. Note that we said liters of solution not just liters of solvent. This means that the total solution volume (including solute volume) is in the denominator ($M = mol/L_{Soln}$).

For our solution, the solute is the potassium permanganate. Since we know the molarity of the solution is supposed to be 0.75 M and the volume is supposed to be 50.0 mL, we can calculate the number of moles of $KMnO_4$ needed:

$$\text{moles}_{KMnO_4} = \frac{0.75\text{mol } KMnO_4}{L_{KMnO_4\ Solution}} \times 0.050\text{L} = 0.0375\text{mol } KMnO_4$$

The hardest part is now done! With the number of moles of $KMnO_4$ known, all we have to do is multiply by the molecular weight of potassium permanganate to get the number of grams.

$$0.0375\text{mol } KMnO_4 \times \frac{158.04\text{g } KMnO_4}{1\text{mol } KMnO_4} = 5.93\text{g } KMnO_4$$

We now need to prepare the flask in which we will make the solution. Solutions are normally made in specially calibrated volumetric flasks. For our solution, we will use a 50.0 mL volumetric flask, filling it about 2/3 full with the 3.0 M HCl we are using as the solvent for our solution. Note that you should never completely fill the flask before adding the solute. This is because addition of the solute will affect the overall volume of the solution and you don't want to exceed the 50.0 mL total volume. Next, we weigh out the 5.93 grams of $KMnO_4$ using the proper weighing technique, and add this amount to the 3.0 M HCl in the flask. After stirring, we can add as much of the remaining HCl required so that the final total solution volume is 50.0 mL. We will use this same process to produce the solution needed to perform this week's experiment.

Volume of H_2 Evolved

As stated above, the reaction we will be observing produces hydrogen gas. In order to determine that the mass of reactants and products has been conserved, we need a method by which we can collect and determine the mass of this hydrogen gas. In order to accomplish this task successfully, you will use a **side-armed Erlenmeyer flask**:

and a balloon. By sealing the end of the balloon to the side-arm, all of the gas released from the reaction will fill up the balloon. We can then determine the volume of the balloon by measuring its radius (r), and height (h) if necessary, and substituting them into one of the following equations:

$$Vol_{Sphere} = \frac{4}{3}\pi r^3 \qquad \text{or} \qquad Vol_{Cyl} = \pi r^2 h$$

Assuming the temperature of the gas is approximately room temperature, 298 K (or 25 °C) and that the atmospheric pressure is 1.00 atmospheres, you can calculate the mass of hydrogen gas by substituting the volume of hydrogen gas in L into the following equation:

$$\text{Mass of } H_2 \text{ gas (in g) evolved} = 8.244 \times 10^{-2} \text{ X (volume in L)}$$

[*Reminder:* 1 cm^3 = 1 mL = 0.001 L]

Procedure

SAFETY NOTES: The hydrochloric acid (HCl) used in this lab is very concentrated and can cause severe burns. If the acid comes in contact with your skin, rinse the area immediately with lots of water.

Part I: Making the Copper Sulfate Solution

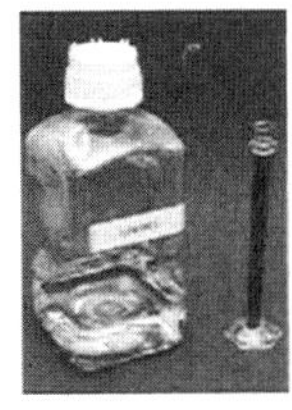

You will need to create 25 mL of a 0.25M solution of $CuSO_4$ in 3M HCl. Using your 100 mL graduated cylinder, collect ~ 25 mL of 3.0M HCl from the front counter and place it in a beaker. Be careful!

Using the weigh boats provided, weigh out the correct amount of $CuSO_4$ you calculated in your pre-laboratory assignment. Be sure to use proper weighing technique.

Fill the volumetric flask about 1/3 full with 3.0 M HCl. Add a small amount of HCl to the weigh boat containing your $CuSO_4$. Slowly pour the HCl/$CuSO_4$ slurry into the volumetric flask. Add HCl to the $CuSO_4$ in the volumetric flask until the volume is 25 mL exactly as indicated by the white line on the flask. Allow the solution to return to room temperature. Check to see that the volume is still correct. Add a little more HCl if necessary.

Part II: Precipitating Cu^{2+} Ions as Copper Metal

Collect and weigh all of the following items and record the masses in your notebook to the nearest 0.0001 g.

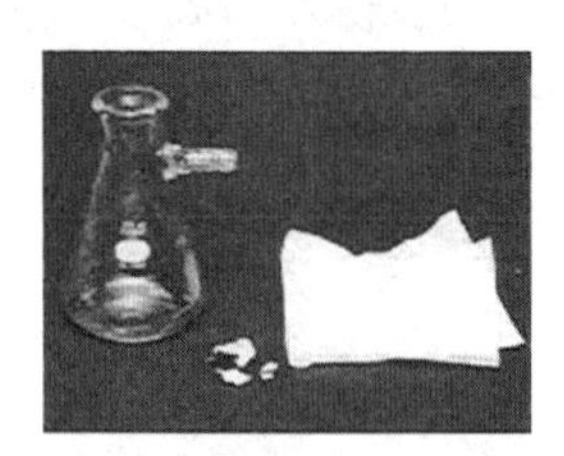

1. **A 100 mL side-arm flask**
2. **4 or 5 pieces of solid zinc metal (weigh together on a clean Kimwipe)**

Place 25 mL of the copper sulfate solution in the sidearm flask and re-weigh. Record the mass in your lab notebook to the nearest 0.0001g.

Collect a balloon from the front counter and fit it onto the sidearm of the pre-weighed flask. Using a rubber band or parafilm, make sure the seal of the balloon to the sidearm is airtight.

Collect a stopper from the front counter. You can add parafilm to the flask before adding the stopper to guarantee a good seal.

You are now ready to start the reaction. Quickly place the pre-weighed zinc metal pieces into the solution and cap with the stopper. You should see bubbles form right away. This is the hydrogen gas being produced.

Make observations about how the solution's appearance changes as the reaction proceeds, noting if there are any color changes. Make sure the balloon stays on the sidearm of the flask as increased gas pressure might push it off.

When the reaction is complete (no more bubbles and the solution should be clear) you will need to carefully measure the balloon. Record the diameter in your notebook.
Make observations about the solution and the metal pellets in the flask. Specifically, detail how these observations may support the claim that a reaction took place.

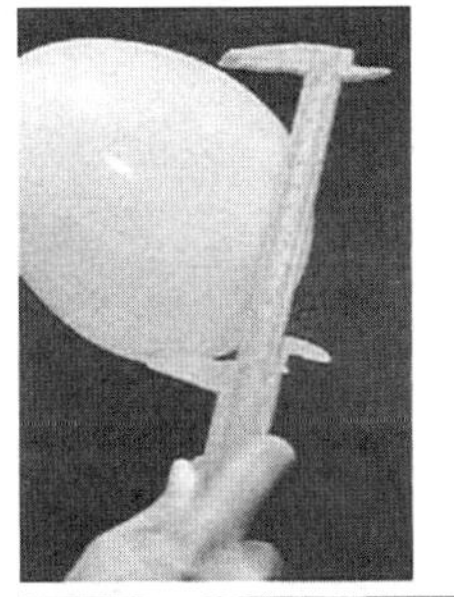

Finally, remove the balloon, stopper and parafilm from the flask. Reweigh the flask and its contents and record the weight in your notebook to the nearest 0.0001g.

Dispose of the solution and pellets in the appropriate waste container(s).

Experiment 1

Pre-Laboratory Questions

Name: ______________________________ **Date:** ___________________

Instructor: ______________________________ **Sec. #:** ____________

Show all work for full credit.

1) Based on your reading of the laboratory Introduction and Background, prepare a hypothesis regarding what results you expect from your experiment. Be detailed.

2) Calculate the number of moles or grams in each of the following quantities of salt.

5.233 g of $Cu(NO_3)_2 \cdot 6H_2O$ = ________________________

7.77 g of zinc nitrate trihydrate = ________________________

6.215×10^{-2} mol of $Na_2SO_4 \cdot 10H_2O$ = ________________________

2.0×10^{-2} mol of zinc nitrate hexahydrate = ________________________

2) You would like to make up 100 mL of a 0.23 M solution of $Na_2SO_4 \cdot 10H_2O$. How many moles of the salt do you need? How many grams of the salt do you need?

You have a 2.5 M solution of $Cu(NO_3)_2 \cdot 6H_2O$. What volume of the solution (in mL) must you measure in order to have 0.50 moles of the salt?

You have a 0.75M solution of $ZnSO_4 \cdot 7H_2O$. What volume of the solution (in mL) must you measure in order to have 32.5 grams of the salt?

3) In your laboratory you will be required to make a solution of 0.25 M $CuSO_4 \cdot 5H_2O$ in a 25.00 mL volumetric flask. This exercise will walk you through the steps involved. After completion, write down the values for your answers in your notebook and take them to the laboratory to use in preparing your solution.

a. What is the formula weight of $CuSO_4 \cdot 5H_2O$?

b. How many moles of $CuSO_4 \cdot 5H_2O$ are required to make 25.00 mL of a 0.25 M solution?

c. What mass of $CuSO_4 \cdot 5H_2O$ will be needed to make this solution?

d. You will add the salt to a weigh boat, which you determine has a mass of 1.9785 g. What will be the reading on the balance when you have put sufficient $CuSO_4 \cdot 5H_2O$ in the boat?

Experiment 1

Laboratory Report

Name: ______________________ **Date:** ______________

Instructor: ______________________ **Sec. #:** __________

PURPOSE: (*The purpose should be several well-constructed sentences describing what your experiment was designed to accomplish and the criteria used to determine success. These sentences should include both concepts and techniques.)*

PROCEDURE: (*The procedure section should reference the lab manual and include any changes made to the procedure during the lab.)*

DATA:

Observations:

Initial Mass Data	
125 mL Side-Arm Flask (g)	
Zinc Metal (g)	
125 mL Side-Arm Flask w/ $CuSO_4$ (g)	
$CuSO_4$ Solution Alone (g)	
Total Initial Mass	

Final Mass Data	
125 mL Side-Arm Flask w/ Solution and Pellets (g)	
Volume of Balloon (L)	
Mass of H_2 Gas (g)	
Total Final Mass	

Percent Recovery = ___________________

CALCULATIONS: (*For each type of calculation an "empty equation" should be described followed by an example use of the equation presented. Significant figures and units should be observed.)*

Average Mass of Side-Arm Flask:

Average Mass of Zinc Metal:

Mass of $CuSO_4$ Solution:

Volume of Balloon:

Mass of H_2 Gas:

CONCLUSION: (*State total mass before the reaction and the total mass after the reaction, then discuss the mass percent recovered. Be sure to explain whether your experiment confirms the Law of Conservation of Mass and provide evidence for your conclusion. Also discuss the mass of* H_2 *gas that the reaction produced. Finally, include a discussion of possible errors in the experiment and how they might have affected the results.*)

Experiment 1

Post Laboratory Questions

Name: ______________________________ **Date:** __________________

Instructor: ______________________________ **Sec. #:** ____________

Show all work for full credit.

1) Based on your experimental results, does the Law of Conservation of Mass apply to this experiment? Explain.

2) If the balloon had expanded more how would the mass of H_2 and the mass percent recovered have changed?

3) How would increasing the molar concentration of $CuSO_4$ have affected this reaction?

4) Zinc metal reacts with acid solution to produce hydrogen gas and zinc ion as follows:

$$Zn(s) + 2H^{+}(aq) \rightarrow Zn^{2+}(aq) + H_2(g)$$

Experiment 1

You add 6.825 grams of Zn metal to 29.0 mL of 1.70 M HCl, and set up the flask to catch the hydrogen in a balloon as in the following picture.

How many moles of Zn have you added to the flask?

How many moles of H^+ are in the solution?

How many moles of Zn will have reacted with the H^+ when the reaction is complete?

How many moles of Zn will remain unreacted?

What mass of Zn will remain unreacted?

How many moles of H_2 gas will be produced?

What mass of H_2 gas will be produced?

Assuming that one mole of H_2 gas would occupy a volume of 22 L at the temperature and pressure in the balloon, what would be the volume of the balloon?

Experiment 2

Atomic Spectra

Introduction

Have you ever wondered how scientists know what elements are present on other planets? Believe it or not, the information regarding the elemental make-up of each of the stars, planets and other heavenly bodies is being broadcast right to us on a daily basis. Each of these bodies has what is called an emission spectrum that can be read from far away using a device called a spectrometer (and a really powerful telescope too). The spectrometer is able to separate the colors of the light being emitted by the star into discrete lines. The origin of these lines is discussed further in both your textbook and in the Background section for this lab. Each element has a unique set of lines; with a sophisticated computer program that can separate the individual spectra from the total array of spectral lines, you can identify the elements present in galaxies far, far away...

In today's lab we will be working closer to home by learning to use spectroscopes and to interpret the meaning of a spectrum. You will experimentally determine the spectra of several unknowns and then, using known "bright line" spectra for comparison, identify unknown ions.

Background

During the past century scientists like Sir J. J. Thomson, Ernest Rutherford, and Niels Bohr studied atomic structure. These studies resulted in the development of several different models, each attempting to further describe the internal structure of the atom. The Bohr model of the atom is especially easy to visualize. Chapter 2 of your textbook describes the development of these models and also describes the calculation of energy and wavelengths.

The Hydrogen Atom

The structure of the hydrogen atom is very simple and consisted of one proton (having a positive charge) and one electron (having a negative charge), a phenomenon known for some time before Bohr. This created some intellectual difficulty for scientists: they did not understand why the electron doesn't spiral into the proton—since negative charges are attracted to positive charges—and completely destroy the atom. But, since hydrogen is a stable atom, this attraction was obviously not the only rule that governed the behavior of the atom.

Understanding why atoms are stable was a very important problem. Even the most advanced physics of the time, Maxwell's Theory of Electromagnetism, predicted the instability of atoms. There was clearly a very fundamental flaw in their understanding of the universe.

Max Planck, Niels Bohr, and Energy

Prior to Niels Bohr's model of the hydrogen atom, Max Planck had postulated that light was composed of photons that carried quanta, discrete packets of energy. Utilizing this notion, Bohr theorized that since there were well defined states in which atoms could exist, then photons with just the right energy could cause transitions between these states. In fact, the reason that the electron did not spiral into the proton was that the atom was actually in one of these stable states. Therefore, Bohr's theory also stated that the electrons were quantized in energy levels.

Quantization

The idea of quantization is often difficult to accept. In the macroscopic world, things are continuous; they can take on any value. For example, a car can travel 17 mph, 18 mph, or 17.5 mph; it is not restricted to only integer values. The speed is perceived on such a large scale that we cannot tell it is quantized, so it appears to be continuous. However, when examined more closely, this continuity no longer holds. The speed of the car is in fact quantized, but since the difference between two adjacent levels is so small, quantization  cannot be observed.

Quantization actually means that only specific values are allowed. For example, consider a set of steps and a ramp. Since potential energy is a function of height, the potential energy is quantized for the steps. An object can rest only on one step or another; it cannot be between steps for a considerable amount of time. Thus, the potential energy increases only in increments of the height of the step. On the other hand, the ramp is not bound by such restrictions. For the ramp, the potential energy is continuous.

The Predictions and Mathematics

Bohr's model predicted that the energy of a hydrogen atom was quantized, and that the energy of the atom was dependent on the principal quantum number (n), which is always an integer. The exact dependence of the energy on this value is complicated to derive, but it turns out that the energy, often abbreviated E_n, is given by the equation shown below where β is a constant based on Planck's constant, the mass of the electron, and the charge of the electron. The value of β is ~2.18 x 10^{-18} Joules.

$$E_n = -\beta\left(\frac{1}{n^2}\right)$$

Note that the energy is negative. If E_n is thought of as a measure of the energy of attraction between the electron and proton, we can see that the larger the value of n, the less negative the value of E_n. This means that there is less attraction between the electron and the proton, so the distance between them increases. When n is very large, the value of E_n approaches zero and there is no attraction between the electron and the proton. Therefore, the electron is free to go wherever it wants—it is no longer bound to the proton.

It should also be noted that the smallest value of n is one; thus the lowest energy state for hydrogen atoms is the value of β, not zero as one might expect. In fact, the only time the energy of the electron is zero is when it approaches an infinite distance from the nucleus.

Rydberg Equation

By itself, the equation given above is not very useful. Because there is no energy parameter used to measure the absolute energy, only energy differences can be measured. Consider two possible states of the hydrogen atom that has quantum numbers n_I and n_F. (Note: n_I and n_F stand for n initial and n final, respectively. These are the initial energy level from which the electron transfers and the final energy level where it ends up.) If we consider a transition from the n_I level to the n_F level, the change in energy (ΔE_{Level}) will be:

$$\Delta E_{Level} = E_{Final} - E_{Initial} = \beta\left(\frac{1}{n_F^2} - \frac{1}{n_I^2}\right)$$

The change in energy values calculated using this equation is positive if absorption is occurring (i.e., when the electron moves from a lower energy state to a higher energy state) and negative if emission is occurring (i.e., when the electron is moving from a higher energy state to a lower energy state.) This energy is equal to the energy of the photon that either was absorbed or released by the atom. The energy of a photon with a frequency ν is simply the value of that frequency multiplied by Planck's constant (h), which is 6.626 x 10^{-34} J-sec.

$$E = h\nu$$

Furthermore, since the frequency is equal to the speed of light (c), 3.00 x 10^{8} m/sec, divided by the wavelength (λ) of the photon in meters:

$$\nu = \frac{c}{\lambda}$$

we can solve for the specific wavelength of light that is associated with a given transition. This is known as the Rydberg equation and is shown below.

$$\frac{1}{\lambda} = -R\left(\frac{1}{n_F^2} - \frac{1}{n_I^2}\right)$$

In this equation, R is known as the Rydberg Constant (1.10 x 10^{-2} nm^{-1}) and takes into account Planck's constant, as well as the speed of light, and has units that are more practical for calculating wavelengths. Although calculations may result in negative values of wavelength, only positive values are reported (a negative length does not exist). Emissions are signified by a negative wavelength and absorptions are signified by a positive wavelength.

This Week's Adventure

The Rydberg equation predicts a measurable value, the wavelength of light emitted by a transition between two states. For this experiment, n_F will always be 2, the value corresponding to the visible spectrum. Therefore, by substituting values for n_I that correspond to the visible spectrum of hydrogen, we can predict the wavelengths we should observe.

In this experiment, we will excite helium and hydrogen atoms by subjecting them to an electrical potential in a lamp. We will also excite a variety of cations by heating them in a flame. Each of these lamps or flames will be marked as unknowns. It will be your job to use your knowledge of their known wavelengths to identify them.

The Spectroscope

The spectroscope used in the experiment has a diffraction grating as its central component. A diffraction grating 'bends' light much like a prism. The angle at which the light is 'bent' depends upon the wavelength of light entering the grating.

Before we can start accurately measuring an unknown spectrum, we must first calibrate the spectroscope. You will use the known spectrum of helium, given in the Table 1 below, to calibrate your spectroscope. Once we plot the measured values of the helium lines against the known values, we can use that graph to determine the actual wavelengths of the unknown lamps and flames provided.

Table 1: Known Wavelengths of Helium	
Color	**Wavelength (nm)**
Violet	**388.9**
Blue	**468.6**
Green	**501.6**
Yellow	**587.6**
Orange	**667.8**
Red	**706.5**

Procedure

SAFETY NOTES: If the flame is somehow extinguished while doing the flame spectra, turn off the gas and call the instructor immediately.

Be careful not to get too close to any of the set-ups. The lamps have enough voltage to give a very serious shock. There is no reason to touch any of the set-ups.

Some of the lamps also emit ultraviolet light, which can damage your eyes. Do not look into a lamp for a long period of time.

GENERAL INSTRUCTIONS: Students work in pairs for this experiment. ***Since you are calibrating the instrument, the same spectroscope must be used in each part.***

Part I: Spectroscope Calibration

Obtain a spectroscope from the front counter.

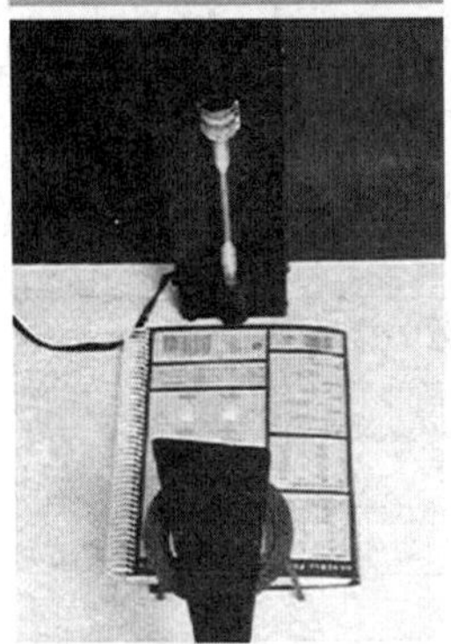

Obtain a ring stand from under the front counter. Set the ring stand up in front of the helium lamp using your lab notebook to space the ring stand from the lamp. (i.e., the legs of the ring stand should be exactly 1 notebook length away from the lamp)

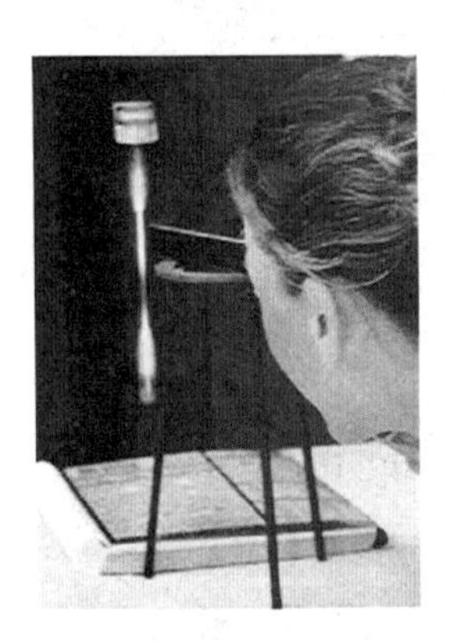

Set your spectroscope up on the ring stand. Focus the spectroscope so that light from the helium lamp comes in directly through the vertical slit. Look through the eyepiece where the grating is located and find the bright visible lines on the scale to the right of the slit. Read the scale accurately. You should be able to read the numbers and marks, as well as estimate between the marks, so each number you record should have three significant figures. Make sure that you do not move the spectroscope once you have aligned it with the light source. Have your partner record your readings. Now let your partner read the spectroscope while you record. He/she should read the spectroscope in the exact same alignment that you did. Check to see if your readings agree. You may decide that several attempts are in order to get the most reliable readings.

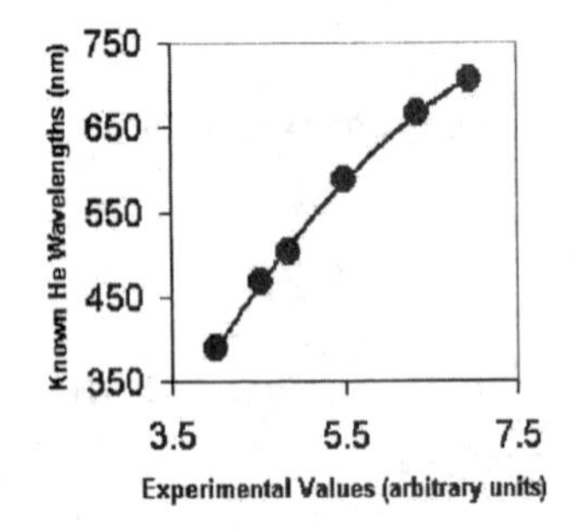

Using graph paper from the back of your lab notebook, create a graph of the known wavelengths of helium versus the wavelengths of the same colors you just observed experimentally. Make a best fit line to your data.

Part II: Identifying the Unknowns

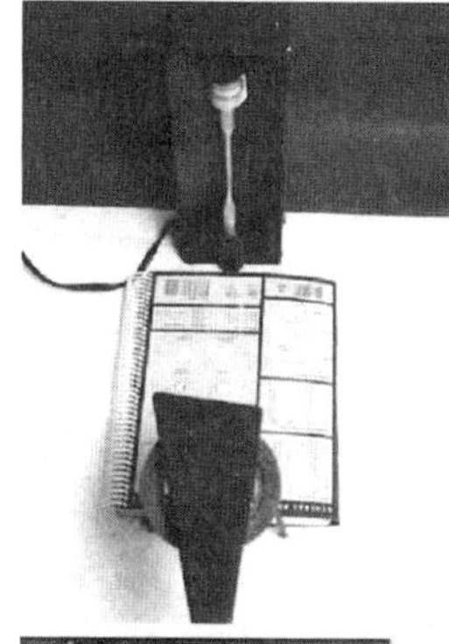

Use the spectroscope that you have just calibrated.

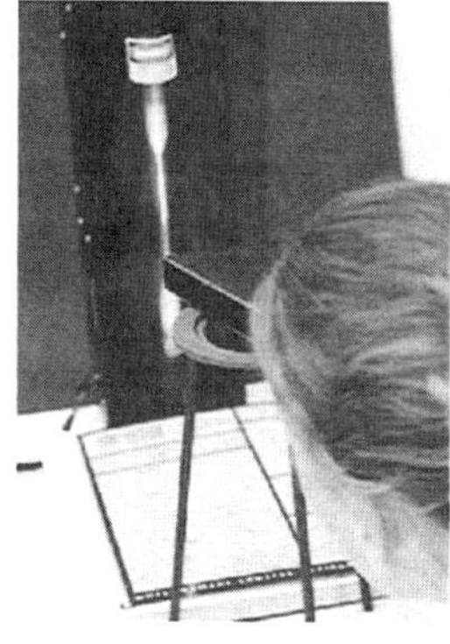

Set up your spectroscope in front of one of the unknown light sources. Record all the colors and wavelengths of each line you observe.

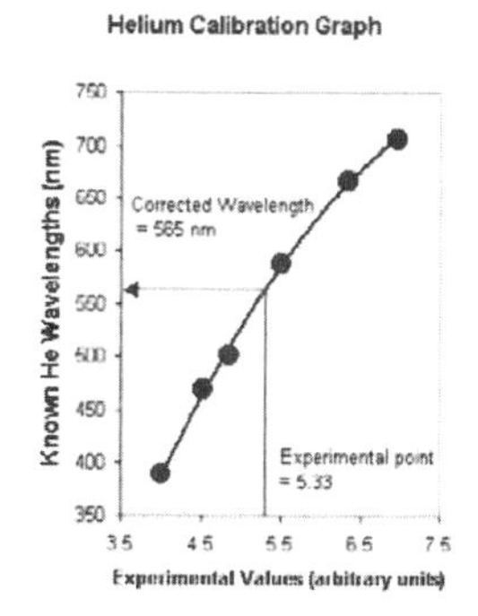

On your calibration graph, on the x-axis find the value of the first of your observed lines. Now using a vertically aligned straight line find the point on your best-fit line that corresponds to that value. Now find the point on the y-axis that also corresponds to that point on the line and record that value in your notebook.

Repeat steps above for the observed lines of all of the unknowns.

Experiment 2

Experiment # 2

Pre-Laboratory Assignment

Name: ______________________________ **Date:** ___________________

Instructor: ______________________________ **Sec. #:** ____________

Show all work for full credit.

1. Some of the elements that may be used in this experiment are listed below. For each element write the complete electron configuration.

    a. Helium

    b. Hydrogen

    c. Mercury

    d. Krypton

    e. Sodium

    f. Lithium

2. What makes the wavelength of the n=4 to n=2 transition for hydrogen different from an n=4 to n=2 transition for mercury? How is this difference related to the spectra we observe? Explain.

3. You have just carried out the calibration of your spectroscope's diffraction grating and you get the values shown in the table for the visible emission lines of helium. Plot these values and produce a calibration graph that can be used to correct the values in the table below:

Color	Spectrometer Reading
Violet	5.3
Blue	5.7
Green	6.1
Yellow	6.8
Orange	7.7
Red	8.3

Spectrometer reading	Corrected Wavelength (nm)	Color of the Emitted Light	Frequency (hz)
6.5			
5.4			
7.9			
4.6			
8.5			

4. The following atomic line spectra display six of the more prominent lines observed in the visible region for the indicated element. The number by each line represents its relative intensity. A photograph of the actual spectrum of the element may show many more lines, or sometimes fewer lines, depending on the overall intensity of the spectrum. If two lines are within 1 nm of each other, they are shown as a single line. You can find a complete listing of the lines for each element at the NIST (National Institute of Standards and Technology) Atomic Spectra Database.

Some spectral lines of Cadmium:

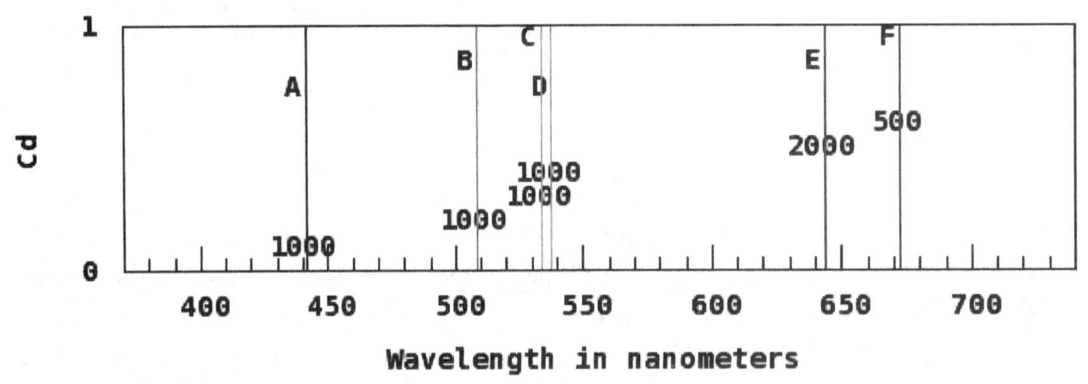

Identify the line in the spectrum of Cd with each of the following characteristics:

a. Light with a photon energy of 2.953×10^{-19} J/photon.

b. Light with a photon energy of 185.8 kJ/mol.

c. Light with a frequency of 5.574×10^{14} s^{-1}.

Experiment 2

Laboratory Report

Name: ______________________________ **Date:** ___________________

Instructor: _______________________________ **Sec. #:** ____________

PURPOSE: (*The purpose should be several well-constructed sentences describing what your experiment was designed to accomplish and the criteria used to determine success. These sentences should include both concepts and techniques.*)

PROCEDURE: (*The procedure section should reference the lab manual and include any changes made to the procedure during the lab.*)

DATA:

Observations:

Calibration Data:

Line Color	He Known Wavelengths (nm)	Spectroscope Reading (AU)

Unknown Data:

Unknown Number: __________
Identity: ___________

Line Color	Spectroscope Reading (AU)	Corrected Wavelength (nm)

Unknown Number: __________
Identity: ___________

Line Color	Spectroscope Reading (AU)	Corrected Wavelength (nm)

Unknown Number: __________
Identity: ___________

Line Color	Spectroscope Reading (AU)	Corrected Wavelength (nm)

Unknown Number: __________
Identity: ___________

Line Color	Spectroscope Reading (AU)	Corrected Wavelength (nm)

Unknown Number: __________
Identity: ___________

Unknown Number: __________
Identity: ___________

Line Color	Spectroscope Reading (AU)	Corrected Wavelength (nm)

Line Color	Spectroscope Reading (AU)	Corrected Wavelength (nm)

CALCULATIONS: (*For each type of calculation an "empty equation" should be described followed by an example use of that equation presented. Significant figures and units should be observed.)*

Calibration Graph:

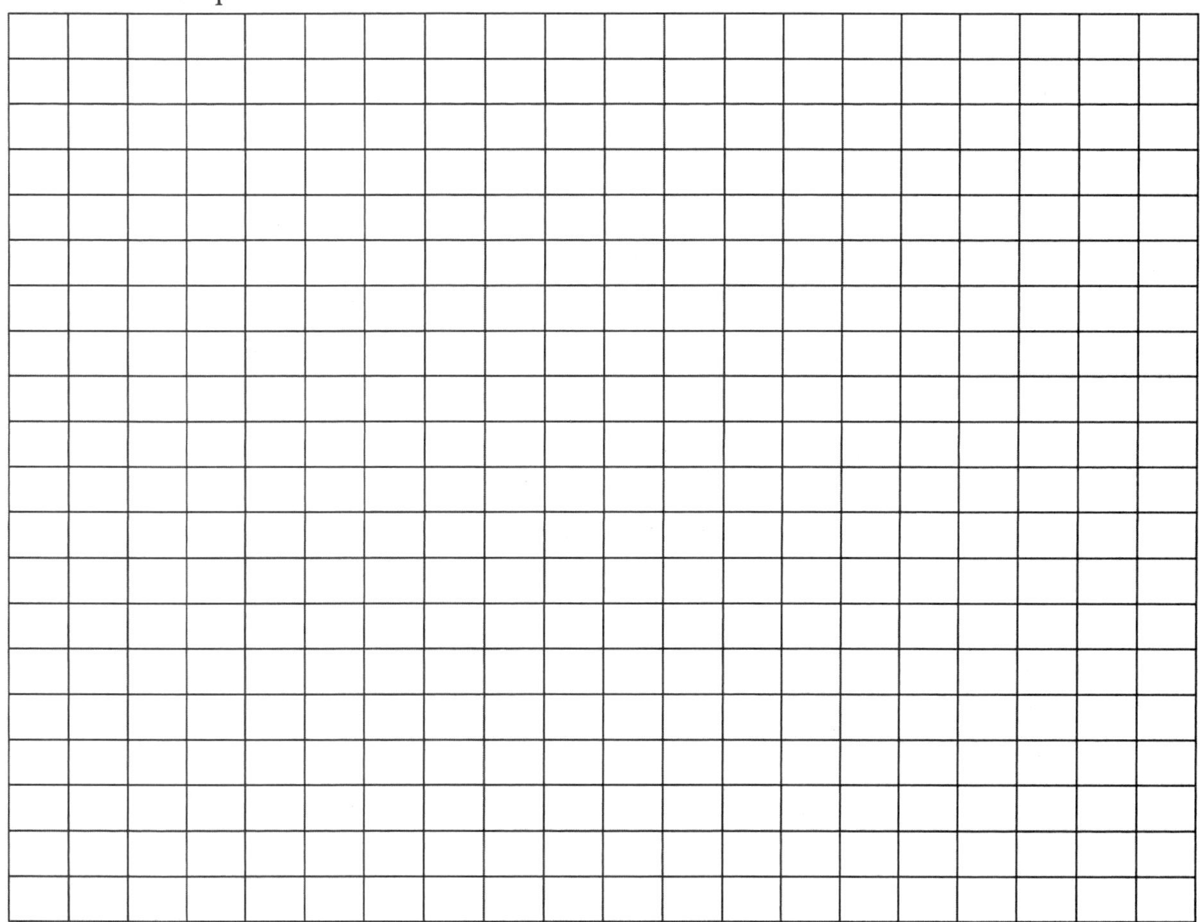

CONCLUSION: (*Include several well-developed paragraphs which include the identity of each unknown light source. Along with the identity, a detailed explanation of how the identity of each light source was determined should be included. A discussion on possible errors should be included.)*

Experiment 2

Post Laboratory Questions

Name: ____________________________ **Date:** __________________
Instructor: _____________________________ **Sec. #:** ___________

Show all work for full credit.

1. What is the difference between an emission and absorption spectrum?

2. In this experiment you compared your findings to known line spectra for the unknowns. The lines given were only the few "strongest" lines in the visible spectrum. Explain what is meant by the "strongest" lines and why some lines are stronger than others.

Experiment 3
Halogen Reactions

Introduction

One of the most basic facets of being a chemist is being able to look at two chemical species and predict whether or not they will react. You will find that most reactions, even the more elaborate organic reactions in your future, are based on the very simple chemical properties that you are learning about in this class.

One of the properties explored in this experiment is electronegativity, a covalently bonded atom's ability to attract electrons towards it. Other concepts illustrated by the reactions of the halogens are polarity of bonds and of molecules, and solubility. These three concepts are interrelated. The presence of an electronegative atom in a molecule is required for a molecule to be polar. A molecule's solubility is dependent on its polarity. Thus knowledge of the simple concept of electronegativity allows you to make predictions about a molecule's solubility.

Because this lab is qualitative in nature rather than quantitative, there are no calculations and no measurements. Rather, we will be collecting colorimetric evidence. Colorimetric simply means you will be recording the colors observed for each of the halogen solutions placed in the test tubes.

Background

One of the most useful things about studying chemistry is discovering the amount of useful information contained in the periodic table. Many of these periodic properties of the elements are discussed in your textbook.

In today's experiment, some of the properties of the Group 7A (Group 17) elements, known as halogens, and their compounds will be explored. The solubility properties of the halogens will be used to observe their reactions.

The more electronegative an element is, the more it attracts electrons. Group 17 atoms that have become ions by gaining an extra electron, such as F^-, Cl^-, Br^-, and I^-, are called *halides*. Note that the chlorine atom in NaCl, sodium chloride, is a halide—specifically a chloride. Also note that anions such as halides must always be paired with cations when found in the formula for a binary ionic compound. Group 17 atoms in their natural diatomic state, such as F_2, Cl_2, Br_2, and I_2, are called *halogens*. In this experiment, the relative *electronegativities* of the halogens will be determined.

If a solution containing halide, X^-, is added to a solution of a different halogen, Y_2, there are two possibilities. When element Y is more electronegative than element X, Y_2 will take the electron from X^-, leaving X_2 as a *halogen*. On the other hand, when Y_2 is less electronegative than X^-, no reaction will take place, and Y_2 remains as the halogen. In terms of balanced equations:

$$2X^- + Y_2 \rightarrow X_2 + 2Y^-$$

Or

$$2X^- + Y_2 \rightarrow \text{No RXN}$$

Polar and Non-Polar Solvents

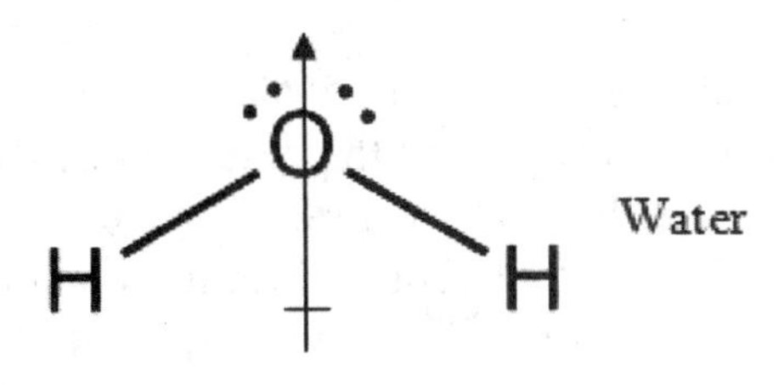

Water

The properties of two solvents, water and hexane, will be useful in sorting out what happens in this type of reaction. A water molecule contains an oxygen atom which is very electronegative and two hydrogen atoms which are not. This difference in electronegativities creates a dipole in water shown as an arrow with a plus sign in its tail. The arrow indicates the direction of flow of the electrons within the water molecule. Because of this permanent dipole, water is a polar solvent and it will solvate *polar* species whether they are ionic or molecular. This means that a polar molecule (one that has a dipole moment) or an ionic compound may dissolve in water. A diatomic molecule is polar if the two atoms have different electronegativities. Identify which of the halogens and halides in the above equations are ionic and which are non-ionic.

Hexane

Non-polar solvents solvate *non-polar* molecules. Hexane is an organic molecule that is non-polar. Note that carbon and hydrogen have very similar electronegativites (2.5 and 2.1 respectively) and that the molecule is very symmetrical. The small difference in the electronegativity values means that each C-H bond in hexane has a slight dipole but the symmetry of the molecule "cancels out" the overall dipole resulting in a non-polar molecule. Since water is polar and hexane is non-polar, the two do not mix. When combined, two distinct, colorless layers are formed with water, the denser liquid, on the bottom.

If colored substances are added to a test tube containing water and hexane, the polarity of the compounds can be determined. If they are non-polar, they will color the hexane layer. If the colored substances are polar, the color is observed in the water layer.

For the first reaction described above, if Y_2 is green before any reaction takes place, the hexane layer is green because non-polar compounds will reside in the hexane layer. As the reaction proceeds, the green disappears from the hexane layer because the Y_2 molecules are reacting and disappearing. The hexane will then take on the color of X_2. In the second case, where X is more electronegative than Y, the more electronegative atom already has the electrons, so no reaction will occur. Since no reaction occurs, the hexane layer will remain green, the color of Y_2.

Whenever a color change occurs, this is a clue that a reaction is taking place. Three halogens in aqueous (water) solutions will be available: chlorine, bromine and iodine. Each of these halogens has a distinctly different color in hexane. Therefore, by observing the color of the hexane layer, the halogen present can be determined.

Procedure

SAFETY POINTS: All halogen solutions are considered somewhat dangerous. Iodine can burn the skin and chlorine and bromine are corrosive. The volatility of chlorine and bromine increases the danger of inhaling fumes. Avoid all contact with these compounds. If a spill of any of the halogen solutions occurs, call the instructor immediately.

Any solution containing bromine must be used only under the hood and should be disposed of in the proper waste container. The only solutions that may be disposed of in the sink are the water solutions of the halide salts. All other solutions must be discarded in the proper waste container.

Although hexane is not considered a toxic substance, it is an organic solvent which cannot be disposed of via the sewer. Therefore, all solutions that contain hexane must be disposed of in the organic waste container.

GENERAL INSTRUCTIONS: At least 12 clean test tubes will be needed for today's experiment.

Part I: Solubility Testing of the Halides

This portion of the experiment tests the solubility of the *halides*. Two sets of halide salts are available, sodium salts (Na^+) and potassium salts (K^+). Choose one of the sets. Get together with another student who will test the other set to share data.

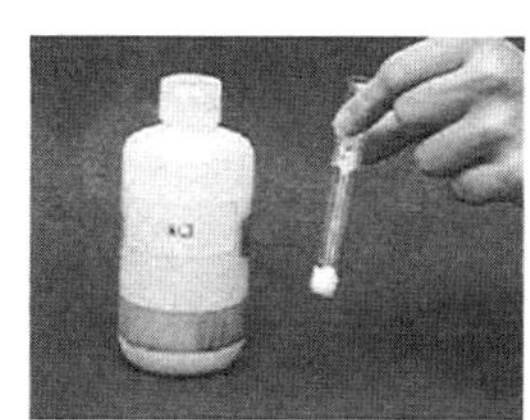

Add a *small* amount (the size of a small pea) of the *solid chloride* salt to a test tube. Note the appearance and other properties in your notebook.

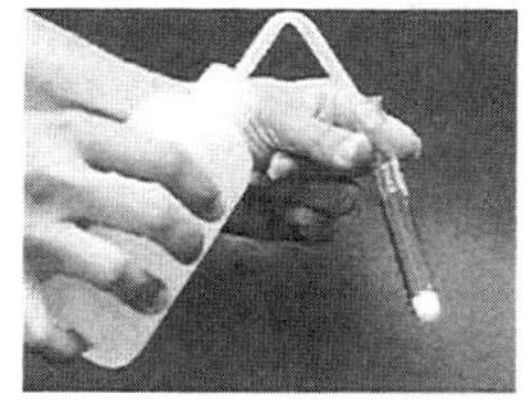

Now add 2 mL of distilled water and agitate the test tube. Record your observations. Repeat for each of the other *solid halide salts* in the set.

Repeat steps 1 and 2 using hexane as the solvent instead of water. NOTE: Any solution that contains hexane should be disposed of in the waste jar under the hood.

Part II: Solubility Testing of the Halogen

The solubility of the *halogens* will be tested next. NOTE: If you are not good at describing subtle differences in colors, you may use the crayons provided to make accurate color notes on the two layers of the test tubes for the remaining tests. Simply draw a diagram of the test tube in your notebook and match the color of the experimental layers using the crayons. Remember, sometimes a picture is worth a thousand words.

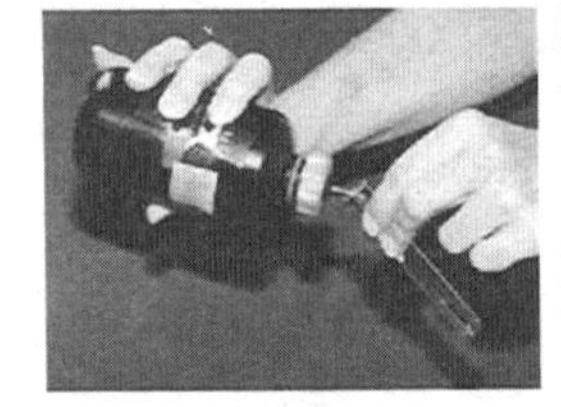

Add 1 mL of the aqueous (water) solution of iodine (I_2) to a test tube. Iodine is a solid at room temperature. Can you hear the solid crystals rattling in the bottom of the dropper bottle?

Do not measure the amount in a graduated cylinder. An approximate amount is good enough, and 20 drops is roughly 1mL. Be careful not to spill any of the halogen solutions on your skin or clothes, as it will stain. Record the appearance of the aqueous solution.

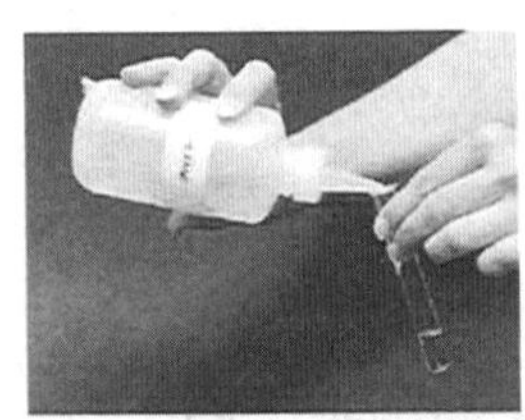

Now add 1 mL hexane and gently tap the bottom of the test tube to mix. Be sure it mixes; it may take a more vigorous shake, but start gently. Don't spill, but *do not* put your finger over the top and shake. Record all observations of both layers. When your observations are complete, put the solutions in the labeled waste jar in the hood. Rinse the test tube and put the rinsing in the jar too.

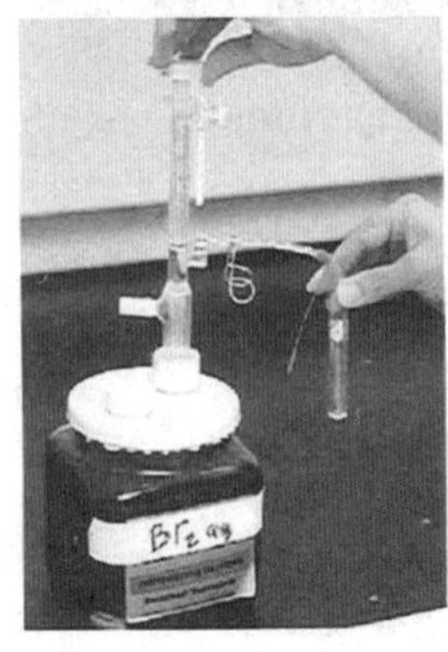

Bromine is a liquid at room temperature, and is quite volatile (that is, it evaporates easily). The aqueous solution of bromine (Br_2) is under the hood. Do not take any test tube containing bromine out of the hood. When finished with the bromine solutions, pour them in the waste jar in the hood, and then rinse the test tube with water from a wash bottle.

Add 1 mL of the bromine solution to a test tube. Record the appearance of the aqueous solution. Just as before, add 1 mL of hexane and mix. Record all observations.

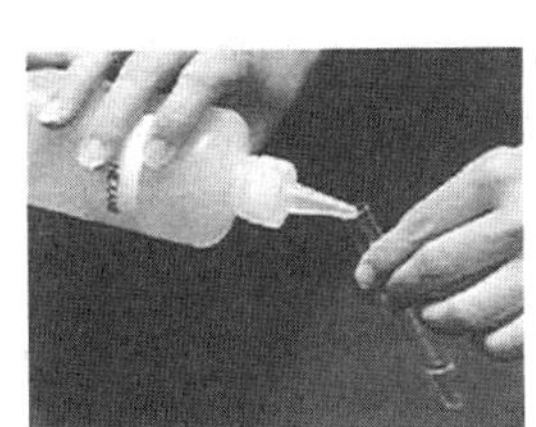

Cl_2 is normally a gas at room temperature. Since aqueous solutions of chlorine are not stable for very long (it's the equivalent of a soda going flat due to the loss of CO_2), each student will make a solution of chlorine (Cl_2).

Add 1 mL hexane to a test tube.

Now add 0.5 mL Clorox and 0.5 mL 1 M HCl to the test tube. Do not breathe the bubbles that are given off–this is chlorine gas. Note the color of each layer. When finished, pour the mixture in the waste jar and rinse the test tube.

Part III: Electronegativity Testing

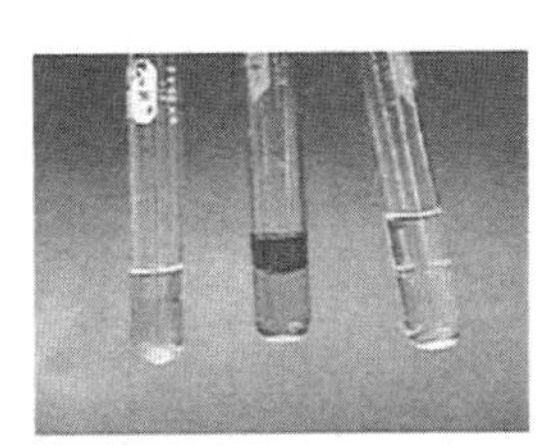

Now that baseline observations of what each of the ions and molecules looks like in water and in hexane, this set of reactions will look at all possible combinations of two reactants. This should allow relative electronegativities of the halogens to be confirmed. Remember that you are now doing a reaction, so you must make sure that your solution layers are thoroughly mixed to make sure the reactants come in contact with each other.

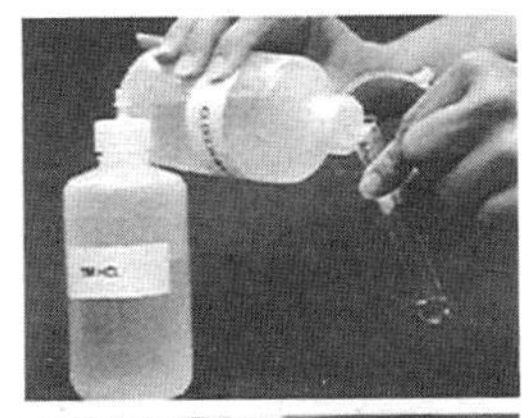

Add 0.5 mL Clorox to 0.5 mL 1M HCl to make $Cl_2(aq)$, and then add 1 mL hexane and mix. To this, add 1 mL of the water solution of I^-. Mix well and record your observations. Dispose of the solution in the waste jar.

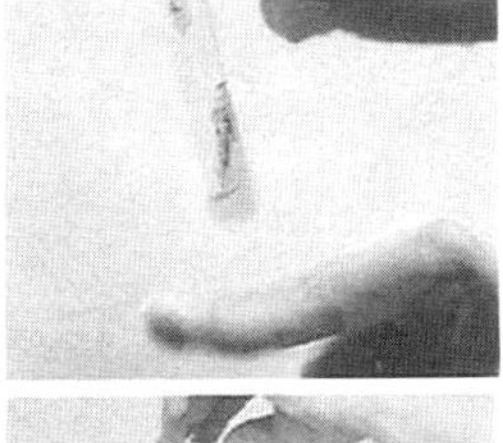

Add 1 mL of the $Br_2(aq)$ to 1 mL hexane, mix, and then add 1 mL of the water solution of I^-. Mix well and record all observations. Again, dispose of the solution in the waste jar.

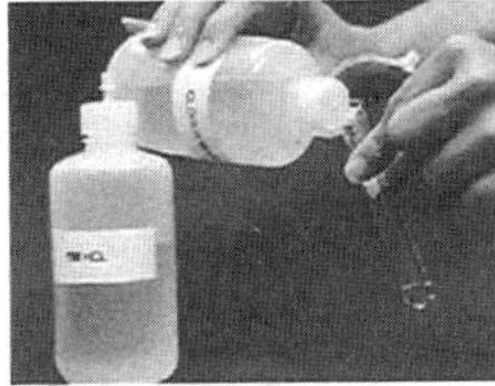

Repeat using the water solution of Cl^- instead of the I^-.

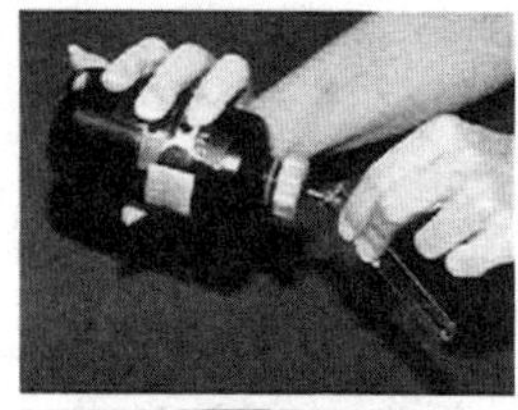

Add 1 mL of the $I_2(aq)$ to 1 mL hexane, mix, then add 1 mL of the water solution of Cl^-. Record all observations. Mix well and record all observations. Again, dispose of the solution in the waste jar.

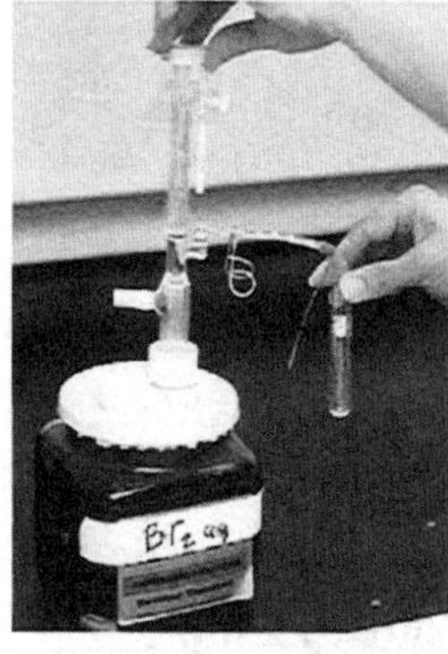

Repeat for the water solution of Br^- instead of the Cl^-.

You have now added each of the *halide* solutions to each of the *halogen* solutions. By noting the color of the hexane layer in each reaction, you can determine whether a reaction took place or not and you can organize an experimental ranking of the electronegativities of the halogens. Before cleaning up your glassware and solutions, ensure that your data makes sense. You should have three reactions and three no reactions. Talk with your instructor about repeating any tests that give contradictory conclusions.

Experiment # 3

Pre-Laboratory Assignment

Name: ______________________ **Date:** ______________

Instructor: ______________________ **Sec. #:** __________

Show all work for full credit.

1) When you go into your laboratory to carry out the halogen displacement reactions, you already know from lecture that the order of reactivity, and electronegativity, is $Cl_2>Br_2>I_2$, so you are able to predict which results to expect from the various combinations of halogen and halide salt.

   Let us suppose, though, that you are a student with a color perception disorder, so that you see the color of the halogens differently from other students. And suppose the halogen samples aren't identified for you, but are given as unknowns, labeled X_2, Y_2 and Z_2 for the halogens, and X^-,Y^-, and Z^- for the corresponding halides. You are to identify the relative reactivity of the unknowns, and hence their identity, by mixing different combinations of halogen and halide, and identifying the halogen product from the color of the hexane layer after extraction.

   Z_2 X_2 Y_2

   As a start, you first add the free halogens to your test tube, extract into hexane, and get the following results:

   Z_2 hexane layer is red; X_2 hexane layer is yellow; Y_2 hexane layer is blue.
   Next, you mix the following combinations of reagents and record the resulting color of the hexane in your test tubes, giving the results shown in the table below.

Reagent Combination	Resulting Color
$Y_2 + Z^-$	red
$X_2 + Y^-$	blue
$Z_2 + X^-$	Yellow

   Using the results of the tests given write the balanced reactions for each of the six possible combinations of X_2, Y_2, Z_2, X^-, Y^-, and Z^-.

2. Complete and balance the following reactions as a "molecular" equation. If no reaction occurs, write NR in place of giving products.

   $NaI + F_2 \rightarrow$

   Complete and balance the following reaction as a "net ionic" equation. If no reaction occurs, write NR in place of giving products.

   $NaBr + At_2 \rightarrow$

3. For each of the following solvents, determine if they are polar or non-polar. Explain your conclusions.

a. 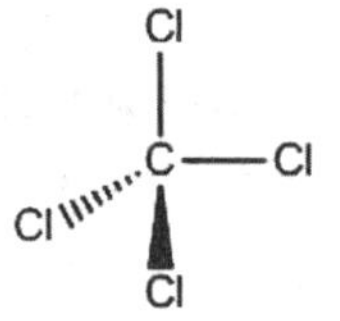 Cl, Cl, C, Cl, Cl

b. 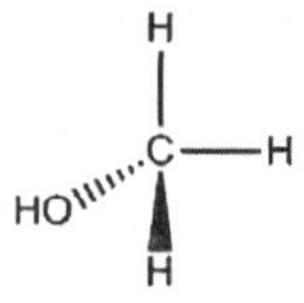 H, HO, C, H, H

Experiment 3

Laboratory Report

Name: ______________________ **Date:** ______________

Instructor: ______________________ **Sec. #:** __________

PURPOSE: (*The purpose should be several well-constructed sentences describing what your experiment was designed to accomplish and the criteria used to determine success. These sentences should include both concepts and techniques.)*

PROCEDURE: (*The procedure section should reference the lab manual and include any changes made to the procedure during the lab. Please note whether you tested potassium or sodium salts)*

DATA:

Observations:

Table 1: Sodium and Potassium Salt Solubility Tests

Salt	Soluble in Hexane (Yes/No)	Soluble in Water (Yes/No)	Observations
KCl			
KBr			
KI			
NaCl			
NaBr			
NaI			

Table 2: Halogen Solubility Tests

Halogen	Color of Hexane Layer	Color of Water Layer	Observations
Cl_2			
Br_2			
I_2			

Table 3: Halogen and Halide Reactions

Reaction	Color of Hexane Layer	Color of Water Layer	Observations
$Cl_2 + Br^-$			
$Cl_2 + I^-$			
$Br_2 + Cl^-$			

$Br_2 + I^-$			
$I_2 + Cl^-$			
$I_2 + Br^-$			

CALCULATIONS: (*The calculation section should have the six balanced equations for the reactions being studied based on your electronegativity data. If no reaction occurred write "no reaction" as the product.*)

CONCLUSION: (*The conclusion should be several well-developed paragraphs that include the following: a discussion of the results, including a list of the halogens in order of electronegativity, identifying the most electronegative, with support from the data section. This discussion should also include a discussion of the agreement between your results and the pre-lab predictions. An explanation of how you decided which reactions occurred and which did not should be included. A final discussion should focus on what the solubility tests indicated. Finally, include a discussion of any errors in the experiment.*)

Experiment 3

Post-Laboratory Questions

Name: ______________________________ **Date:** ___________________

Instructor: ______________________________ **Sec. #:** ____________

Show all work for full credit.

1) This experiment uses a qualitative colorimetric process. What is meant by this statement?

2) All of the halogens with the exception of fluorine make strong acids. Using electronegativity, explain why hydrofluoric acid is a weak acid.

Experiment 4

Paper Chromatography: M&M's® True Colors

Introduction

Of all the analytical techniques you will learn in this class, the most powerful will be chromatography. Chromatography is used to separate components in a mixture. In a real chemistry research lab, chromatography may be used to separate ions, proteins, DNA or a host of other molecules. It is an invaluable technique that regardless of style is based on the same general principles that you will be learning in this experiment.

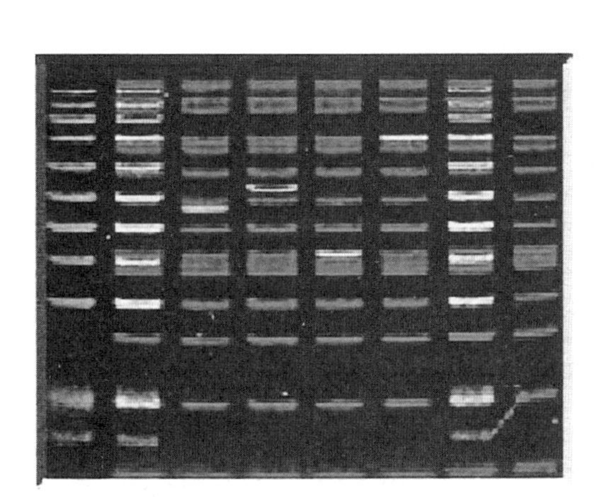

Figure 1a: Bands of DNA Separated by Chromatography

Figure 1b: Bands of Color Dyes Separated by Chromatography

Paper chromatography is a version of thin layer chromatography. Thin layer chromatography is often used in forensics to separate and identify inks and other dyes found in trace evidence. In this experiment, you will be using paper chromatography to identify the food color dyes Mars™ company uses to color their M&M candy shells.

Background

Separations

All separation techniques utilize differences in the chemical and physical properties of various components in the mixture to be separated. For example, filtration is used to separate solids and liquids. Specifically, the filter paper used in filtration allows the liquid to flow through it while trapping the solids. Another separation technique you will encounter several times in this lab is liquid-liquid extraction which utilizes the constituent's solubility properties. In this technique two immiscible solvents are used to distribute the components of a mixture between the two solvents according to their solubilities in those particular solvents. In Proof of Alcohol, the more non-polar ethanol migrates to the non-polar organic layer and in Halogen Reactions the covalent forms of the halogens migrate to the hexane layer of the separation while the ionic halides migrate to the polar water layer. In both of these examples you should notice that the polarity of the molecules plays a big role in the direction of their migration.

While these forms of separation are good they do share a common problem. The overall efficiency of these separations is not 100%. In filtration, separation efficiency is complicated by the solubility of the solid in the liquid. Likewise, in two-layer (liquid-liquid) solution separation, the components of the mixture are partially soluble in both layers. Therefore, in order to obtain an acceptable degree of separation using these techniques, it may be necessary to repeat the separation several times.

A more elegant approach is to use a differential technique in which the composition does not change in discrete steps, but rather continuously from the time the experiment begins. In other words, the degree of separation will then depend on the length of time that the mixture is processed.

Chromatography

Chromatography is probably the most powerful, and widely used, differential separation technique available today. Though many forms of the technique exist, every one of them involves the same basic principles. In these methods, a sample mixture migrates through a solid, porous material under the influence of fluid flow. The solid material, stationary phase, either provides a surface to which the sample material will adhere, or it might be coated with a liquid that can dissolve the sample material but is immiscible to the mobile phase. The mobile phase might be a liquid (liquid chromatography) or a gas (gas chromatography). The migration rate of each component in the material depends upon its relative affinity for the stationary or mobile phase, just as in liquid-liquid extraction the distribution of a substance depends on its relative affinity for the two solvents.

Individual components of the sample mixture will interact with the stationary and mobile phases to differing degrees, and as a result, will migrate through the system at varying rates. Specifically, components which have a strong affinity for the stationary phase but only a slight affinity for the mobile phase will move at a slower rate than those which have a weak affinity for the stationary phase and a large affinity for the mobile phase. The consequence of this division between stationary and mobile phases is a mixture that is separated into zones, or scientifically referred to as bands. Therefore, each band represents an individual component of the mixture.

In more detail, in paper chromatography, both the ***molecular weight*** and the ***polarity*** of the structures being separated play a role in the separation. The molecular weight, size and shape of the molecule cause its retention, while its affinity for the polar solvent produces its migration. Another phenomenon involved in the differing rates of migration is the formation of intermolecular bonds, a topic you will become all too familiar with next semester. All in all, the point at which one force overcomes the other is the point at which the dye will stop moving up the paper.

Rf

In order to use the results of a chromatography experiment a measurement of the bands needs to be completed. The Rf value is the height or length the component travelled or eluted from the starting point divided by the total length the mobile phase or solvent travelled.

$$R_f = \frac{Distance\ Compound\ Moved}{Distance\ Solvent\ Moved}$$

Comparison of the resulting Rf values of the bands of each separated component to the Rf value(s) of a reference allow the scientist to determine the components of an unknown sample.

Food Dyes

The Federal Drug Administration certifies only seven food coloring additives (other than those from natural sources, which do not require certification). They are the first seven dyes listed in the table below. Two additional dyes are certified for restricted use, Orange B in hot dogs and sausage casings, and Citrus Red No. 2, a carcinogen, for use on the skin of some Florida oranges (so don't eat orange peels).

Dye Name	Reflected Color	Absorbed Color
Blue No. 1	Royal Blue	Lt. Orange
Blue No. 2	Periwinkle Blue	Lt. Orange
Red No. 3	Red	Lt. Green
Red No. 40	Orange	Lt. Blue
Green No. 3	Blue-Green	Red-Orange
Yellow No. 5	Yellow	Purple
Yellow No. 6	Lt. Orange	Royal Blue
Orange B	Lt. Orange	Royal Blue
Citrus Red No. 2	Red	White

These nine dyes are complex organic molecules with extensive conjugated double bond systems where the electrons can be promoted to higher molecular orbital energy levels by light in the visible region. A dye with a particular color reflects light of that color, absorbing photons of light from the remainder of the visible spectrum. The table shows the approximate color of the dye and the corresponding color of the "complementary" light that is being absorbed.

Both the behavior of the dyes on chromatographic separation and the wavelength of visible light absorbed are determined by structural features of the dye molecule. The food colors readily available in most grocery stores are FD&C Yellow 5, FD&C Red 40, FD&C Blue 1, and FD&C Red 3. Let's examine some of the features of these dyes:

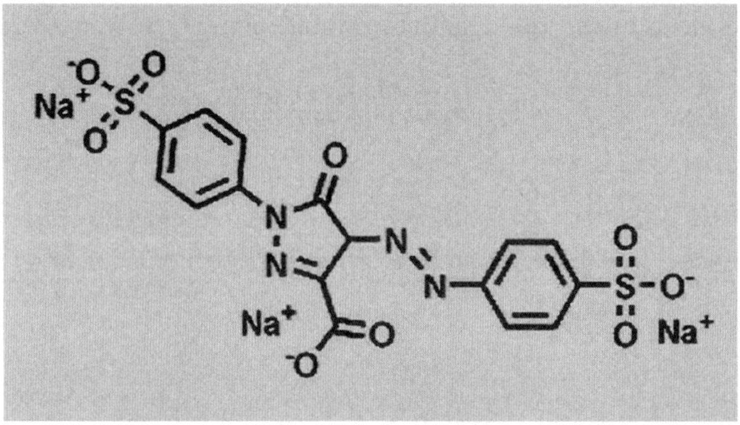

Dye Name: Yellow No. 5 Formula: $C_{16}H_9N_4Na_3O_9S_2$

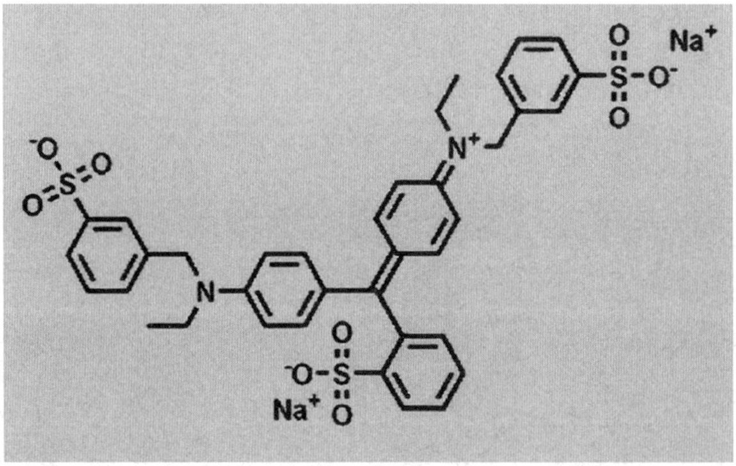

Dye Name: Blue No. 1 Formula: $C_{37}H_{34}N_2Na_2O_9S_3$

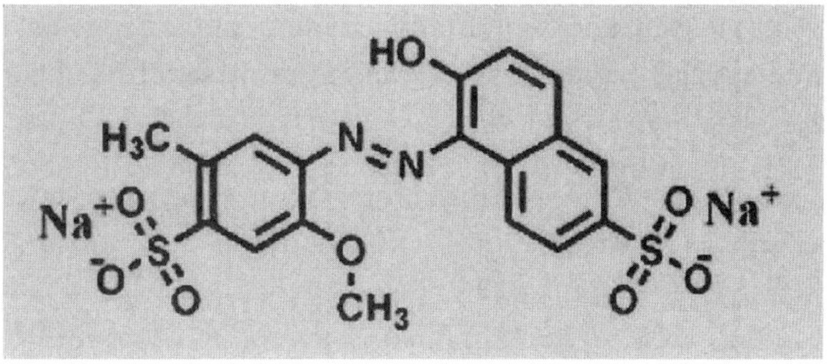

Dye Name: Red No. 40 Formula: $C_{18}H_{14}N_2Na_2O_8S_2$

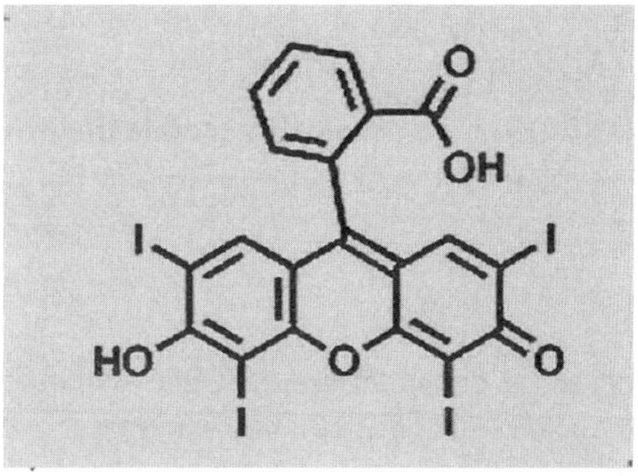

Dye Name: Red No. 3 Formula: $C_{20}H_8I_4O_5$

The Experiment

The mixtures being separated in this experiment are the dye components of 'standard' food colors. In order to resolve which dye components are part of each food color, a chromatography paper will be spotted with each of the food colors shown above. This technique is quite easy and reliable, provided that directions are followed carefully.

In addition to spotting the chromatography paper with the food color standards, you will also spot the paper with the dyes used to color the candy shell portion of M&M candies. The resulting bands of the standard food colors will be compared to the bands of the M&M candies in order to determine which dyes are used in their making.

A Couple of Hints

Don't put too much coloring on the chromatography paper, as doing so will overload the separation capability of the paper and the dyes will not separate well.

Insure that all the spots are exactly 1.5-cm from the bottom of the paper and higher than the solvent level in the developing beaker. If the solvent is above the spots, the dyes will leach out into the solvent.

Avoid touching or creasing the surface of the chromatography paper when you are handling it. Pick up the paper only by its edges!

Procedure

SAFETY NOTES: The solvent used in the chromatography tank generates significant fumes that can be corrosive to contact lens material. No contacts should be worn during this lab experiment. Persons with asthma should notify their TA so that they can work at the hood to avoid any irritation that the fumes might cause. Gloves should be worn when handling the chromatogram.

GENERAL INSTRUCTIONS: Each student will generate their own chromatogram.

Part I: Preparing the Chromatography Tank

Prepare the chromatography tank by pouring 40 mL of the solvent mixture (3:2 ethanol/water) into a clean 1000 mL beaker. Cover the top of the beaker with plastic wrap securely to allow the solvent to saturate the air in the beaker with solvent vapor.

Part II: Preparing the Chromatogram

Obtain from the instructor a pre-cut 11 x 20 cm sheet of chromatography paper. The width of the paper should be a little less than the depth of the chromatographic tank and the length should allow the formation of a cylinder that will fit into the tank with adequate clearance from the sides.

Avoid contamination of the paper by handling it only along the narrow ends, and use a sheet of clean paper to protect it from the bench.

Make a pencil line across one side 1.5 cm from the edge. In increments of 1.5 cm, mark off the line using a pencil. Label the marks in pencil with abbreviations of the color standards and M&M colors that will be used. See Figure 1 for clarification.

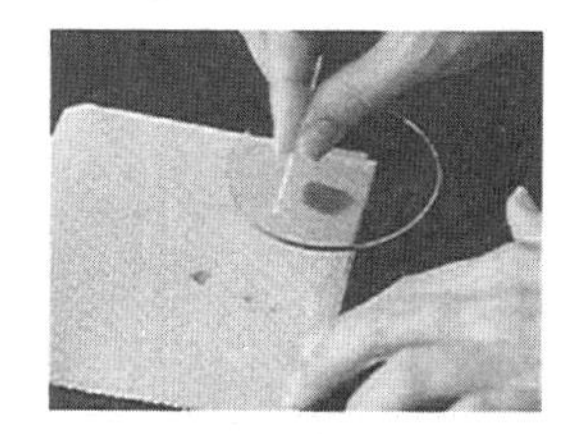

Spotting requires practice and patience. Obtain several toothpicks from the front counter. Using your large watch glass, place a drop of one of the food colors in the center of the glass. Use one of the toothpicks to practice making several spots on a scrap of filter paper. The spots should be no more than 2 mm in diameter. When you are confident you can make good spots, continue on to the next step.

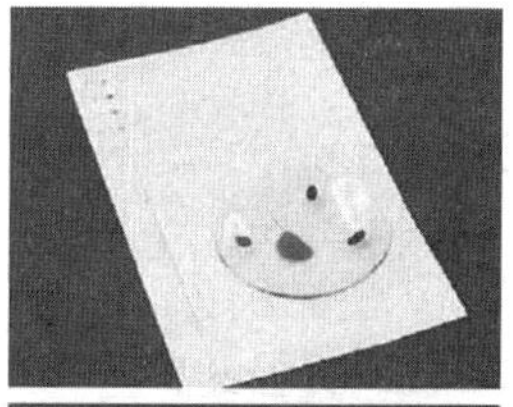

Carefully apply the four food colorings to your chromatography paper. Allow the spots to dry completely.

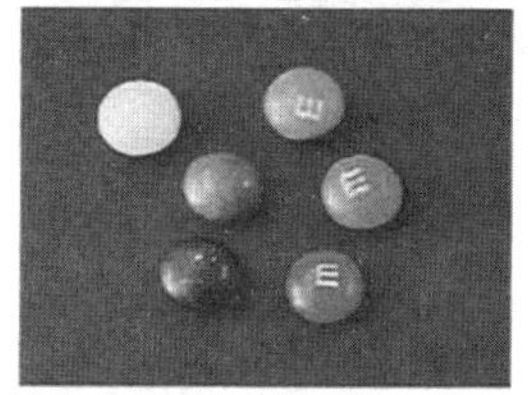

Obtain 1 M&M of each color from the front counter.

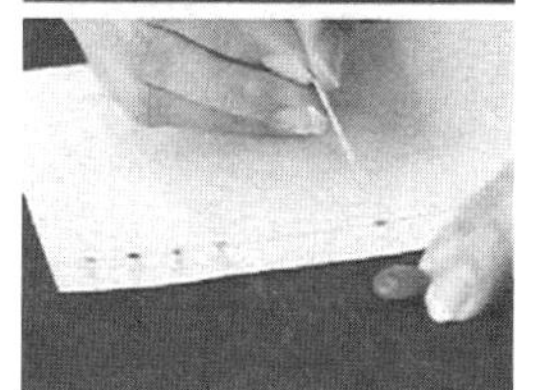

Dip a toothpick in DI water to moisten the tip and then rub it on one of the M&Ms until you see enough color transfer to spot the chromatography paper. Spot the paper as before. You may need to make several transfers to the same spot so that it is as dark as the standard food color spots. Repeat the process until each M&M color has a spot. (Return the used M&M's to the front counter for others to use.)

Part III: Running the Chromatogram

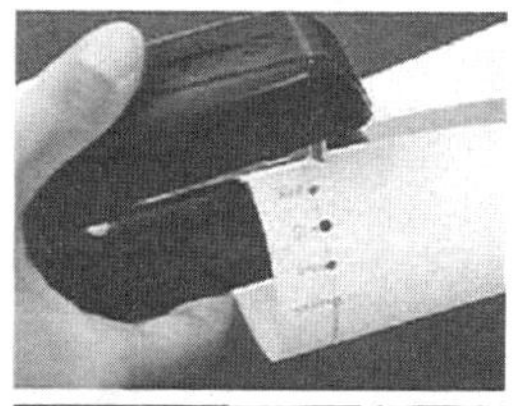

Coil the chromatography paper into a cylinder and fasten with three staples, making sure that the stapled ends do not touch or overlap as touching ends will cause an uneven flow of solvent up the paper.

Insert the cylinder into the beaker. Cover the beaker with the Saran wrap during the development. Record the time when development started. Make sure your set-up looks like the one pictured here.

Allow the beaker to remain undisturbed for one hour.

At this time you may leave the lab. Be prepared to return exactly 55 minutes after starting your chromatogram. You will need the five minutes to prepare to take the chromatographic tank apart and prepare a place to store the chromatogram while drying.

Part IV: Removing the Chromatogram

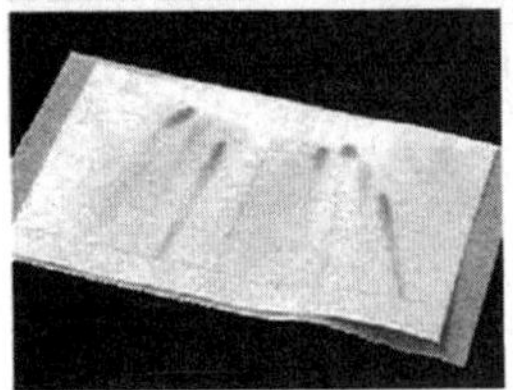

At the end of one hour, the solvent front should be about 2 cm from the top of the upper edge of the cylinder. Remove the cylinder and immediately mark the position of the solvent front with a pencil line. Allow the cylinder to dry standing on a clean piece of paper.

When dry, remove the staples, flatten the cylinder out and examine the chromatogram. Measure and record the distance each component in the food colorings traveled from the original spot 1.5 cm from the bottom. Record the distance the solvent front traveled beginning at the line 1.5 cm from the bottom.

Experiment # 4

Pre-Laboratory Assignment

Name: ______________________ **Date:** ______________

Instructor: ______________________ **Sec. #:** __________

Show all work for full credit.

1) **Below is a paper chromatogram comparing four standard dyes (in lanes 1-4, Red, Green, Blue, Yellow, respectively) with six dye mixtures (lanes 5-10). The position of the solvent front is noted by the wavy blue line. (A number of factors can affect the mobility of the solvent, including non-uniformity in the paper and temperature variations within the chamber while the chromatogram is being run. Ideally the front should be horizontal, but this problem illustrates a case where it is not.)**

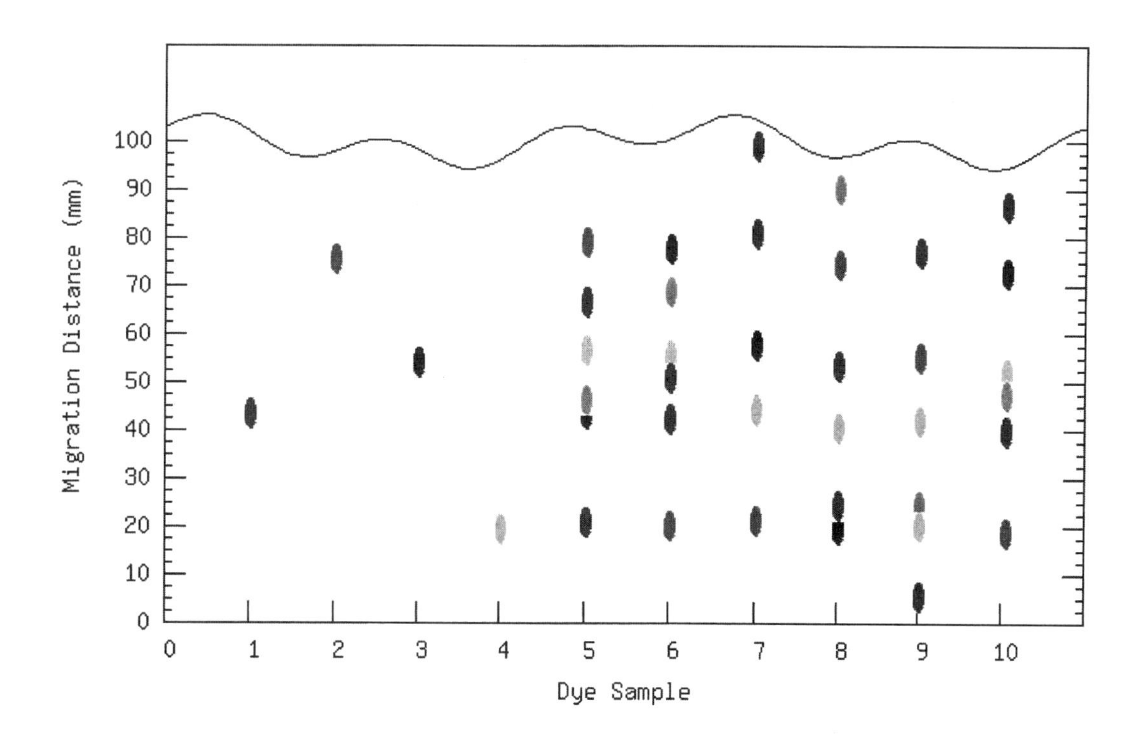

The relative mobility of substances in thin layer and paper chromatography is recorded as an Rf value. The Rf value is simply the ratio of the distance the substance has moved from the origin (measured to the front of the spot), divided by the distance the solvent front has moved. It is important to record the solvent front position before the chromatogram dries completely. When the solvent front is irregular, as shown in the example above, the solvent front distance should be measured above the spot. Identical substances should have the same Rf value even if they do not line up exactly on a horizontal line.

Measure the Rf values of the the four standard dyes in lanes 1-4 to 2 significant figures.

Rf of the red spot in lane 1: ________ **Rf of the blue spot in lane 3: ________**

Rf of the green spot in lane 2: _______ **Rf of the yellow spot in lane 4: ________**

Each of the dye mixtures in lanes 5 through 9 contains only one of the four standard dyes. Identify the standard dye in the mixture by comparing its Rf .

Which of the four standard dyes is contained in the mixture in lane 7? ________

Which of the four standard dyes is contained in the mixture in lane 10? ________

Which of the four standard dyes is contained in the mixture in lane 9? ________

Which of the four standard dyes is contained in the mixture in lane 5? ________

2) **For each of the dyes shown in the background, answer the following questions:**
    a) **What is the molecular weight? (Unit=g/mol)**

    b) **How many double bonds are present? (Include the S=O double bonds.)**

    c) **What is the total number of charges on Blue No. 1? (Do not count the Na+ counterions.)**

    d) **How many sulfonate (-SO_3^-) groups are there?**

e) Some dyes are classified as "azo" dyes because they have the azo group (-N=N-) as part of the structure. Indicate which of the dyes are azo dyes.

3) Using the information from 2) above and your knowledge of polarity, make a prediction of the rankings of the food dyes from largest to smallest Rf values and explain your reasoning. Copy your predictions into your lab notebook for later use.

Experiment 4

Laboratory Report

Name:______________________________ **Date:**__________________

Instructor: ______________________________ **Sec. #:** ____________

PURPOSE: *(The purpose should be several well-constructed sentences describing what your experiment was designed to accomplish and the criteria used to determine success. These sentences should include both concepts and techniques.)*

PROCEDURE: *(The procedure section should reference the lab manual and include any changes made to the procedure during the lab.)*

DATA:

Observations:

Experiment 4

Attach Chromatogram Here.

Dye Components Data

Food Color Bands	Reference #1 ______	Reference #2 ______	Reference #3 ______	Reference #4 ______
Rf/Color				
Rf/Color				
Rf/Color				
Rf/Color				
Rf/Color				
Rf/Color				
Rf/Color				

Food Color Bands	M&M Color ______	M&M Color ______	M&M Color ______	M&M Color ______	M&M Color ______	M&M Color ______	M&M Color ______
Rf/Color							
Rf/Color							
Rf/Color							
Rf/Color							

Rf/Color							
Rf/Color							
Rf/Color							

CALCULATIONS: (*The calculation section should include a sample calculation for determining the Rf values of the food colors.)*

CONCLUSION: (*The conclusion section should be several paragraphs addressing the following: a) the identity of the dye components of each of the M&M candy shells, b) a supported discussion of the identity of the dye components based on the food coloring Rf values (be sure to use Rf values to support your conclusions), c) a discussion of any dyes that do not appear to be any of the four food coloring colors tested (make sure to use the Rf values to support this discussion also), d) an explanation of why the components moved up the paper at different rates and e) any errors in the experiment.)*

____

____

____

____

____

____

Experiment 4
Post-Laboratory Questions

Name: ______________________________ **Date:** __________________
Instructor: ______________________________ **Sec. #:** ____________

Show all work for full credit.

1) Using the internet or a chemistry textbook find at least three other types of chromatography techniques and list them.

2) Explain the role of the solvent in this experiment. Why was a 3:2 ethanol water mixture used?

3) Now that you have completed the experiment, compare your actual results to the predictions you made regarding the Rf values of the food dyes. Explain any discrepancies.

Experiment 5

Building Molecular Models

Introduction

In any subject there is always a language to learn. In chemistry the language uses elements as the alphabet and compounds as the words. Reactions will eventually be the sentences, but for now, you will need to learn the proper way to write the words, and then to develop a mental picture of the structures represented by these words. The use of Lewis structures gives us a step-by-step way to construct molecular models of any species: covalent or ionic. Lewis structures are based on the idea that every atom is most stable when its electron configuration matches that of the nearest noble gas. This is often called "the octet rule" because many common atoms are most stable when they have eight electrons in their outermost, or valence, shell.

Because Lewis structures are conceptualized at the molecular level, this experiment will be used as a workshop to practice drawing and otherwise visualizing molecular structures. In the lab itself you will complete a worksheet that requires you to draw and analyze models of several inorganic compounds. Additionally, you will use a protractor to determine the bond angles of the resulting molecular structures.

Background

The premise behind Lewis structures is the octet rule: that all atoms would like to be surrounded with an octet of electrons. Of course, there are some exceptions: very small atoms (H, Be and B) have less than an octet, and some main group atoms with low energy d orbitals (P, S, Cl, Br, and I) may have more than an octet. This is especially true when these atoms are central atoms and combined with highly electronegative atoms.

Drawing correct Lewis structures takes practice but the process can be simplified by following a series of steps:

Step 1. Count all the valence electrons for each atom. Add or subtract electrons if the structure is an anion or cation, respectively.

Step 2. Determine which atoms are bonded to one another. Draw a skeletal structure.

Step 3. Connect the atoms with a pair of electrons in each bond. Subtract the bonding electrons from the total valence electrons.

Step 4. Add electron pairs to complete octets for all peripheral atoms attached to the central atom. Beware of hydrogen – hydrogen ***never*** has more than one bond or one pair of electrons.

Step 5. Place remaining electrons on the central atom, usually in pairs. The octet rule may be exceeded for P, S, Cl, Br, or I.

Step 6. If the central atom does not have an octet, form double or triple bonds by moving electron pairs from one or more peripheral atoms to achieve an octet.

Step 7. Look for resonance structures by rearranging bonds. The structure with the lowest total formal charges will be the most likely form to be found in nature. (See below for explanation.)

Drawing Lewis Structures

Let's look at an example of how this works using a real molecule. Consider the molecule most responsible for the greenhouse effect, carbon dioxide (CO_2).

To draw the Lewis structure:

Step 1. Count all the valence electrons for each atom:

Carbon 1 x 4 valence electrons = 4 electrons
Oxygen 2 x 6 valence electrons = 12 electrons
Total = 16 electrons

Step 2. Determine which atoms are bonded to one another. Generally the least electronegative atom is the central atom. However, if the only choice is between a more electronegative atom and hydrogen, the more electronegative atom will be the central atom (e.g., water). Hydrogen NEVER makes more than one bond and thus can NEVER be the central atom.

For CO_2, carbon is the less electronegative atom so it should be the central atom.

O:C:O

Step 3. Connect each atom with a single pair of electrons or single bond: (16 valence electrons – 4 bonding electrons = 12 electrons left.)

O-C-O

Step 4. Add electron pairs to peripheral atoms for octets:

:Ö—C—Ö:

Step 5. No electrons are left over, but the central atom doesn't have an octet!

Step 6. Move electrons from peripheral atoms, forming double bonds to give the central atom an octet:

:Ö=C=Ö:

Step 7. Look for resonance structures and identify the one with the smallest formal charges:

:Ö=C=Ö: ⇔ :Ö—C≡O: ⇔ :O≡C—Ö:

0 0 0 +1 0 -1 -1 0 +1

For some molecules, more than one structure can be drawn. Note that a Lewis structure for carbon dioxide can be written using a carbon-oxygen single bond on one side, and carbon-oxygen triple bond on the other. How can these two possibilities be distinguished? How can the most important structure be chosen, or are they all equally likely? When several structures can be drawn, they are called resonance structures.

Resonance Structures

In resonance structures, all the atoms are in the same relative position to one another, but the distribution of electrons around them is different. To evaluate the importance of each structure, the formal charge on each atom must be determined.

Formal Charge

Formal charge is a somewhat arbitrary way of describing how many electrons an atom seems to have in a particular compound. Electron pairs in bonds between atoms are assumed to be split equally between the two atoms. Non-bonding electron pairs are counted as belonging to the atom on which they reside. This can be put into an equation:

$$\textbf{Formal Charge} = \textbf{\#Valence } e^{-}\textbf{s} - (\textbf{\#non-bonding } e^{-} + \tfrac{1}{2}\textbf{ bonding } e^{-})$$

or

$$\textbf{Formal Charge} = \textbf{\#Valence } e^{-}\textbf{s} - (\textbf{\#non-bonding } e^{-} + \textbf{\# of Bonds})$$

The most stable resonance structure is the one in which:

1. There is a minimum number of formal charges;
2. If there are formal charges, like charges are separated; and
3. Negative formal charges are on the more electronegative atoms and positive formal charges are on the less electronegative atoms.

For the CO_2 structure with two double bonds, the formal charges can be calculated as follows:

Oxygens: Formal Charge = 6 - (4 + 1/2(4)) = 0
Carbon: Formal Charge = 4 - (0+ 1/2(8)) = 0

For the CO_2 structure with a single and triple bond:

Oxygen (single): Formal Charge = 6 - (6 + 1/2(2)) = -1
Oxygen (triple): Formal Charge = 6 - (4 + 1/2(6)) = +1
Carbon: Formal Charge = 4 - (0+ 1/2(8)) = 0

So, while both structures work as Lewis structures, the one which results in zero formal charges for any of the atoms is more stable and thus more likely to exist in nature than the one having charges on the two oxygen atoms.

Oxidation Numbers

Formal charges need to be distinguished from oxidation numbers (which can also be determined from Lewis structures). Oxidation numbers are used to indicate whether a molecule is neutral, electron-rich or electron-poor. The rules for determining oxidation numbers are found in your textbook. A short summary of these rules is given here:

1. The oxidation number for an element in its elemental form is 0 (holds true for isolated atoms and for molecular elements, e.g., Cl_2 and P_4).

2. The oxidation number of a monatomic ion is the same as its charge (e.g., oxidation number of $Na^+ = +1$, and that of S^{2-} is -2).

3. In binary compounds the element with greater electronegativity is assigned a negative oxidation number equal to its charge if found in simple ionic compounds (e.g., in the compound PCl_3 the chlorine is more electronegative than the phosphorus. In simple ionic compounds Cl has an ionic charge of 1-, so its oxidation state in PCl_3 is -1.)

4. The sum of the oxidation numbers is zero for an electrically neutral compound and equals the overall charge for an ionic species.

5. Alkali metals exhibit only an oxidation state of +1 in compounds.

6. Alkaline earth metals exhibit only an oxidation state of +2 in compounds.

Once Lewis structures are drawn successfully, they can be used to predict the electron cloud geometry, molecular shape and polarity of molecules and ions. For a thorough discussion, refer to your textbook. In particular, look at the three-dimensional representations for all the geometries and shapes.

Electron Cloud Geometry and Molecular Geometry

The electron cloud geometry around a central atom is determined by the number of electron groups surrounding it. Each set (2, 3, 4, 5, and 6) has a different name and arrangement in three-dimensional space. Electron clouds, all being negative, are most stable when separated as far from one another as possible. This is called the valence shell electron pair repulsion theory (VSEPR). While electron cloud geometry describes the orientation of the electrons around the central atom, the molecular geometry describes the arrangement of peripheral atoms.

The Experiment

In the lab you will be presented with six molecular models as unknowns. It will be your job to name them. You will also be asked to determine their electron pair and molecular geometries by measuring their bond angles using a protractor. A worksheet is provided containing other questions that should be completed for each of the molecules. You should make five additional copies of the worksheet to use during class. These worksheets will then be used as the data section of your lab report.

Procedure

SAFETY NOTES: Since no chemicals will be used today, goggles are not required.

GENERAL INSTRUCTIONS: Each student works independently in this lab. Print out 6 worksheets to use for data collection during the experiment.

Part I: Identifying Lewis Structures

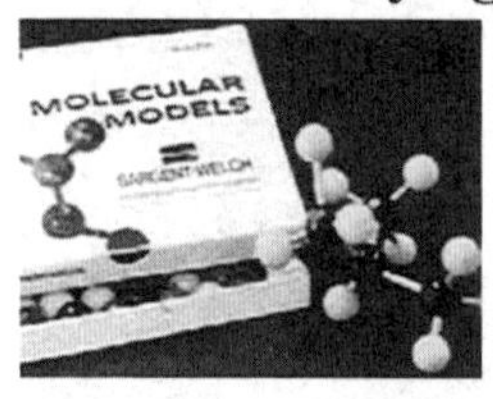

Draw a good Lewis Structure for each of the six molecular models present.

Part II: Analyzing Models

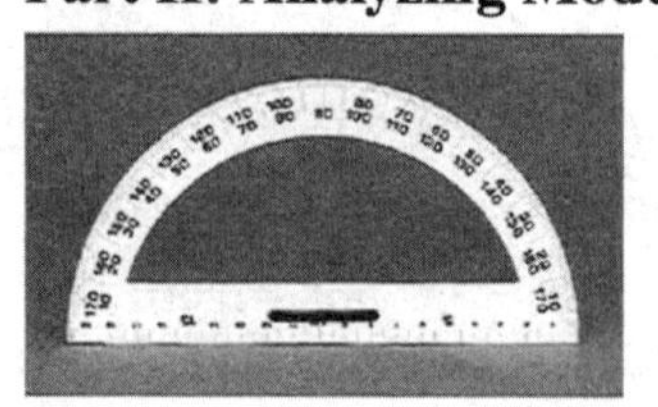

Collect a protractor from the front counter.

Use the protractor to measure the angles between the bonds of the molecular models and record them on your worksheet.

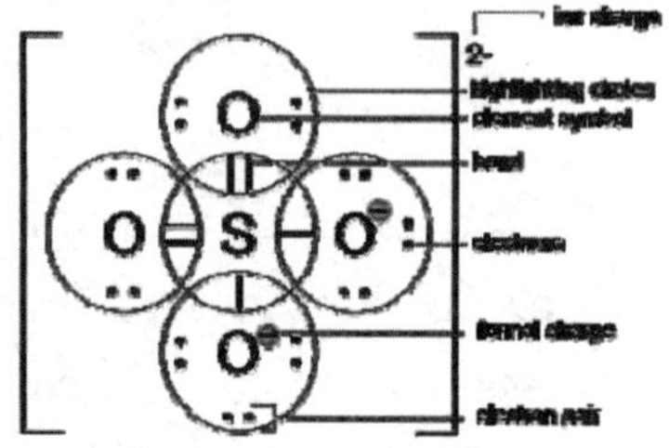

For each Lewis Structure you have drawn, answer the remaining questions posed by the worksheet.

Molecular Structure Worksheet

Unknown ID: ______________
Formula: ______________
MW: ________________

Atom Identity	Ox. Number	Formal Charge

Bond Angles:

Angle #	Identity Ex. C-H	Measured Angle
1		
2		
3		
4		
5		
6		
7		

Molecule Name	
Resonance (Y/N)	
Electron Cloud Geometry	
Shape	
Polar (Y/N)	

Experiment 5

Experiment # 5

Pre-Laboratory Assignment

Name: ______________________ **Date:** ____________

Instructor: ______________________ **Sec. #:** ________

Show all work for full credit. FC = # val e⁻'s − (# non-bonding + (½) bonding e⁻'s)

1) Make six or more copies of the worksheet provided at the end of the procedure section and bring them with you to lab.

2) Sulfur is a Period 3 element and as such can create structures that exceed the octet rule. Because of this ability, there are 2 resonance structures for the sulfuric acid molecule. a) Draw the two possible structures; b) Calculate the formal charges; and c) Determine the best structure and explain your choice. H_2SO_4

H − O − S(=O)(=O) − O − H ⟷ H − O − S(−O)(=O) − O − H

S FC = 6 − 0 − (½) 12 = 0

H FC = 1 − 0 − (½) 2 = 0

O between S & H FC = 6 − 4 − (½)4) = 0

=O FC = 6 − 4 − (½)4 = 0

net FC = 0

stable

S FC = 6 − 0 − (½)(10) = 1

H FC = 1 − 0 − (½) 2 = 0

O between S & H same as ← = 0

=O same as ← = 0

:Ö: FC = 6 − 1 − (½)(2) = 4

net FC = 5

not stable

3) For each of the molecules listed below, complete the following:
   a. Calculate the number of valence electrons
   b. Draw a good Lewis structure and any resonance structures
   c. Calculate the formal charges for each atom
   d. Calculate the oxidation numbers for each atom
   e. Determine the electron cloud and molecular geometries for the central atom(s)
   f. List the bond angles
   g. Determine if the structure is polar or non-polar

Molecules: Dichromate ion, Iron II Hydroxide, Disulfur Decafluoride

Experiment 5

Laboratory Report

Name: ______________________ **Date:** ______________

Instructor: ______________________ **Sec. #:** __________

PURPOSE: (*The purpose should be several well-constructed sentences describing what your experiment was designed to accomplish and the criteria used to determine success. These sentences should include both concepts and techniques.*)

PROCEDURE: (*The procedure section should reference the lab manual and include any changes made to the procedure during the lab. Please note whether you tested potassium or sodium salts.*)

DATA:

Observations:

****Input 6 worksheets here****

CALCULATIONS: (The calculation section should include sample calculations for determining the formal charges and molecular weight (MW) for each molecule.)

CONCLUSION: (*The conclusion section should be several well-developed paragraphs and should cover the following information: a) identification of each of the molecular model unknowns; b) a discussion of the polarity of each molecule and how it was determined; and c) a discussion of which molecules have resonance forms and why.)*

Experiment 5

Experiment 5
Post-Laboratory Questions

Name: ______________________________ **Date:** ___________________
Instructor: ______________________________ **Sec. #:** ____________

Show all work for full credit.

1) What physical forces create the three-dimensional structure of a molecule?

2) Now that you know more about molecular structure, discuss how the structure of a molecule influences its polarity.

3) Oxidation numbers and formal charges are often confused by first year chemistry students. If you were teaching the course, explain how you would differentiate the two types of values so that the concepts would be clear to your students:

Experiment 6

Limiting and Excess Reagents

Introduction

In this experiment we expand our knowledge of stoichiometry by working with a reaction between copper(II) nitrate and potassium iodide, a reaction that is not a one-to-one reaction.

$$2Cu(NO_3)_2(aq) + 4KI(aq) \rightarrow 2CuI(s)\downarrow + I_2(aq) + 4KNO_3(aq)$$

In addition to stretching our ability to calculate product and reactant amounts using reaction stoichiometry, we will also be exploring the concept of limiting and excess reagents. It is not always obvious from gram amounts which reactant will run out first when a reaction occurs. Two factors complicate the prediction: 1) chemical equations are balanced in moles not grams; and 2) moles are based on molecular weights. Thus 100 grams of O_2 is a lot fewer moles than 100 grams of H_2.

In addition to practicing stoichiometry, we will also learn about two new devices: the centrifuge and the dessicator.

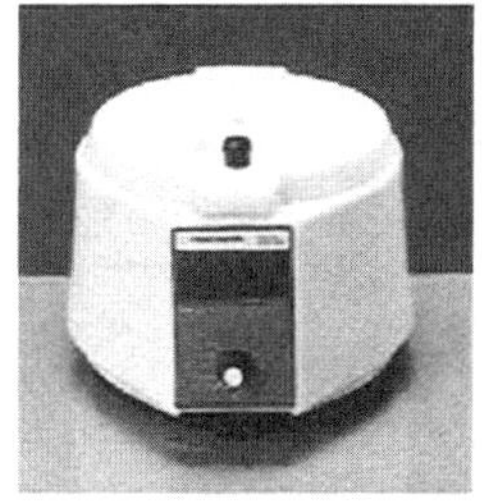

The centrifuge is a device that uses centrifugal force to separate solids (often called precipitates) from solution.

A desiccator is a piece of laboratory equipment used to protect chemicals from moisture. Dessicators use compounds called dessicants to remove moisture from the air. A hydrate is a compound that has water molecules incorporated into its crystal structure. A desiccant is a hydrate that has been stripped of all its water molecules, making it very "hungry" for water. If a desiccant is placed in a sealed container, the container is then called a dessicator, since anything placed inside it will be stripped of its moisture by the desiccant.

Background

Many of the concepts used in this experiment, including limiting and excess reagents, moles, and theoretical yields are covered extensively in your textbook. Since the treatment is extensive, another deep discussion here would be redundant, so we will just hit the highlights.

Limiting and Excess Reagents

The concept of limiting and excess reagents deals with how much product results when two or more reactants are mixed. This concept has very practical applications in the real world.

For example, in the business of manufacturing a chemical product of some kind, it is important to do it as efficiently and with as little waste as possible. This is also true in the laboratory, where you want to use only the amounts of reactants necessary to produce the largest amount of product, as anything more would be simply wasted.

Consider a simple reaction where one mole of reactant A reacts with one mole of reactant B:

$$A + B \rightarrow C$$

What happens if too much reactant A is added? When all of reactant B is used up, there will be some reactant A left over. In scientific terms, we say that reactant B is the limiting reagent. In other words, no matter how much A is added, no more product is made when B is consumed—reactant B limits how much product is obtained. On the other hand, reactant A is called the excess reagent because there is more than enough of reagent A.

In the reaction above, if we add more of the limiting reagent, reactant B, more of product C is formed. Why, you may ask? Well, reactant B limits how much product is obtained, so when more B is added, the reaction will resume until one of the reactants is used up again. Therefore, the final question is “What is the most efficient mixture of these two reactants?” In fact, the best mixture consists of correctly matched amounts of reagent A and reagent B that would allow for both reactants to be completely used. Note that unless the mole ratios are one to one, “correctly matched” does not mean “evenly matched.”

Finding the Limiting Reagent

For practice, let’s look at the following example and see if we can determine which reagent is limiting and which is in excess. Let's assume we have time-warped to the early 1970s where we all are employed by NASA as rocket scientists. We are working with a fuel mixture composed of dinitrogen tetraoxide (N_2O_4) and hydrazine (N_2H_4). These two reagents react to produce nitrogen gas (N_2) and water vapor (H_2O), as shown below. We have been assigned to figure out which reactant is limiting when 1.40 kg of N_2H_4 and 2.80 kg of N_2O_4 are allowed to react.

$$N_2H_4(l) + N_2O_4(l) \rightarrow N_2(g) + H_2O(g)$$

The first step is to balance the equation:

$$2N_2H_4(l) + N_2O_4(l) \rightarrow 2N_2(g) + 4H_2O(g)$$

Next, we have to use stoichiometry to see how many moles of one of the products are produced based on the initial amounts of each reactant. Let’s find the number of moles of H_2O.

Based on moles of Hydrazine (N_2H_4)

$$1.40\text{kg } N_2H_4 \times \left(\frac{1000\text{ g}}{1\text{ kg}}\right) \times \left(\frac{1\text{ mol } N_2H_4}{32.06\text{ g } N_2H_4}\right) \times \left(\frac{4\text{ mol } H_2O}{2\text{ mol } N_2H_4}\right) = 87.3\text{ mol } H_2O$$

Based on moles of Dinitrogen Tetraoxide (N_2O_4)

$$2.80\text{kg } N_2O_4 \times \left(\frac{1000\text{ g}}{1\text{ kg}}\right) \times \left(\frac{1\text{ mol } N_2O_4}{92.02\text{ g } N_2O_4}\right) \times \left(\frac{4\text{ mol } H_2O}{1\text{ mol } N_2O_4}\right) = 122\text{ mol } H_2O$$

Since hydrazine produces fewer moles of water, it must be the limiting reagent!

As you can now see, it is very important to calculate exactly how much product can be made based on the stoichiometry of each and every reactant. Another point to note is that in any reaction there is always a portion of the product that is lost to human error, incomplete mixing of reagents, etc. Therefore it is important in any chemical preparation to take this loss into account so that enough product is made.

Precipitation

In this experiment, the reaction of the two solutions, copper(II) nitrate, $Cu(NO_3)_2$, and potassium iodide, KI, forms a precipitate. A precipitate is a solid formed during a reaction between two aqueous compounds. (This is the same word meteorologists use when talking about falling rain or snow.) In order to better separate the solid and liquid phases, test tubes containing precipitates are placed in a centrifuge. By spinning them around at an extremely high velocity, the centrifuge forces the heavier precipitate to the bottom of the test tube. After centrifugation, a solution that was cloudy has become clear with a solid accumulated at the bottom of the tube.

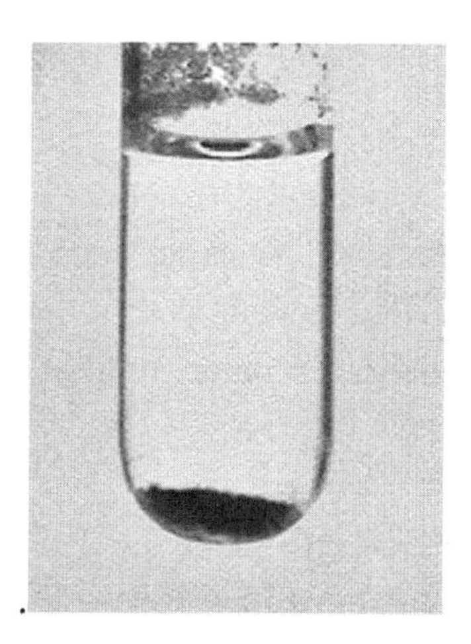

The clear (but not necessarily colorless) liquid found above the precipitate is called the supernatant. In this experiment, we will test the supernatant to see which reactant remains after the reaction is complete. The reactant that is present in the supernatant when the reaction stops is obviously in excess, while the reactant not detected in the supernatant was completely used and is in fact our limiting reagent. This can be verified by adding more of the two solutions we are studying. Using our example earlier, if more solution A is added and additional precipitate forms, then reagent A is the limiting reagent. Similarly, if more solution B is added and additional precipitate forms, then reagent B is the limiting reagent.

The Reaction in the Experiment

Specifically, this experiment involves mixing aqueous solutions of $Cu(NO_3)_2$ and KI, which will react as shown:

$$2Cu(NO_3)_2(aq) + 4KI(aq) \rightarrow 2CuI(s)\downarrow + I_2(aq) + 4KNO_3(aq)$$

Since we are dealing with solutions, the concentration is expressed in molarity, a term used in our Conservation of Matter experiment. Recall that molarity (M) is defined as the number of moles of solute in one liter of solution. If we know the volume of solution we can then calculate the number of moles present by multiplying the volume in liters by the molarity of the solution:

$$L_{Solution} \times \frac{\text{moles}}{L_{Solution}} = \text{moles}$$

A reaction between 0.50-M $Cu(NO_3)_2$ and 0.50-M KI now has a little more meaning to us. In the experiment, each student will be assigned different volumes of reactants to investigate. Based on the volume of reactants assigned, a theoretical amount of product can be calculated. After the solutions react, the precipitate is recovered and its mass determined. The final step in the process is to evaluate the overall efficiency of the reaction by reporting the percent yield of the reaction based on the experimental results.

Procedure

SAFETY NOTES: The supernatant in this experiment contains I_2. The I_2 will leave a yellow stain on your skin if you come in contact with it. I_2 will also stain your clothes, so be careful. ***If you know you are allergic to iodine, notify your instructor immediately so that further precautions may be taken.***

GENERAL INSTRUCTIONS: In order to make the most of time in the lab, the class will pool all results. Be sure to fill in the table of the class results when you are finished your experiment. Since some students will finish earlier than others, bring something to read or study along with you to class so that you will not waste valuable time while waiting for the rest of the class to finish. You must have these results in order to complete the lab report.

****NOTE** The pictures below are for demonstration purposes only and may show improper technique with respect to handling of the test tubes in order to produce a clearer pictures. You, however, should only handle weighed test tubes with Kimwipes or clamps.**

Part I: The Precipitation Reaction

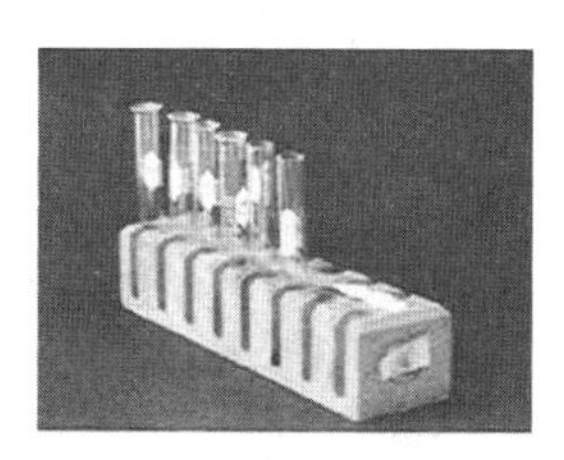

Collect six small test tubes (clean and dry) as well as clamps or tongs. Label the test tubes 1, 1a, 1b, 2, 2a, and 2b. The test tubes marked 1 and 2 should be Pyrex since they will be heated later in the experiment. Wipe off any fingerprints or other smudges from tubes 1 and 2, and then weigh them carefully to the nearest 0.0001g. Do not touch them with your fingers from now on. Test tubes 1a, 1b, 2a and 2b will not be weighed and can be touched as needed.

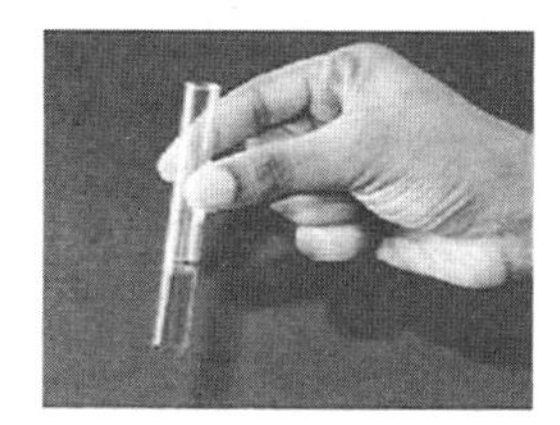

Each student in the lab section will be assigned certain volumes of the 0.50M $Cu(NO_3)_2$ and 0.50M KI. Tubes 1 and 2 will have identical mixtures so that if any error is made there is a duplicate. Add the assigned volume of $Cu(NO_3)_2$ to the assigned volume of KI in tubes 1 and 2. Be sure to record any observations of the reaction in your lab notebook.

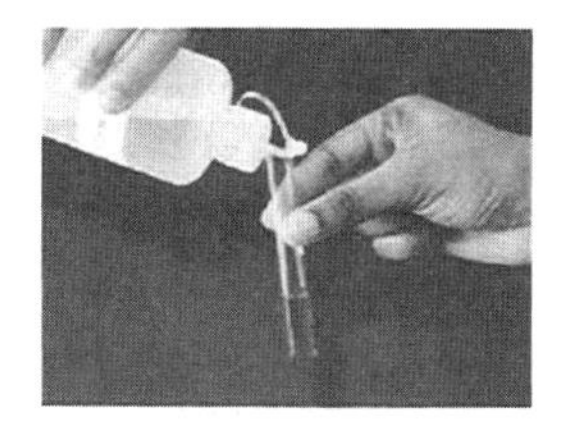

Holding the top of the test tube gently but firmly between the thumb and forefinger of one hand, tap the bottom of the test tube gently with the forefinger of your other hand to thoroughly mix the chemicals. Set the tubes in the test tube rack for five minutes to allow the reaction to go to completion. Remember not to touch test tubes 1 and 2 with your hands. Use a clamp, tongs or a Kimwipe.

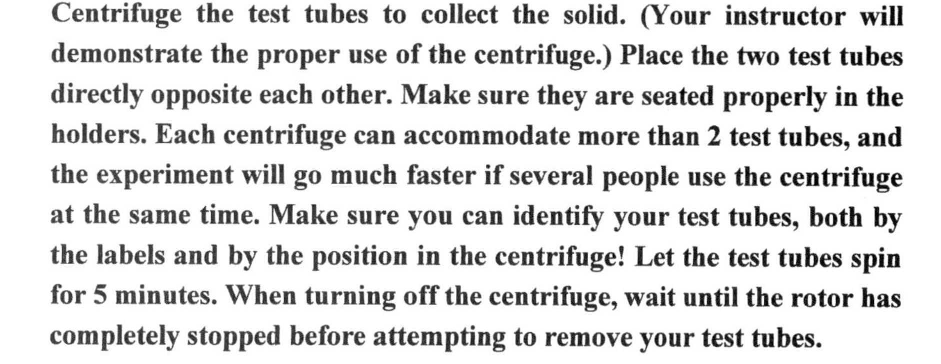

Centrifuge the test tubes to collect the solid. (Your instructor will demonstrate the proper use of the centrifuge.) Place the two test tubes directly opposite each other. Make sure they are seated properly in the holders. Each centrifuge can accommodate more than 2 test tubes, and the experiment will go much faster if several people use the centrifuge at the same time. Make sure you can identify your test tubes, both by the labels and by the position in the centrifuge! Let the test tubes spin for 5 minutes. When turning off the centrifuge, wait until the rotor has completely stopped before attempting to remove your test tubes.

*****Note: Be aware of any unusually loud noises or "walking" of the centrifuge, which might indicate that the tubes are off balance. If this occurs turn it off immediately.*****

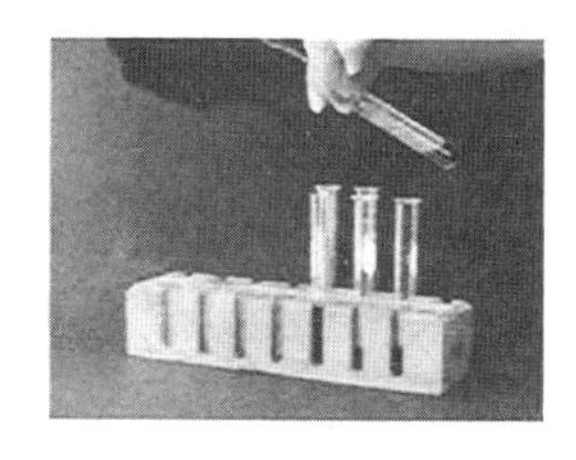

Remove the test tubes, being careful not to touch them with your fingers. They should contain a clear brown-green colored liquid (supernatant) and a gray solid at the bottom (precipitate). Put the test tubes in the test tube rack.

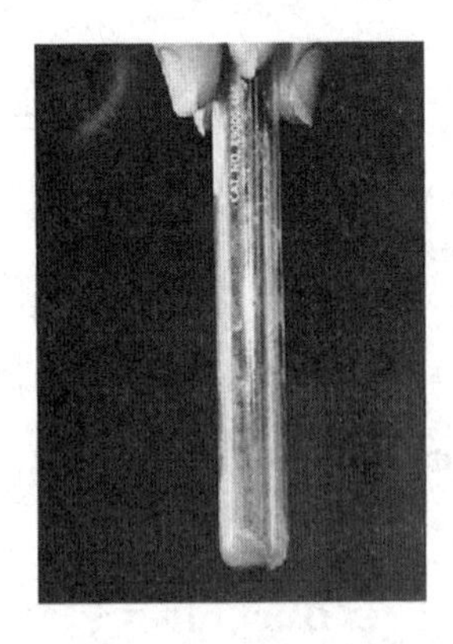

The next step is to draw off the supernatant, leaving the precipitate in the test tube. It is not essential that every last drop of supernatant is removed from the test tube, but it is important that the precipitate be undisturbed, so follow these directions carefully. Obtain two disposable pipettes. Squeeze the air out of the pipette before putting it into the liquid, and then carefully draw up the supernatant.

Split the supernatant from test tube 1 between the test tubes labeled 1a and 1b. Discard the pipette in the sharps disposal box at the front of the room. Use the second pipette for test tube 2 and follow the same procedure.

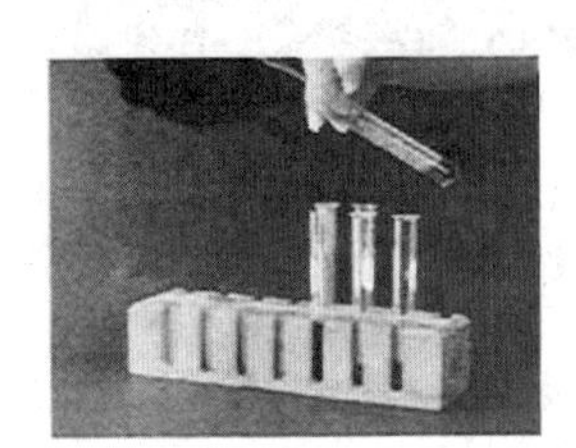

Test tubes 1a, 1b, 2a and 2b contain an excess of either $Cu(NO_3)_2$ or KI. The supernatants will be tested later so don't throw them away. Test tubes 1 and 2 contain the precipitates. This precipitate is wet and must be dried before it can be weighed. Any unreacted reagents that may be trapped must also be removed.

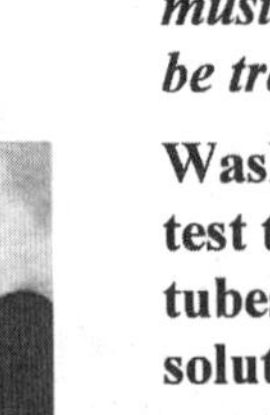

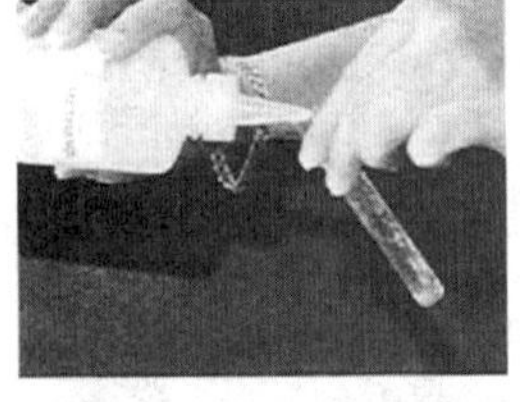

Wash the precipitate by adding approximately 3 mL methanol to each test tube. It is important that the same amount be added to both test tubes. Gently shake the tubes so that the precipitate starts to cloud the solution. Remember not to touch the tubes with your fingers. Also, be careful not to spill any methanol on the outside of the test tubes because it will dissolve the marker.

Spin the test tubes in the centrifuge for 5 minutes. The supernatant should be clear once again.

Without disturbing the precipitate, remove the supernatant with a disposable pipette. Since this supernatant will not be tested, you can use the same pipette for both test tubes. Discard the liquid in the proper container.

To dry the precipitate, place test tubes 1 and 2 in a 100 mL beaker. Write your name on a label or a piece of paper and place it on the inside of the beaker. Put the beaker and test tubes into the drying oven for at least 20 minutes. If there is more time left in the period, leave them in longer, keeping in mind that the test tubes must be cooled before weighing.

*****Note: Methanol is used to wash the precipitate rather than water because methanol is much more volatile. This means that it evaporates very quickly, making the precipitates dry much faster. Methanol is also flammable, so no flames in the lab today!*****

Part II: Testing the Supernatants

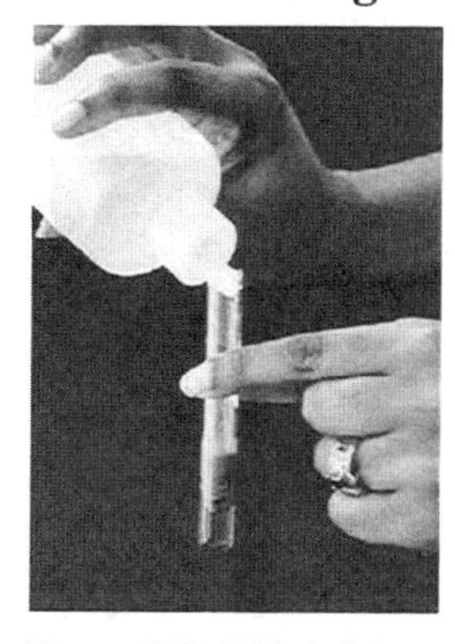

While the test tubes are in the oven, test the supernatants. Add 1 mL of the $Cu(NO_3)_2$ solution to test tubes 1a and 2a. Record any observations.

Now add 1 mL of the KI solution to test tubes 1b and 2b and record observations.

Part III: Weighing the Precipitates

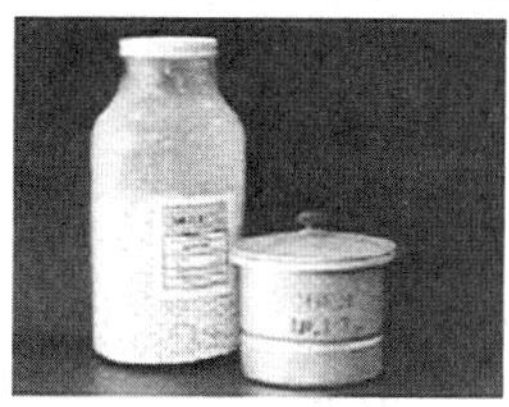

After the test tubes have been in the oven at least 20 minutes, remove them and place the test tubes in a desiccator to cool for at least 10 minutes. Remember not to touch the test tubes with your fingers (the beaker will be hot anyway).

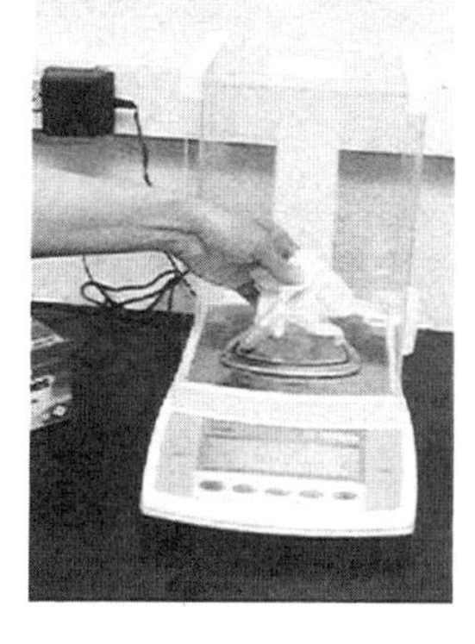

Weigh each test tube to the nearest 0.0001g.

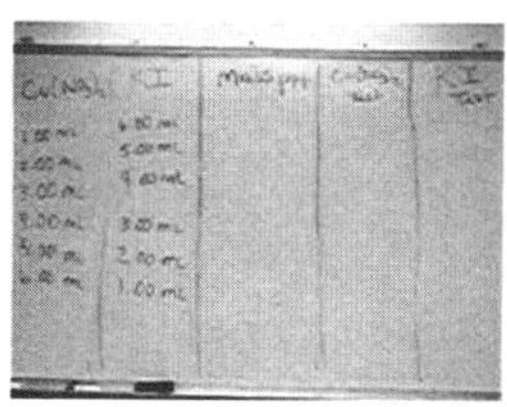

Calculate the mass of precipitate for each test tube. Give your instructor the results to post for the whole class. Make sure you have a copy of everyone's results before leaving the lab.

Be sure that precipitates, supernates, and methanol wash are all disposed of in the proper containers.

Experiment 6

Pre-Laboratory Assignment

Name:______________________________ **Date:**__________________

Instructor: ______________________________ **Sec. #:** ___________

Show all work for full credit.

Your laboratory experiment on "Limiting and Excess Reagents" involves the reaction between aqueous solutions of $Cu(NO_3)_2$ and KI to form CuI and I_2. The CuI precipitates from the reaction as a solid, and the other products (KNO_3 and I_2) remain in aqueous solution. In the lab you will vary the relative amounts of each reactant, determining the amount of solid product in each case. In this pre-lab you will explore some features of this reaction.

1. This reaction is an oxidation-reduction (redox) reaction in which one reactant loses electrons (is oxidized) and another reactant gains electrons (is reduced). To recognize a redox reaction you need to check the oxidation states of all the elements in both the reactants and products. If any change their oxidation state, the reaction is a redox. An increase in oxidation state is an oxidation; a decrease in oxidation state is a reduction. (Note: If an element is a reactant or product, converted to or formed from a compound, the reaction must be a redox reaction.)

Indicate the oxidation state of the elements in each of the following compounds:

Reactants		Products		
$Cu(NO_3)_2$	KI	CuI	KNO_3	I_2
Cu______	K______	Cu______	K______	I______
N______	I______	I______	N______	
O______			O______	

Which reactant is oxidized, and which is reduced?

2) Calculating Molarity:

What is the formula weight of CuI?

What quantity of CuI is present in 50.0 mL of a 0.50 M solution?

What is the formula weight of $Cu(NO_3)_2$?

What mass of $Cu(NO_3)_2$ is present in 225 mL of a 0.30 M solution?

What is the formula weight of KI? (Enter unit as g/mol)

What mass of KI is needed to add to a 1000 mL volumetric flask to produce a solution that is 0.75 M?

What is the formula weight of I_2?

What is the molarity of a solution produced by dissolving 53 g of I_2 in a 250 mL volumetric flask?

3) Limiting Reagent Calculations:

You mix 3.5 mL of 0.50 M $Cu(NO_3)_2$ with 2.0 mL of 0.50 M KI, and collect and dry the CuI precipitate. Calculate the following:

Moles of $Cu(NO_3)_2$ used:

Moles of KI used:

Moles of CuI precipitate expected from $Cu(NO_3)_2$:

Moles of CuI precipitate expected from KI:

What mass of CuI do you expect to obtain from this reaction?

If the actual mass of the precipitate you recover is 0.133 g, what is the percent recovery of the precipitate?

Experiment 6

Laboratory Report

Name: ______________________ **Date:** ______________

Instructor: ______________________ **Sec. #:** __________

PURPOSE: *(The purpose should be several well-constructed sentences describing what your experiment was designed to accomplish and the criteria used to determine success. These sentences should include both concepts and techniques.)*

PROCEDURE: *(The procedure section should reference the lab manual and include any changes made to the procedure during the lab. Please note whether you tested potassium or sodium salts.)*

DATA: *(The data section should include your own personal data, as well as the table below filled in with class data. For your personal data include all observations, the masses, and the supernatant test results, using a "+", [more precipitate formed] or "-", [more precipitate did not form].*

Observations:

Balanced Chemical Equation:

____

Class Data Table

	Vol. $Cu(NO_3)_2$ (0.50M)	Vol. KI (0.50M)	Masses of ppt. in test tubes	$Cu(NO_3)_2$ Test	KI Test
A	1.0 mL 1.0 mL 1.0 mL 1.0 mL	6.0 mL 6.0 mL 6.0 mL 6.0 mL			
B	2.0 mL 2.0 mL 2.0 mL 2.0 mL	5.0 mL 5.0 mL 5.0 mL 5.0 mL			
C	3.0 mL 3.0 mL 3.0 mL 3.0 mL	4.0 mL 4.0 mL 4.0 mL 4.0 mL			
D	4.0 mL 4.0 mL 4.0 mL 4.0 mL	3.0 mL 3.0 mL 3.0 mL 3.0 mL			
E	5.0 mL 5.0 mL 5.0 mL 5.0 mL	2.0 mL 2.0 mL 2.0 mL 2.0 mL			
F	6.0 mL 6.0 mL 6.0 mL 6.0 mL	1.0 mL 1.0 mL 1.0 mL 1.0 mL			

	Moles $Cu(NO_3)_2$ Added (mol)	Moles KI Added (mol)	Theoretical Moles CuI (mol)	Theoretical Mass CuI (g)	Experimental Average Mass CuI (g)	% Error	Limiting Reagent
A							
B							
C							
D							
E							
F							

CALCULATIONS: (*The calculation section should have sample calculations for the following:* ***a)*** *moles of* $Cu(NO_3)_2$*,* ***b)*** *moles of KI,* ***c)*** *moles of product (CuI),* ***d)*** *average number of grams observed ,* ***e)*** *determination of the limiting reagent and* ***f)*** *percent error in mass of CuI.)*

Graph of the number of moles of CuI expected vs. the number of moles of KI added. All graphing guidelines apply here.

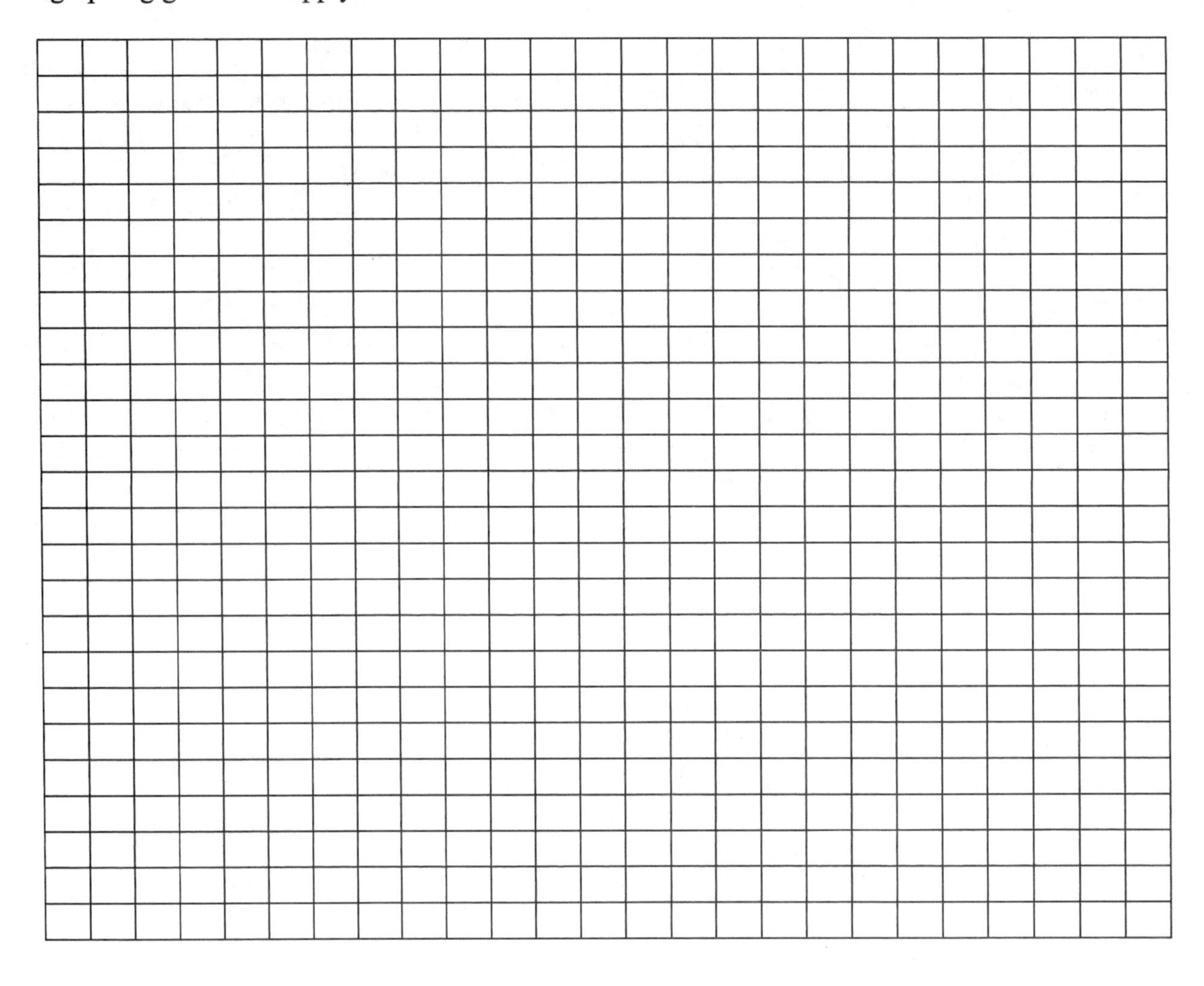

CONCLUSION: (*The conclusion should be in paragraph form and include a summary of the information determined by performing the experiment. This section should include the findings for each set of solutions. For solutions A-F include the test results, experimental product yield, theoretical yield, the limiting reagent and the percent error. Compare the results of your predicted LR with the results of the tests on the supernatant solutions. Be sure to discuss the variation in the different solutions and what caused the differences. Also discuss what information can be determined from the graph. Finally discuss all sources of error and how they might have affected the results.*)

Experiment 6

Post-Laboratory Questions

Name: ____________________ **Date:** ______________

Instructor: ____________________ **Sec. #:** __________

Show all work for full credit.

1. How would a decrease in the concentration of the reactants affect the moles of reactant?

2. Explain the role of limiting and excess reagents in an experiment.

3. Copper compounds are fairly common. Using the internet or any other reliable source, look up at least five other copper compounds. List the name, chemical formula, and uses of each compound. Be sure to cite your source in acceptable format.

Experiment 7

Redox Reactions in Voltaic Cells: Construction of a Potential Series

Introduction

Have you ever wondered how a battery worked? We use them every day for our cell phones, our cars, our mp3 players and even some important things like pacemakers. But have you ever really thought about how or why they work? What's inside of them? For one second, just imagine that there was no power, no electricity coming from the plug in the wall, but you are desperately in need of power to make something work and your life depends on it. What would you do? What (as a chemist) could you do?

Electrochemistry is the study of electrochemical reactions. These are reactions that create a flow of electrons from one place to another. A flow of electrons is "electricity." A battery is a small holding cell of electrochemical energy. In each battery there are two reservoirs of chemicals that will allow a transfer of electrons if connected by a conductive material. To use this flow of electrons (electricity) to power our devices we simply need to wire them into the loop:

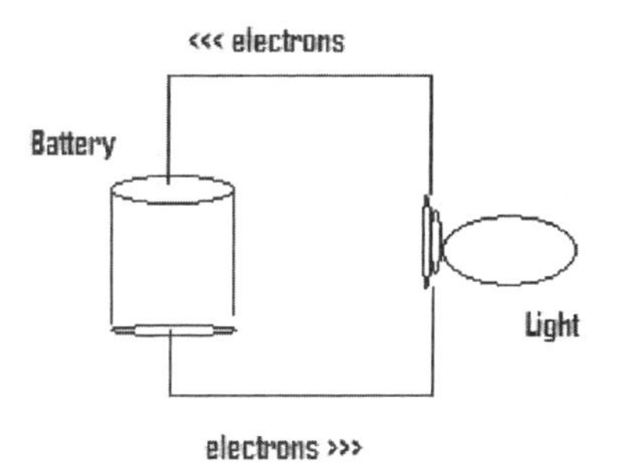

As a chemist, you should know the basic design of a common battery and also the chemistry behind why it works. In this lab, we are going to create some small voltaic cells out of various chemicals and measure their potentials (e.g., how strong the flow of electrons is from one metal to another). The greater the potential difference (electronegativity difference) between two metals, the stronger the electron flow from one to the other and the stronger the electron flow, the more electricity is generated. A lot of research has been devoted to the production of a better battery. The lithium batteries that power our laptops are the result of some of that research, and are approximately twice as strong as a normal zinc-carbon battery.

Development of a potential series of metals is therefore a valuable tool for chemists, allowing them to select metals/chemicals that will produce the most power.

In addition to learning about redox reactions, potentials, and the chemistry behind a battery, you will also learn to use a new device called a volt- or multi-meter. We will use the volt/multi-meter in the lab to measure the potentials of the cells that we construct.

Background

Redox

In both the Conservation of Matter and Limiting and Excess Reagents experiments you are given a brief introduction to redox reactions. Recall that redox reactions are also known as oxidation-reduction reactions. In particular, oxidation and reduction are the terms used to describe the transfer of electrons in a chemical reaction. Atoms or ions that accept electrons in a reaction have undergone reduction, while species that lose electrons undergo oxidation. These two definitions can be remembered by either of these two mnemonics:

1. "LEO the lion goes GER"—Loses Electrons, Oxidized (LEO); Gains Electrons, Reduced (GER); or
2. "OIL RIG"—Oxidation Is Loss; Reduction Is Gain.

When an oxidation reaction and a reduction reaction are paired together, electrons can flow from the oxidized species (losing electrons) to the reduced species (gaining electrons). The reduced species is often referred to as the oxidizing agent, or oxidant, as it is taking electrons away from another atom or ion, causing oxidation. Conversely, the species being oxidized is referred to as the reducing agent or reductant, because it is donating electrons, thus causing reduction. The oxidation or reduction of an atom or ion can be followed by observing the species' oxidation number on the reactant side (left) and seeing if it has changed on the product side (right). A brief summary of the rules for assigning oxidation numbers are provided here:

Rule 1: The oxidation number of an element in its natural state is zero.

Examples: $Fe(s)$, $F_2(g)$, $S_8(s)$

Rule 2: The oxidation number of a monatomic (one-atom) ion is the same as the charge on the ion.

Examples: $Na^+ = +1$; $N^{3-} = -3$

Rule 3: The sum of all oxidation numbers in a neutral compound is zero. The sum of all oxidation numbers in a polyatomic (many-atom) ion is equal to the charge on the ion.

Examples: PO_4^{3-} , P = +5, O = -2 so 5 + 4(-2) = -3 same as charge

Rule 4: The oxidation number of an alkali metal (IA family) in a compound is +1; the oxidation number of an alkaline earth metal (IIA family) in a compound is +2.

Examples: K = +1, Mg = +2

Rule 5: The oxidation number of oxygen in a compound is usually –2. If, however, the oxygen is in a class of compounds called peroxides (for example, hydrogen peroxide), then the oxygen has an oxidation number of –1. If the oxygen is bonded to fluorine, the number is +1.

Examples: H_2O_2, H = +1, O = -1.

Rule 6: The oxidation state of hydrogen in a compound is usually +1. If the hydrogen is part of a binary metal hydride, then the oxidation state of hydrogen is –1.

Examples: NaH, Na = +1, H = -1

Rule 7: The oxidation number of fluorine is always –1. Chlorine, bromine, and iodine usually have an oxidation number of –1, unless they're in combination with an oxygen or fluorine.

Examples: $HClO_4$, H = +1, O = -2, Cl = +7

Electricity and Voltaic Cells

In general, the transfer of electrons from one place to another is referred to as electricity. In fact, the electricity that you use to power your iPod® is simply the flow of electrons along a metal wire. Another way to produce electricity is through the spontaneous redox reactions that occur in voltaic cells. The oxidation and reduction processes are separated, so that the transfer of electrons occurs through an external wire. These 'separated' parts of the cell are called half-cells, one for oxidation and one for reduction.

The construction of a voltaic cell takes advantage of the spontaneous nature of a redox reaction. For example, let's look at the following reaction where zinc metal is oxidized and copper ion is reduced.

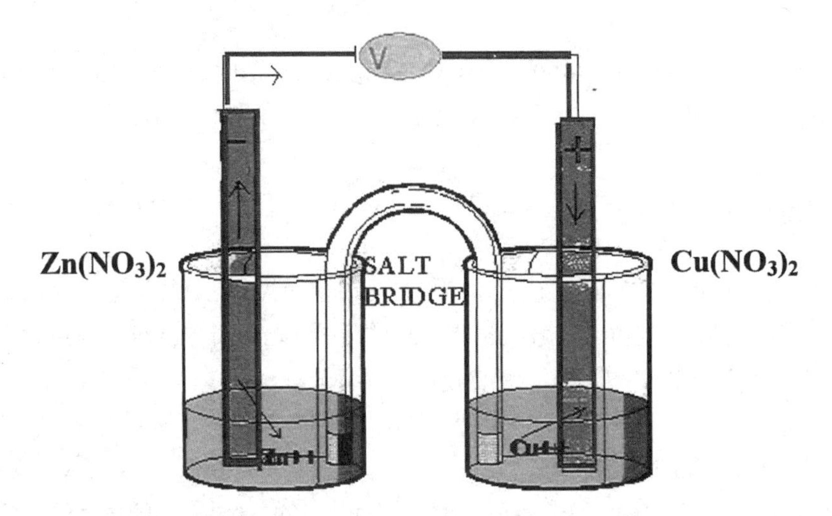

If we were to construct the characteristic voltaic cell, we would need 1.0-M solutions of copper nitrate ($Cu(NO_3)_2$) and zinc nitrate ($Zn(NO_3)_2$). Additionally, we would have to suspend a piece of copper metal in the container holding the $Cu(NO_3)_2$ solution and a piece of zinc in the $Zn(NO_3)_2$ as shown in the diagram above. Both metals serve as electrodes, the conductor used to allow contact with the ions in solution. The solutions are then connected by a third solution which acts as a salt bridge.

The piece of zinc metal is the negative electrode, referred to as the anode, while the piece of copper metal is the positive electrode, termed the cathode. Finally, when the circuit is completed by connecting the two electrodes with an external metal wire, usually a voltmeter, the flow of electrons can be observed.

When a voltaic cell is constructed correctly, oxidation (loss of electrons) always occurs at the anode, while reduction (gain of electrons) always occurs at the cathode. A little hint to help you differentiate these two components is that oxidation and anode both start with vowels, while cathode and reduction both start with consonants. Thus, the flow of electrons should always have a positive potential value since they are flowing from negative (anode) to positive (cathode). In the case that you do obtain a negative potential value, which indicates that your electrons are flowing in the opposite direction, all you have to do is switch the electrode connections on the voltmeter.

In this example, the electrons flow away from the zinc and towards the copper. This means that the copper ions are being reduced to copper atoms (gaining electrons) while the zinc atoms are being oxidized to zinc ions (losing electrons). Furthermore, since copper is being reduced, we say that it is the stronger oxidizing agent (more electronegative), and by default, zinc has to be the stronger reducing agent. The direction of the flow of electrons depends on the potential of each metal ion.

In order to further understand what we have been talking about thus far, let's look at the reactions taking place at the individual electrodes. Recalling that oxidation occurs at the anode and reduction occurs at the cathode, we can break down our reaction into two half-reactions:

Anode: $\mathbf{Zn^0(s) \rightarrow Zn^{2+}(aq) + 2e^-}$

Cathode: $\mathbf{Cu^{2+}(aq) + 2e^- \rightarrow Cu^0(s)}$

We can then "add" these two half reactions together in order to generate the overall reaction:

$$\mathbf{Zn^0(s) + Cu^{2+}(aq) \rightarrow Zn^{2+}(aq) + Cu^0(s)}$$

The potential of this cell is 1.1 V as measured by the voltmeter.

Standard Cells

A cell in which all the materials are in their standard states is termed a standard cell. The standard states of elements and compounds are the most stable form found naturally at 25 °C. For gaseous substances the standard state is 1.00 atm of pressure and 25 °C, while for solutions the standard state is 1.00 M and 25 °C.

Voltaic Cell Representation

Rather than writing out long chemical equations to represent redox reactions occurring in voltaic cells, a short-hand form has been devised using the following rules.

1. The negative electrode (anode) is written at the left hand side and the positive electrode (cathode) on the right hand side.

   Zn **Cu**

2. All materials involved in the cell are represented with symbols and formulas, including the physical state.

   $\mathbf{Zn^0_{(s)}\ Zn^{2+}_{(aq)}}$ $\mathbf{Cu^{2+}_{(aq)}\ Cu^0_{(s)}}$

3. Direct contact is indicated by a single vertical line.

   $\mathbf{Zn^0_{(s)}|Zn^{2+}_{(aq)}}$ $\mathbf{Cu^{2+}_{(aq)}|Cu^0_{(s)}}$

4. Indirect contact through a salt bridge is indicated by double vertical lines ||.

   $\mathbf{Zn^0_{(s)}|Zn^{2+}_{(aq)}\ ||Cu^{2+}_{(aq)}|Cu^0_{(s)}}$

5. The concentrations of the electrolytes are given in parentheses.

   $\mathbf{Zn^0_{(s)}|Zn^{2+}_{(aq)}(1.0M)\ ||Cu^{2+}_{(aq)}(1.0M)|Cu^0_{(s)}}$

Calculations of a Cell's Potential

A term left unmentioned in the previous sections was the overall cell potential (E^0_{Cell}), which is simply the sum of the standard reduction potential (E^0_{Red}) and the standard oxidation potential (E^0_{Ox}). Mathematically, the overall cell potential is expressed as:

$$\mathbf{E^0_{Cell} = E^0_{Red} + E^0_{Ox}}$$

The standard oxidation potential (E^0_{Ox}) has the same value as the standard reduction potential (E^0_{Red}) but the sign is reversed to show an opposite flow of electrons. You will often see the equation above written as:

$$\mathbf{E^0_{Cell} = E^0_{Red\ Anode} - E^0_{Red\ Cathode}}$$

Because you are now subtracting the potential of the sign of the potential if negative will change and the net result will be the same as the first equation given.

Using the example shown in our diagram above, let's make a reasonable hypothesis about the overall potential for this cell. In the figure, we see that Zn is being oxidized at the anode, while Cu^{2+} is being reduced at the cathode, as shown in the half reactions provided below:

Anode: $\mathbf{Zn^0(s) \rightarrow Zn^{2+}(aq) + 2e^-}$
Cathode: $\mathbf{Cu^{2+}(aq) + 2e^- \rightarrow Cu^0(s)}$

If you look in Appendix D, you will find a list of reduction potentials for a great many elements, ions and compounds. Because these are reduction potentials, in order to determine the oxidation potential, we must reverse the sign of the potential value. Remember that oxidation (the loss of electrons) is simply the reverse of reduction (the addition of electrons) for an element, ion or compound. If we use Appendix D to look up the potentials for copper and zinc, we find that the E^0_{Red} for Cu^{2+} is +0.34 V, while the E^0_{Red} for Zn^{2+} is -0.76 V. Therefore after reversing the sign of the potential the E^0_{Ox} for Zn is +0.76 V, and our predicted overall cell potential is:

$$\mathbf{E^0_{Cell} = 0.34V + 0.076V = +1.10V}$$

However, when you construct this cell in the experiment, the reading from your voltmeter may not be exactly +1.10 V. This is because you are probably not working at completely standard conditions. Still, this is not the most important thing that needs to be considered—the most important condition that needs to be met is that the ratio of concentrations of the corresponding salts is 1:1. If this ratio is met, any differences in the readings could be explained by other factors such as resistance in the salt bridge, lack of calibration of the voltmeter, or non-standard conditions.

Procedure

SAFETY NOTES: Use the same precautions required when working with any chemical solution.

GENERAL INSTRUCTIONS: Each student should complete a series of measurements for three metals and share the results with the class as directed by your instructor.

Part I: Building the Cells

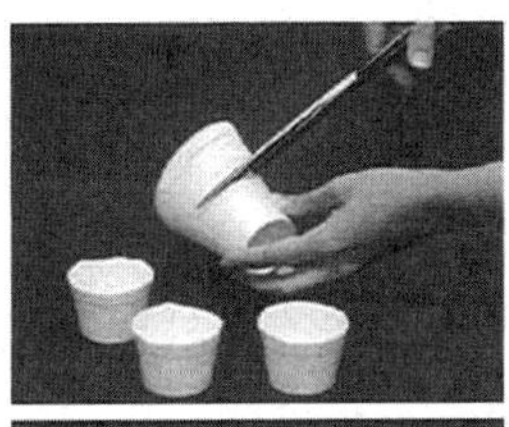

Obtain 4 styrofoam cups from the front counter. Cut the tops off the cups so that the cup that remains is ~5 cm deep.

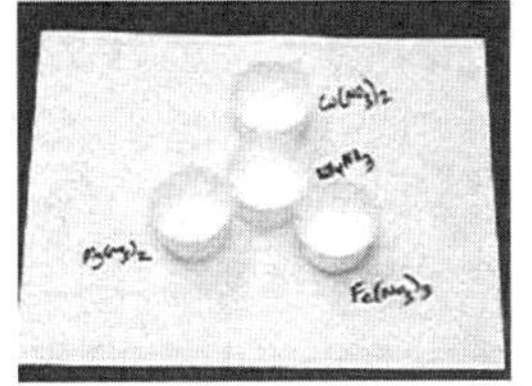

Obtain a sheet of white cardstock from the front counter and tape the cups to the paper according to the picture to the left.

Fill the cup in the middle about half full with 0.5M NH_4NO_3. Label the paper above the cup with the name and concentration of what is in the cup.

Fill each of the other three cups half full with one of your assigned metal nitrate solutions. Label the paper to the side of the cup with the name and concentration of the contents of the cup.

Obtain 3 filter papers from the front counter and fold them into 1 cm strips.

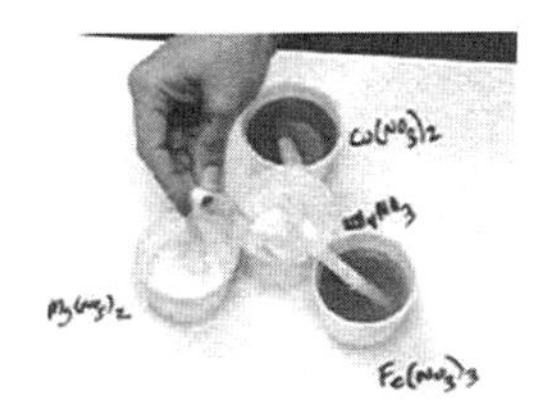

Bend one of the strips into a "U" shape. Place one end of the "U" into the center cup making sure that it contacts the solution and then place the other end into any of the other three cups. The filter paper will act as a salt bridge between the three solutions you are measuring. You may need to pre-soak the filter paper strips in NH_4NO_3 to make sure the salt bridge is functioning properly. Repeat until there is a salt bridge between the center cup and all three of the other cups.

Part II: Measure the Potentials

Turn on the voltmeter. Set the voltmeter to read volts (V dc). Attach the positve end (red) of the voltmeter to a piece of metal and place it in the correct (matching) nitrate solution so that the metal piece is partially submerged. (DO NOT SUBMERGE THE ALIGATOR CLIPS IN THE SOLUTION.) Use tape to adhere the clips and metal in place if necessary. It is easiest if you simply hold both clips in place.

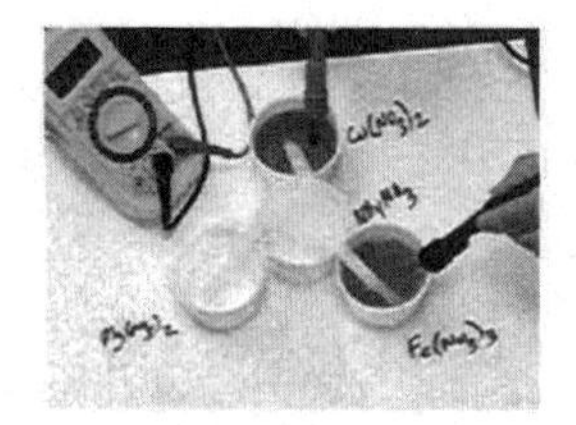

Attach the negative end (black) of the voltmeter to another of your assigned pieces of metal and place it in the correct (matching) nitrate solution so that the metal piece is partially submerged. (DO NOT SUBMERGE THE ALIGATOR CLIPS IN THE SOLUTION.) Use tape to adhere the clips and metal in place if necessary. It is easiest if you simply hold both clips in place.

Measure the potential and record it in your lab notebook. If the potential is negative, switch the aligator clips between the metals.

Repeat the potential measurements until a potential is established for each metal combination. Report your findings to the class.

Experiment 7

Pre-Laboratory Assignment

Name: ______________________________ **Date:** ____________________

Instructor: ______________________________ **Sec. #:** ____________

Show all work for full credit.

Following is a diagram of a voltaic cell.

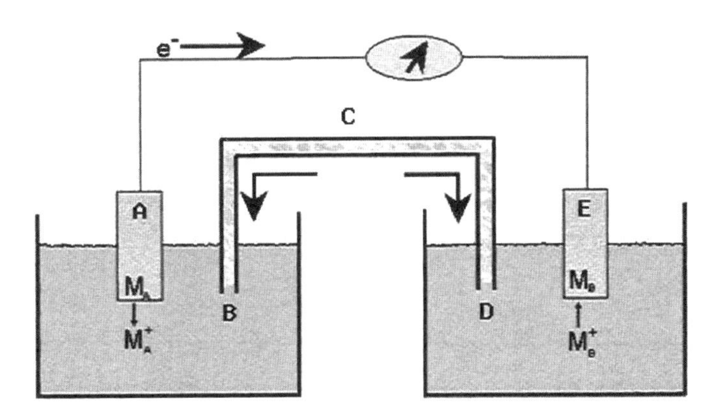

1) Identify the parts of the cell shown above by selecting the appropriate letter from the figure.

_____This is the positive electrode.

_____Reduction occurs at this electrode.

_____This electrode is the cathode.

_____Anions depart from the salt bridge here.

_____Oxidation occurs at this electrode.

2) Write the shorthand notation for the following voltaic cell reaction:

$$Fe(s) + Ni^{2+}(aq) \rightarrow Fe^{2+}(aq) + Ni(s)$$

Write the reaction represented by the following voltaic cell notation:

$$Mn(s)|Mn^{2+}(1M)||Pb^{2+}(1M)|Pb(s)$$

3) The list below shows the reduction potentials for ten different metal ions.

(a) Calculate ΔE for the redox reaction: $Ni(s) + Mn^{2+}(aq) \rightarrow Ni^{2+}(aq) + Mn(s)$

Will this reaction occur spontaneously?

(b) Calculate ΔE for the redox reaction: $Co(s) + Sn^{2}+(aq) \rightarrow Co^{2}+(aq) + Sn(s)$

Will this reaction occur spontaneously?

Half-Reaction	**E°(V)**
$Cu^{2+}(aq) + 2e^{-} \rightarrow Cu(s)$	+0.34
$Pb^{2+}(aq) + 2e^{-} \rightarrow Pb(s)$	-0.13
$Sn^{2+}(aq) + 2e^{-} \rightarrow Sn(s)$	-0.14
$Ni^{2+}(aq) + 2e^{-} \rightarrow Ni(s)$	-0.25
$Co^{2+}(aq) + 2e^{-} \rightarrow Co(s)$	-0.28
$Cd^{2+}(aq) + 2e^{-} \rightarrow Cd(s)$	-0.40
$Fe^{2+}(aq) + 2e^{-} \rightarrow Fe(s)$	-0.44
$Zn^{2+}(aq) + 2e^{-} \rightarrow Zn(s)$	-0.76
$Mn^{2+}(aq) + 2e^{-} \rightarrow Mn(s)$	-1.18
$Mg^{2+}(aq) + 2e^{-} \rightarrow Mg(s)$	-2.37

Experiment 7

Laboratory Report

Name: ______________________________ **Date:** ___________________

Instructor: ________________________________ **Sec. #:** ____________

PURPOSE: *(The purpose should be several well-constructed sentences describing what your experiment was designed to accomplish and the criteria used to determine success. These sentences should include both concepts and techniques.)*

PROCEDURE: *(The procedure section should reference the lab manual and include any changes made to the procedure during the lab. Please note which three metals you used for your cells.)*

DATA: (*The data section should include your own personal data and observations.)*

Observations:

Cell Data Table: *(An example for the cell representation and half reaction is included in this data table. Use this example to complete the cells you did. The E_{cell} theoretical can be determined using the half cell potential table found in your textbook. The E_{cell} experimental is the voltage you measured with your voltmeter.)*

Cell Representation	Anode half-reaction	E^0_{anode}	Cathode half-reaction	$E^0_{cathode}$	Overall reaction
$Fe^o \vert Fe^{2+} \Vert Cu^{2+} \vert Cu^o$	$Fe^o \rightarrow Fe^{2+} + 2e^-$		$Cu^{2+} + 2e^- \rightarrow Cu^o$		$Fe^o + Cu^{2+} \rightarrow Fe^{2+} + Cu^o$

E_{cell} Theoretical	E_{cell} Experimental	% error

CALCULATIONS: *(The calculation section should include sample calculations for determining E_{cell} theoretical.)*

CONCLUSION: *(The conclusion should be in paragraph form and include a summary of the information determined by performing the experiment. A discussion on the relationship of the weakest reducing metal to the strongest oxidizing metal should also be included. Another discussion should be included on the comparison of the E_{cell} theoretical and E_{cell} experimental along with the percent error values. Finally, discuss any errors that might have led to the discrepancies in the E_{cell} values.)*

Experiment 7

Post-Laboratory Questions

Name: ______________________________ **Date:** ___________________
Instructor: ______________________________ **Sec. #:** ____________

Show all work for full credit.

1) What is a reactivity series and how is it related to this experiment?

2) Create a potential series ranking all of the metals tested by the class arranged from strongest oxidizing agent to weakest oxidizing agent.

3) Explain your reasoning behind the rankings you list in problem 2).

Experiment 8

Reactions in Aqueous Solutions: Strong Acids and Bases

Introduction

Strong acids and bases are a part of our daily lives even though we seldom think about them. The gastric acid in your stomach is a strong (pH 2–4) hydrochloric acid that hydrolyses the bonds in the carbohydrates and the proteins that make up your food. The drain cleaner that unclogs your sink is made up of concentrated sodium hydroxide that, when combined with grease or oils, converts the compounds into soap, which is then soluble in water and can be washed away. In this experiment we will test the relationship between pH and acid or base concentration. We will also discuss how strong acids and bases react to neutralize each other, forming predictable products of salt and water.

Background

Acid and Base Strength

The strength of an acid or base is determined by its ability to dissociate in water.

Acid Dissociation Reaction: $\mathbf{HA \Leftrightarrow H^{+} + A^{-}}$

Base Dissociation Reaction: $\mathbf{BOH \Leftrightarrow B^{+} + OH^{-}}$

An acid that completely dissociates to form its component ions H^+ and A^- when placed in water is a strong acid. The same can be said for a base that completely dissociates into B^+ and OH^-.

Acid and base strengths are characterized quantitatively by a value called the K_a and K_b (acid/base dissociation constants), respectively. These dissociation constants are calculated as a ratio of the amount of acid or base that is separated into ions over the amount of acid or base that is still together in acid or base form.

$$K_a = \frac{[H^+][A^-]}{[HA]} \qquad \text{or} \qquad K_b = \frac{[B^+][OH^-]}{[BOH]}$$

Therefore, strong acids and bases will have K_a and K_b values greater than one since they are completely dissociated. Weak acids and bases have values less than one since the concentration of the acid or base that remains together will be greater than the concentration of the separated ions.

pH

In addition to the K_a or K_b indicating the strength of an acid or base, we often use the pH of the solution to describe whether a solution is acidic or basic. Solutions with a pH of less than 7 are considered acidic and solutions with a pH of greater than 7 are considered basic. Below are some more of the relationships between acid and base concentration and pH:

$$\mathbf{pH = -log\,[H^+] \quad and \quad [H^+] = 10^{-pH}}$$
$$\mathbf{pOH = -log\,[OH^-] \quad and \quad [OH^-] = 10^{-pOH}}$$
$$\mathbf{pH + pOH = 14 \quad and \quad [H^+]*[OH^-] = K_w = 1 \times 10^{-14}}$$

The determination of the pH and concentration of a weak acid or base solution is complicated by the less than 100% dissociation and will be discussed further in later chapters. For now, we will only worry about calculating the concentration and pH of solutions containing strong acids and bases.

The pH of a strong acid solution is equal to the –log of the hydrogen ion concentration of that solution. Because the strong acid completely dissociates into ions, the acid concentration equals the hydrogen ion concentration.

For example: A 2.5M HCL solution has a $[H^+]$ = 2.5M and the pH = -log(2.5) = -0.40

The same relationship can be described for the pOH of a strong base. The hydroxide ion concentration will equal the concentration given for the strong base:

For example: A 2.5 M NaOH solution has a $[OH^-]$ = 2.5M and the pOH = -log(2.5) = -0.40
So the pH = 14 –(-0.40) = 14.4

A couple of things should be noted about the examples above. First, students often think that pH values cannot be above 14 or below 0. As you can see from the examples given this is simply not true. When dealing with strong acids and bases the pH values are often very large or even negative. Second, you have to be careful when calculating the concentration of a base solution. The negative log of that solution gives you the pOH, not the pH and students often forget this very important second calculation.

The Reaction of a Strong Acid with a Strong Base

In this experiment, the strong acid, HCl, and the strong base, NaOH, are used to create serial dilutions that are combined to form salt solutions of NaCl according to the reaction below:

$$\mathbf{HCl(aq) + NaOH(aq) \rightarrow NaCl(aq) + H_2O(l)}$$

These products (a salt and water) are typical of the strong acid/base reaction and can be used to predict the products of other strong acid and base reactions.

$$\mathbf{Ex.\quad HNO_3(aq) + KOH(aq) \rightarrow KNO_3(aq) + H_2O(l)}$$

Although the reactants have changed, the basic products remain the same: a salt and water.

Prediction of the amount of salt produced is based on the stoichiometry of the reaction and the amount of acid and base provided. In the case of HCl and NaOH, the reaction is "one-to-one." Therefore, whichever reactant is present in the smallest amount of moles will limit the product moles that can be formed. This amount of moles will be the *theoretical yield* of the reaction. The *actual yield* will be determined experimentally.

Serial Dilutions

Serial dilution is an important technique in experimentation. The process allows the researcher to create controlled, systematic dilutions from a stock solution. Each successive solution is diluted so that the concentration of solute is changed by a desired ratio. For example, in 1/10 serial dilutions starting with a stock solution of 5M solute concentration, the serial dilution would be comprised of 5M, 0.5M, 0.05M, 0.005M… Each solution is reduced to one tenth of the previous solution's concentration. For another example, 1/25 serial dilutions starting from a stock solution of 12M solute concentration would be comprised of 12M, .48M, 0.0192M, 0.000768M… The greater the ratio of solute concentration change, the greater the difference in concentration for each successive serial dilution. Serial dilutions also work to conserve supplies since they use the previously made solution as the "stock" solution for the next more dilute solution.

For example:
Using a 0.5M stock solution, create 4 solutions of 1/25 serial dilutions. Calculate the resulting concentration for each diluted solution using the following equation:

$$\mathbf{M_1V_1 = M_2V_2}$$

where M_1 is the concentration of solution one, V_1 is the volume of solution one, M_2 is the concentration of solution two, and V_2 is the volume of solution two.

$$(1.0 \text{ mL})(0.5\text{M}) = (25 \text{ mL})(X)$$
$$X = 0.02 \text{ M}$$

The results indicate that 0.02M is the concentration of the new diluted solution.
This process is repeated in order to make the next 1/25 diluted solution. But the stock solution used this time is the newly made 0.02M solution. Repeating the calculations above:

$$(1.0 \text{ mL})(0.02\text{M}) = (25 \text{ mL})(X)$$
$$X = 8.0 \times 10^{-4}\text{M}$$

The process continues using each newly made solution as the stock for the next lower concentration until all of the dilutions needed are produced.

As you can see, making 1/25 dilutions rapidly drops the H^+ concentration to very small amounts. At some point, the concentration of the acid will actually fall below the concentration of H^+ found in pure water. (The pH of pure water is 7 so the $[H^+] = 1.0 \times 10^{-7}$M.) Once you get to this point you must take into consideration not only the ions from your dilute acid but those from the water dissociation as well.

The 100% acid dissociation means that the $[H^+] = [Cl^-]$ when HCl dissociates. We also know that the dissociation of water in terms of concentration means that for every $[H^+]$ a $[OH^-]$ is produced so that $[H^+] = [OH^-]$. When combined the charge-balanced relationship shows that:

$[H^+]_{water}$ + $[H^+]_{acid}$ combine to make the total $[H^+]$ in the solution, so

$$[H^+] = [Cl^-] + [OH^-]$$

You can substitute for $[OH^-]$ from the water ionizatiion constant ($K_w = [H^+][OH^-]$; $K_w/[H^+] = [OH^-]$;) :

$$[H^+] = [Cl^-] + K_w/[H^+],$$

and on rearrangment you get the quadratic equation:

$$[H^+]^2 - [Cl^-][H^+] - K_w = 0$$

This equation can be solved for $[H^+]$ using the quadratic formula.

Percent and Theoretical Yield

To properly report the information you collect in this laboratory, you will need to not only measure the pH using a pH meter but also calculate the theoretical pH. The theoretical pH can be determined for the acid by using pH = -log [H+], the [H+] is the hydrogen ion concentration which is equal to the concentration of acid you calculate. For the pH of the base you must first determine the pOH, using pOH= -log [OH-], where [OH-] is the concentration of the base you determine. Once you have determined the pOH you need to subtract it from 14 to obtain the pH. (pH=14 - pOH). The experimental error between the experimental pH you determine and the theoretical pH you calculate can be determined with the following formula:

$$\% \text{ error} = \frac{\text{Theoretical Value} - \text{Experimental Value}}{\text{Theoretical Value}} \times 100\%$$

In order to determine if the combined acid and base used in the experiment generates the expected amount of salt, the balanced chemical equation must be used to determine the theoretical yield of salt. To do this you will need to determine the moles of acid or base (pick one) used in the mixture. You can determine the number of moles used by multiplying the volume used by the concentration of the acid or base (moles = M x volume in L). Using your balanced chemical equation, the reaction's stochiometric ratios, and the moles of acid or base you start with, you can calculate the moles of salt that should be produced. Finally you will need to convert the moles to grams using the molecular weight of the salt to determine the percent yield of the reaction. The percent yield is calculated by dividing the mass of the salt you measure experimentally by the theoretical yield (mass) calculated in the manner above. High percent yields normally indicate a fully mixed and well run reaction. Percent yields over 100% may indicate errors in measurement or possibly excess water in the 'dried' salt.

$$\% \text{ yield} = \frac{\text{Experimental Yield (g)}}{\text{Theoretical Yield (g)}} \times 100\%$$

Procedure

SAFETY NOTES: Acids and Bases can cause burns if they come in contact with your skin. Wash your hands immediately if you feel a burning or itching sensation. Use all necessary precautions when handling the acid and base containers.

GENERAL INSTRUCTIONS: This experiment is to be done in pairs; one partner will do the acid dilutions and the other the base dilutions. Be sure to get all of the information from your partner before leaving the lab.

Part I: Preparing Salt Solution

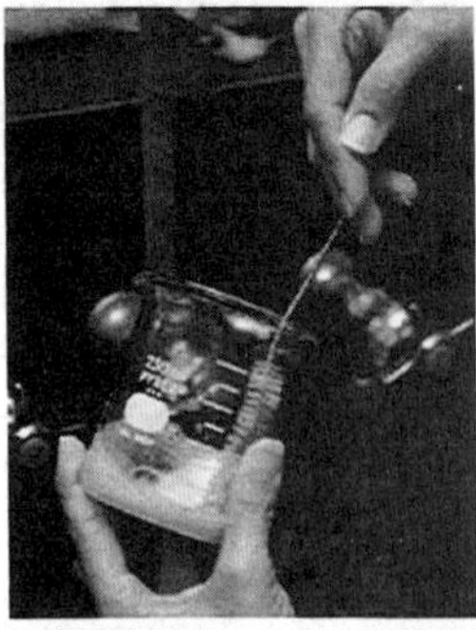

Obtain approximately 15 mL of 0.5M HCl (red solution) in a clean beaker and15 mL 0.5M NaOH (blue solution) in a clean beaker.

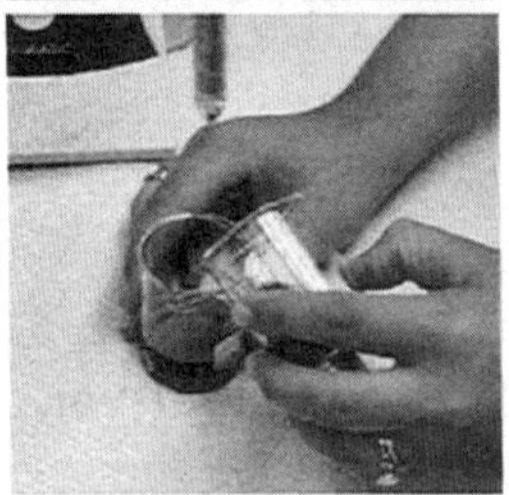

Using a graduated cylinder, measure out and combine exactly 10 mL of NaOH with 10mL of HCl in a clean 250mL beaker.

Place the beaker with the acid-base mixture on a hot plate and set to 100 °C. Allow the mixture to boil gently until all liquid has evaporated.

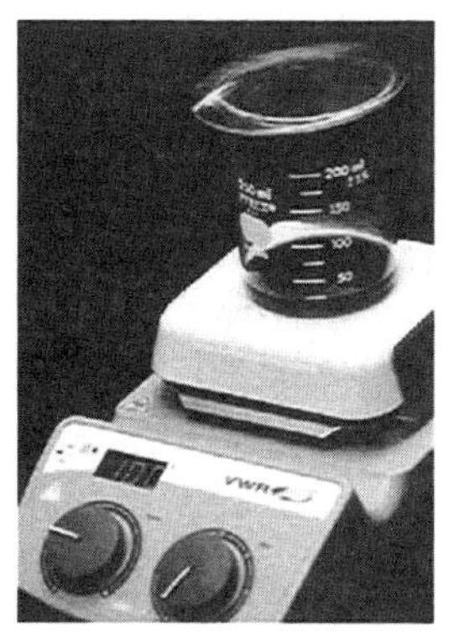

Place a watch glass over the mouth of the beaker if the salt starts to "spit" out.

While the solution is boiling off, continue on to part 2.

Part II: Acid-Base Reactions and pH

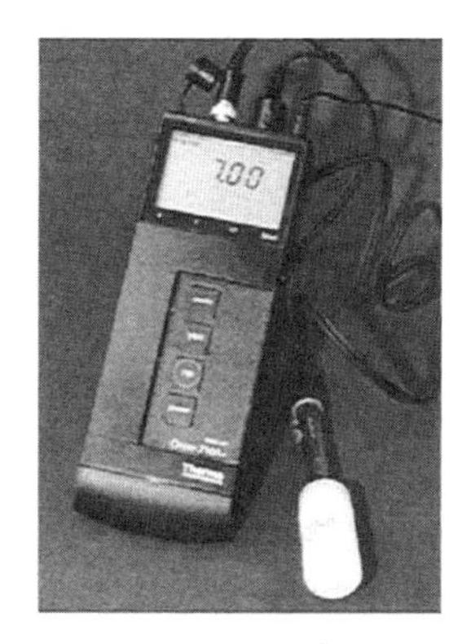

Calibrate the pH meter using the three way calibration procedure provided by your TA.

Collect 100 mL of D.I. water and measure the pH. Record the pH in your lab notebook.

Collect a 25 mL volumetric flask from the front counter.

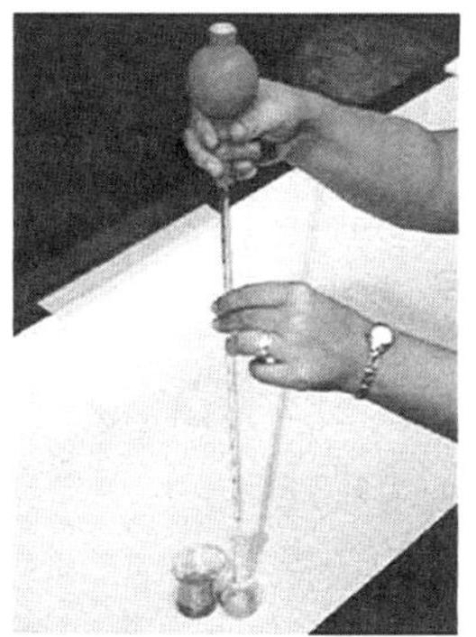

Using a graduated 1.0 mL pipette, the 25mL volumetric flask and either the acid or base, perform four serial dilutions using 0.5 mL of stock solution at each step.

For example: Dilution one, pipette, using the three way bulb, 0.5 mL of 0.5M HCl OR 0.5 mL of 0.5M NaOH into the 25mL volumetric flask and dilute to the 25mL mark with DI water. Pour the solution into a clean beaker and then repeat until all dilutions are made.

Calculate the resulting NaCl, H^+ and OH^- concentrations as you did in your pre-lab.

Record the pH of all 4 HCl dilutions and then of all 4 NaOH dilutions.

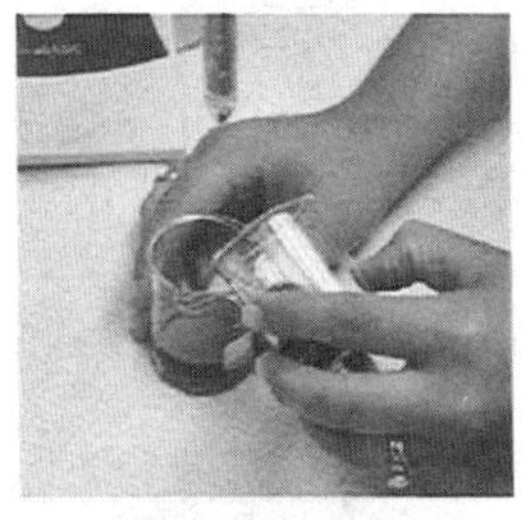

Combine the solutions of acid and base by adding the 0.5M solution of NaOH with the 0.5M solution of HCl. Record the resulting pH. Repeat the combination of solutions for the other four dilutions. Record the pH for each combination.

Part III: Experimental Yield

Carefully remove the beaker and watch glass from the hotplate using the oven gloves provided. Allow the beaker to cool to room temperature.

Separately weigh the beaker and watch glass containing the salt residue to the nearest 0.001g and record the value in your lab notebook.

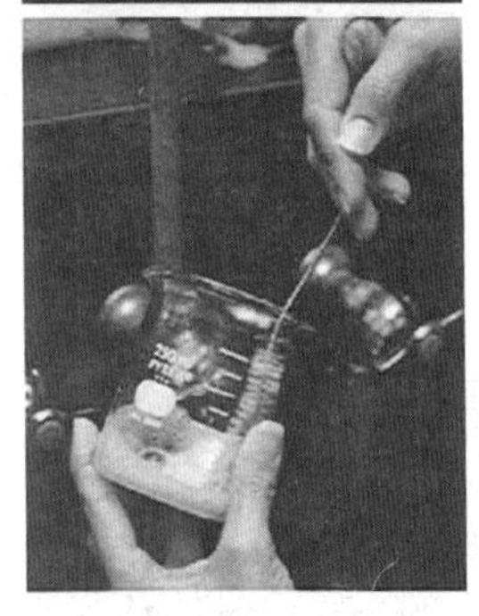

After weighing, clean and dry the beaker and watch glass thoroughly. Reweigh the now-empty beaker and watch glass to the nearest 0.001g and record the value in your lab notebook.

Experiment 8

Pre-Laboratory Assignment

Name: ______________________________ **Date:** ________________

Instructor: ______________________________ **Sec. #:** ____________

Show all work for full credit.

1. Constant boiling HCl has a concentration of 11.6 M. Your laboratory assistant is going to dilute this acid for you to create a stock solution of HCl to use in this experiment. What volume of the 11.6 M HCl must the assistant add to a one liter volumetric flask so that the concentration of the stock solution will be 0.78 M after dilution to the mark with water?

   What is the pH of this stock solution? (Remember the significant figure rules for logarithms!)

   You will now use this 0.78 M HCl solution to carry out a series of dilutions, measuring the pH of each diluted sample. Calculate the $[Cl^-]$ and $[H^+]$ concentrations and the pH of each sample. (Remember the significant figure rules for logarithms!)

   Sample A will be produced by pipetting 1.00 mL of the stock solution into a 25 mL volumetric, and diluting to volume with distilled water.

   [Cl-]=
   [H+]=
   pH=

   Sample B will be produced by pipetting 1.00 mL of sample A into a 25 mL volumetric, and diluting to volume with distilled water.

   [Cl-]=
   [H+]=
   pH=

   Sample C will be produced by pipetting 1.00 mL of sample B into a 25 mL volumetric, and diluting to volume with distilled water.

   [Cl-]=
   [H+]=

pH=

Sample D will be produced by pipetting 1.00 mL of sample C into a 25 mL volumetric, and diluting to volume with distilled water.

[Cl-]=
[H+]=
pH=

Sample E will be produced by pipetting 1.00 mL of sample D into a 25 mL volumetric, and diluting to volume with distilled water.

[Cl-]=
[H+]=
pH=

Sample F will be produced by pipetting 1.00 mL of sample E into a 25 mL volumetric, and diluting to volume with distilled water.

[Cl-]=
[H+]=
pH=

2) Your laboratory assistant is going to prepare a stock solution of sodium hydroxide (NaOH) for you. What mass of NaOH must the assistant add to a one liter volumetric so that the stock solution has a concentration of 0.46 M after dilution to the mark with water?

What is the pH of this stock solution? (Remember the significant figure rules for logarithms!) You will now use this 0.46 M NaOH solution to carry out a series of dilutions, measuring the pH of each diluted sample.

Calculate the [Na+] and [OH-] concentrations and the pH of each sample. (Remember the significant figure rules for logarithms!)

Sample A will be produced by pipetting 1.00 mL of the stock solution into a 25 mL volumetric, and diluting to volume with distilled water.

[Na+]=
[OH-] =
pH=

Sample B will be produced by pipetting 1.00 mL of sample A into a 25 mL volumetric, and diluting to volume with distilled water.

[Na+]=
[OH-] =

pH=

Sample C will be produced by pipetting 1.00 mL of sample B into a 25 mL volumetric, and diluting to volume with distilled water.

[Na+]=
[OH-] =
pH=

Sample D will be produced by pipetting 1.00 mL of sample C into a 25 mL volumetric, and diluting to volume with distilled water.

[Na+]=
[OH-] =
pH=

Sample E will be produced by pipetting 1.00 mL of sample D into a 25 mL volumetric, and diluting to volume with distilled water.

[Na+]=
[OH-] =
pH=

Sample F will be produced by pipetting 1.00 mL of sample E into a 25 mL volumetric, and diluting to volume with distilled water.

[Na+]=
[OH-] =
pH=

3) Write the balanced molecular equation for the neutralization reaction between $Rb(OH)_2$ (a strong base) and HNO_3 (a strong acid).

The lab assistant has prepared for you a 0.125 M solution of $Rb(OH)_2$ and a 0.399 M solution of HCl. What volume of HCl will be required to exactly neutralize 14.0 mL of the Rb(OH)2 solution?

You neutralize 17.0 mL of the $Rb(OH)_2$ solution with this required volume of the HCl solution in a 250 mL beaker which has a mass of 102.7789 g. You heat the solution to drive off all the water, leaving only solid $Rb(NO_3)_2$. What mass of $Rb(NO_3)_2$ do you expect to find in the residue?

You find that the beaker plus the residue has a mass of 104.3628 g.
What is your percent yield of $Rb(NO_3)_2$?

Experiment 8

Laboratory Report

Name: ______________________________ **Date:** __________________

Instructor: ______________________________ **Sec. #:** ____________

PURPOSE: (*The purpose should be several well constructed-sentences describing what your experiment was designed to accomplish and the criteria used to determine success. These sentences should include both concepts and techniques.)*

PROCEDURE: (*The procedure section should reference the lab manual and include any changes made to the procedure during the lab.)*

DATA: *(Complete the tables below using the data you collected during the lab.)*

Observations:

ACID DILUTION DATA:

Dilution	$[H^+]$	Experiment pH	Theoretical pH	% Error
Stock Concentration				
1				
2				
3				
4				

BASE DILUTION DATA:

Dilution	$[OH^-]$	Experiment pH	Theoretical pH	% Error
Stock Concentration				
1				
2				
3				
4				

COMBINED ACID AND BASE:

Dilution	Experiment pH	Theoretical pH (Hint: Think DI Water)	% Error
Stock Concentration			
1			
2			
3			
4			

Write the balanced chemical equation for this experiment:

SALT YIELD:

Experimental Yield	Theoretical Yield	% Error

CALCULATIONS: (*The calculation section should have sample calculations for the following: a) concentration of dilution, b) pH, c) pH from pOH, d) experimental error, e) theoretical yield of salt, and f) percent yield of salt.*)

Graph the pH vs. concentration for both the acid and base. Remember all graphing guidelines apply here.

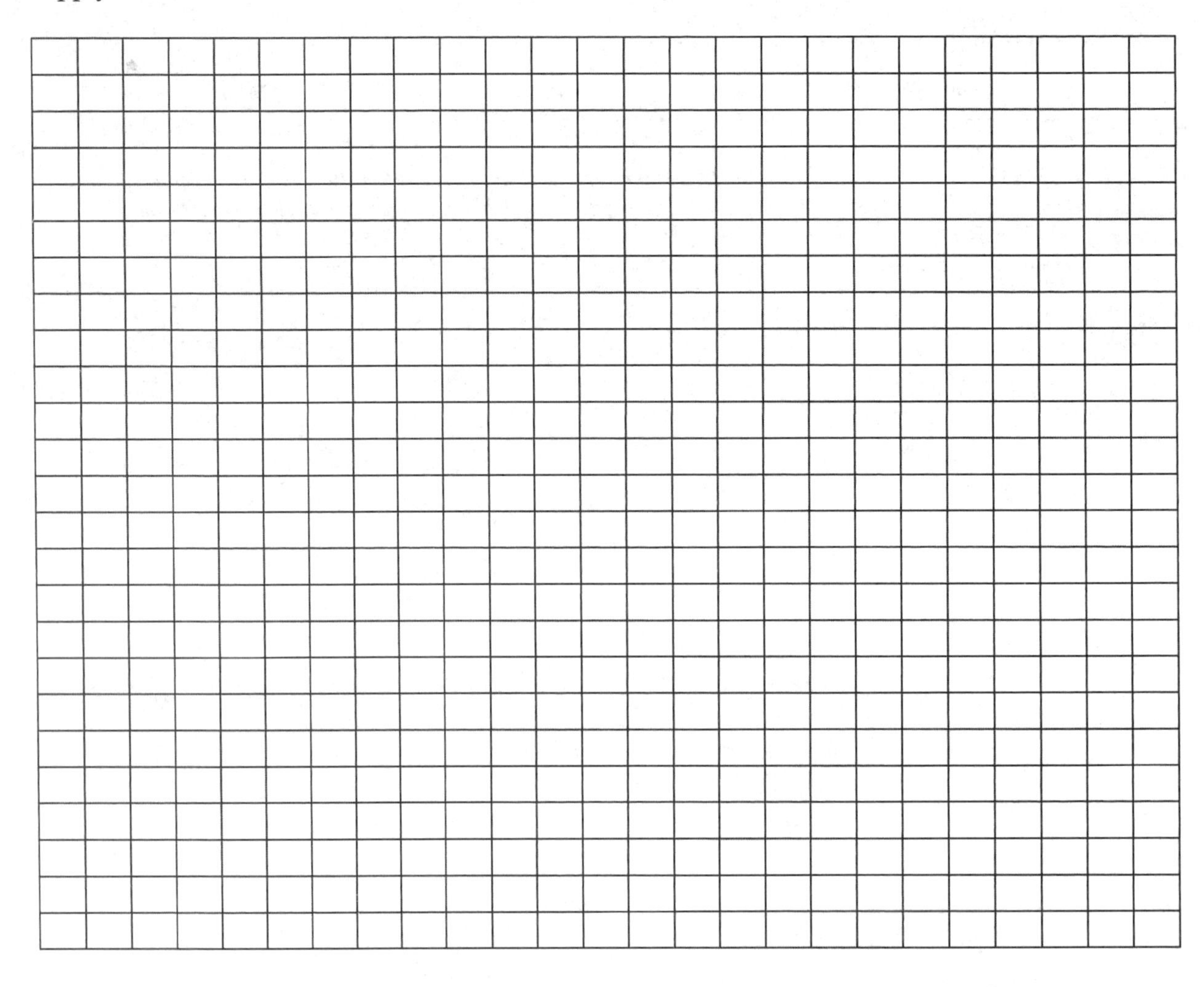

CONCLUSION: (*The conclusion section should be several paragraphs in length and include the following information: a discussion on the experimental and theoretical pH as well as the percent error for both the acid and the base; be sure to discuss in detail any discrepancies in the values. Within this discussion explain the trends seen in the graphs you made. It should also include a discussion on the pH of the combined solutions and the error in the values: be sure to include a statement about what is expected from these mixtures, and why that was not achieved. Finally discuss the percent yield of salt and any errors which may have caused a yield less than or greater than 100 percent.*)

Experiment 8

Post-Laboratory Assignment

Name: ______________________________ **Date:** ____________________

Instructor: ______________________________ **Sec. #:** ____________

Show all work for full credit.

1) Explain how this experiment can be related to limiting and excess reagent problems.

2) The acid used in this experiment is monoprotic. What does this mean and how would the experimental results differ if a diprotic acid had been used instead?

Experiment 9

Calorimetry & Hess's Law

Introduction

It is likely that you have started discussing the concepts of thermodynamics in class. In this lab we will illustrate some of those concepts. Specifically, why in the world do you need to know about processes like calorimetry and Hess's Law?

The purpose of this lab is to determine the enthalpy of reaction for the burning of magnesium in oxygen:

$$Mg(s) + \frac{1}{2}O_2(g) \rightarrow MgO(g) \quad \Delta H_{Rxn} = \text{Experimental Goal}$$

The reaction is quite exothermic. Therefore, trying to measure the enthalpy of this reaction would be difficult and possibly even dangerous. But we can still determine the ΔH_{Rxn} by using calorimetry to collect experimental data and then we can use Hess's Law to manipulate that data to yield the answer we need.

We will construct and use a constant pressure calorimeter to measure the ΔH_{Rxn} of two reactions related by their reagents to our target reaction. These values of ΔH_{Rxn} are much easier to measure as they are much less exothermic. The ΔH_{Rxn} of magnesium with hydrochloric acid and of magnesium oxide with hydrochloric acid will be determined and then used in conjunction with Hess's Law to calculate the ΔH_{Rxn} for our target reaction.

Hess's Law Reactions:

$$Mg(s) + 2HCl(aq) \rightarrow MgCl_2(aq) + H_2(g) \quad \Delta H_{Rxn} = \text{Experimental B}$$

$$MgO(s) + 2HCl(aq) \rightarrow MgCl_2(aq) + H_2O(l) \quad \Delta H_{Rxn} = \text{Experimental C}$$

$$2H_2(g) + O_2(g) \rightarrow 2H_2O(l) \quad \Delta H_F = -285.840 \ {}^{kJ}/_{mol}$$

Background

In studying chemical processes, attention is usually centered upon the properties of the substances involved. However, energy changes due to changes in the state of the system are associated with these processes. A *system* is that region of the universe under consideration. The *surroundings* are everything other than the system. Thus, the state of the system is specified by a number of variables, including temperature, pressure, volume, and chemical composition. A more extensive discussion of these energy relationships can be found in your textbook.

When a system undergoes any chemical or physical changes, the First Law of Thermodynamics requires that the accompanying change in the system's *internal energy* (ΔE) is equal to *heat* (q) plus *work* (w):

$$\Delta E = q + w \qquad \text{Equation 1}$$

Thus the energy of a system increases when heat is added (q is positive) and/or work is done on the system (w is positive). For systems under constant pressure (note subscript p), Equation 1 becomes:

$$\Delta E = q_p - P\Delta V \qquad \text{Equation 2}$$

The value for the heat (per mole) absorbed or given off by a system at constant pressure (q_p) is called the *heat of reaction* or *enthalpy of reaction* (ΔH). By substituting ΔH for q_p and rearranging, Equation 2 becomes:

$$\Delta H = \Delta E + P\Delta V \qquad \text{Equation 3}$$

If there is no change in volume (no ΔV), then ΔH is the change in energy of the system at constant pressure. Therefore, ΔH is a useful quantity to measure for reactions in solution (at constant pressure) where no gaseous products are formed (no change in volume). This describes a lot of chemistry. (Note: Although one of the reactions we are using in this experiment does produce a gas, the change in volume is negligible. So we will maintain our assumption of constant volume. But as with all assumptions, we should acknowledge it as a possible source of error in the conclusion.)

The amount of heat (q) necessary to raise the temperature of a system is an extensive property. That is, it depends on the amount of material, as well as what the material is and how much the temperature changes. This can be expressed in two ways:

$$q = n\, C_m \Delta T \qquad \text{Equation 4}$$

where n is the number of moles of material, C_m is the molar heat capacity and ΔT is the change in temperature (either in K or °C; since it's a *change*, it doesn't matter). Alternatively, this can be expressed for the mass of the material as:

$$q = m\, c_p\, \Delta T \qquad \text{Equation 5}$$

where m is the mass of the material in grams, c_p is called the specific heat of the material and ΔT is the change in the temperature (either in K or °C). If the conditions are restricted to constant pressure, these expressions become:

$$q_p = n\, C_m\, \Delta T \quad \text{and} \quad q_p = m\, c_p\, \Delta T$$

The formal definitions of the constants described above are: *Molar heat capacity* (C_m) is the amount of heat required to raise the temperature of one mole of material by one degree Kelvin or Celsius. *Specific heat* (c) is the amount of heat required to raise the temperature of one gram of material by one degree Kelvin or Celsius. The value of C_m or c_p depends on the identity of the material, its state (gas, liquid, solid), and its temperature. Below is a table of the values these "constants" take on for various states and temperatures of water. Note that for most of the large range of liquid water temperatures, the value does remain fairly constant (to two significant figures).

Table 1. Molar Heat Capacities and Specific Heats for Water

State	Temperature (°C)	Temperature (K)	Heat Capacity (J/mol * K)	Specific Heat (J/g * K)	Specific Heat (cal/g * K)
Solid	-34.0	239.2	33.30	1.846	0.4416
Solid	-2.0	271.2	37.78	2.100	0.5024
Liquid	0.0	273.2	75.86	4.218	1.007
Liquid	25.0	298.2	75.23	4.180	0.9983
Liquid	100.0	373.2	75.90	4.216	1.007
Gas	110.0	383.2	36.28	2.01	0.481

An older (English system) unit used to express heat is the *calorie.* It is defined as the amount of heat necessary to raise 1 gram of water from 14.58 °C to 15.58 °C at one atmosphere pressure. Obviously, since much work is done in aqueous solutions, this is still a useful unit, even though it is not an SI unit. (Note: The food Calorie, abbreviated Cal, is actually 1000 calories or one kilocalorie.)

In order to determine the state function enthalpy (ΔH), the heat (q_P) of a particular reaction is divided by the number of moles of material involved in the reaction.

$$\Delta H = \frac{q_P}{n} \qquad \text{Equation 6}$$

The Calorimeter Cup

The term calorimetry refers to the measurement of heat released or absorbed during a chemical or physical process. The ideal calorimeter is well insulated so that its contents do not gain or lose heat to the surroundings, and is constructed of a material of low heat capacity so that only a small amount of heat is exchanged between the contents and the calorimeter. For many processes a simple, unsealed, insulated cup can be used as a calorimeter since it has a low heat capacity and excellent insulating properties. Constant pressure on the reaction is maintained by the atmosphere.

In reality, no calorimeter is ideal. Thus, to obtain reliable results the calorimeter must be calibrated to determine how much heat is exchanged with the calorimeter cup (and/or other surroundings). This correction factor is called the *calorimeter constant* (heat capacity of the calorimeter in J/K) and fits into an equation similar to the ones above. The temperature change of the cup can be assumed to be the same as that of the solution in it. Since the ΔT of the cup will change with each experiment, it is useful to have a calorimeter constant which can be used in all experiments as follows:

$$q_{cal} = C_{cup}\Delta T \qquad \text{Equation 7}$$

The heat capacity of the calorimeter cup (C_{cup}) is determined by performing a separate experiment in which no chemical reaction takes place. Instead, hot water is added to the cold water in the calorimeter. The amount of heat lost by the hot water (q_{hw}) must be equal to the heat gained by the cold water (q_{cw}) plus the heat gained by the calorimeter cup (q_{cal}). (We can't lose energy, so it only has two places to go.) This can be written as:

$$-q_{hw} = q_{cw} + q_{cal} \qquad \text{Equation 8}$$

Note that the signs of these amounts of heat are opposite because the hot water is losing energy and the cold water and cup are gaining energy.

The amount of heat (q_{cal}) absorbed by the cup is easily found, since q_{hw} and q_{cw} can be calculated using Equation 5. To determine q_{hw}, substitute the mass of the hot water, the specific heat of water (Table 1), and the temperature change for the hot water (ΔT) into the equation. The q_{cw} is found the same way. Once q_{cal} is found, C_{cup} can be calculated from Equation 7.

To clarify the principles of calorimetry, consider the following hypothetical experiment. Using calorimetry methods, a determination of the heat of reaction of NaOH in HCl is made.

$$NaOH(aq) + HCl\ (aq) \rightarrow NaCl\ (aq) + H_2O\ (l)$$

The heat capacity for the calorimeter cup is given as 6.80 J/K. In the experiment, 75 mL (75 g, assume a density of 1.0 g/mL) of a 0.6 M HCl solution is placed in the calorimeter and allowed to equilibrate to room temperature. Then 75 mL (75 g) of a 0.6 M NaOH is added to the calorimeter with stirring. Temperature readings are taken before and after addition of the NaOH for 10 minutes.

Plotting temperature vs. time gives the following plot.

Temperature vs. Time

Time (sec)	Temp. (°C)
0	22.8
30	22.8
60	22.9
90	23.6
120	28.2
150	27.8
180	27
210	26
240	25.5
270	24.8
300	24.1
330	23.7
360	23.6

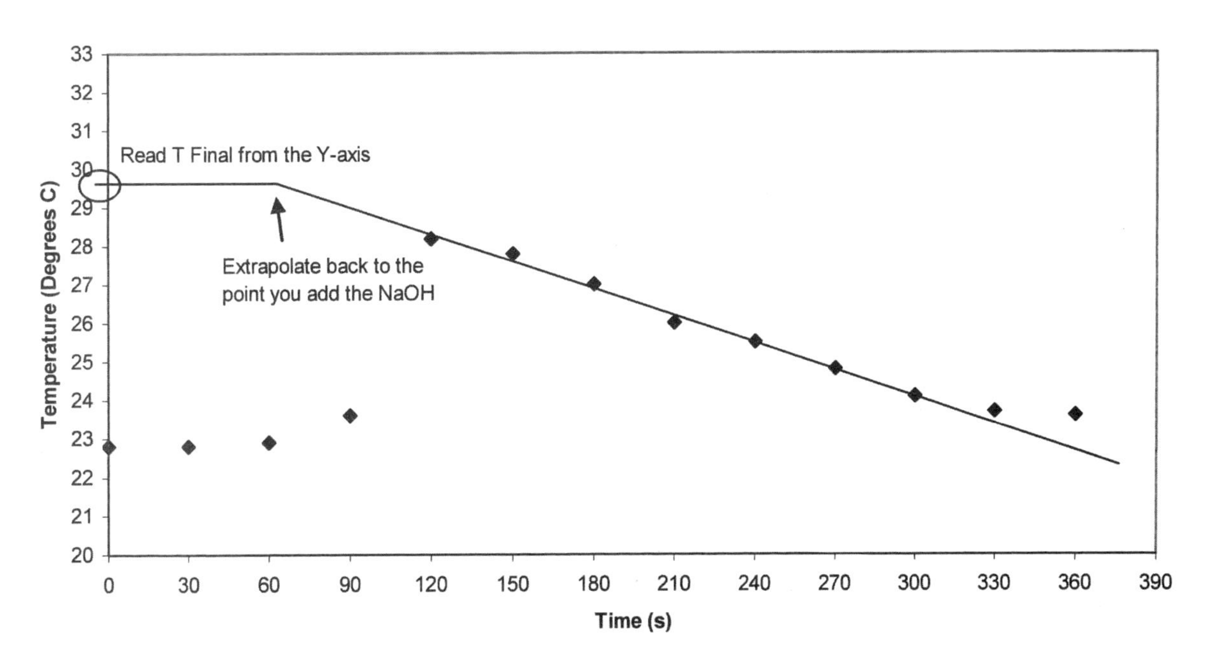

Note that this is a cooling curve, and the data of interest is on the left of the graph. The initial temperature (T_i) of the acid in the cup is 22.8 °C. Temperature readings are started one minute before addition of second substance. Upon addition of NaOH, the acid temperature increases to a maximum of 28.2 °C in one minute and then slowly decreases. The precise determination of T_f is more complicated, since heat exchange is occurring between the contents and the cup during and after the reaction. Evaluation of T_f is accomplished by extrapolation as shown above. This gives a theoretical temperature representing the temperature obtained if instantaneous equilibrium was achieved within the system. By extrapolation, $T_f = 29.7$ °C. The heat capacity for the calorimeter cup is given as 6.80 J/K.

Calculations

NOTE: Since these calculations require only the use of ΔT, it becomes unnecessary to convert all Celsius temperatures to Kelvin, just to arrive at ΔT. (Convince yourself this is true!)

$$\Delta T = T_f - T_i = 29.7\ °C - 22.8\ °C - 6.9\ °C \text{ (also 6.9 K if a } \Delta T\text{)}$$

To determine the heat of reaction, use Equation 5 and Table 1. (Assume that the heat capacity for a dilute solution of acid in water is very similar to that of water.)

$$q_{HCl} = m\ c_p\ \Delta T = 75.0\ g * 4.180\ J/gK * 6.9\ K$$
$$q_{HCl} = 2163\ J = 2.2\ kJ$$

We also need to calculate the heat absorbed by the calorimeter cup.

$$q_{cal} = C_{cup} * \Delta T = 6.80\ J/K * 6.9\ K$$
$$q_{cal} = 46.9\ J$$

The final heat of reaction value is calculated as follows:

$$q_{Rxn} = q_{HCl} + q_{cal} = 2163\ J + 46.9\ J$$
$$q_{Rxn} = 2.2\ kJ$$

Note that the calorimeter constant is very small compared to the q_{HCl}.

Hess's Law

If a reaction is carried out in a series of steps, ΔH for the reaction is equal to the sum of the enthalpy changes for the individual steps.

The overall enthalpy change for the process is independent of the number of steps or the particular nature of the path by which the reaction is carried out. Thus we can use information tabulated for a relatively small number of reactions to calculate ΔH for a large number of different reactions.

In this experiment we want to know the heat of reaction for burning magnesium metal in oxygen:

$$2Mg(s) + O_2(g) \rightarrow 2MgO(s)$$

This is a very exothermic reaction where the ΔH of reaction would be extremely difficult to measure using our calorimetry set-up. However, by experimentally measuring the heat of reaction for two other reactions:

$$Mg(s) + 2HCl(aq) \rightarrow MgCl_2(aq) + H_2(g)$$
$$\Delta H_{Rxn} = \text{(Experiment B)}$$

and

$$MgO(s) + 2HCl(aq) \rightarrow MgCl_2(aq) + H_2O(l)$$
$$\Delta H_{Rxn} = \text{(Experiment C)}$$

and using the heat of formation reaction for water:

$$2H_2(g) + O_2(g) \rightarrow 2H_2O(l)$$
$$\Delta H_f = -285.840 \text{ kJ/mol}$$

we can use Hess's Law to calculate the heat of reaction for magnesium burning in oxygen. Refer to your textbook for further explanation on how to use Hess's Law.

Procedure

SAFETY NOTES: The thermometers are fragile! If one is broken, inform your instructor immediately. Thoroughly wash off any acid or base which comes in contact with your skin. Acids and bases can cause burns!

GENERAL INSTRUCTIONS: Students will work in pairs with one person swirling the cup and reading the thermometer, while the other keeps time and records the data in both lab notebooks.

Experiment A: Determination of the heat capacity of the calorimeter (calorimeter constant)

The calorimeter consists of an insulated cup covered with a cardboard lid. The lid has a hole to accommodate the thermometer (see picture). A clamp may be used to support the thermometer. After assembling the calorimeter, make sure there is room to swirl the calorimeter without bumping the thermometer.

In *each* part of the experiment, measure the volume of solution put in the calorimeter with a graduated cylinder. Record these values to be used in calculations later. Allow the calorimeter and solution to come to room temperature. Equilibrium is reached when the temperature remains constant for 2 to 3 temperature readings at 30 second intervals. Once equilibrium is reached, you will add the second material to the calorimeter and continue recording the temperature at 30 second intervals for the amount of time specified. This will provide sufficient data to determine T_F by extrapolation. The calorimeter must be swirled constantly during the temperature reading phase of the experiment to insure complete mixing of the contents.

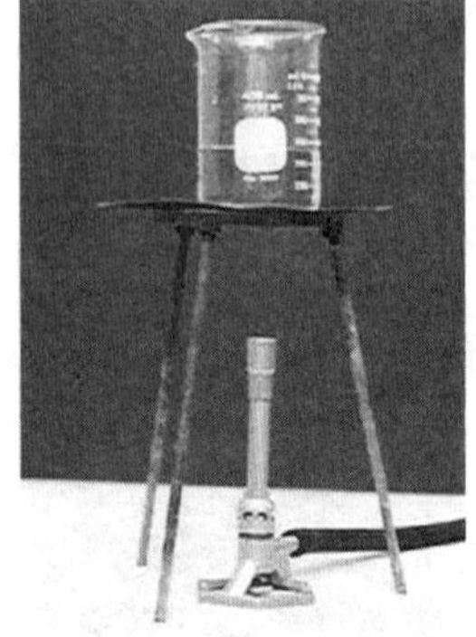

Place 150 mL of water in a 400 mL beaker and heat to 50–60 °C.

While the water is heating, measure out 50 mL of cold water in a graduated cylinder and place it in your calorimeter cup.

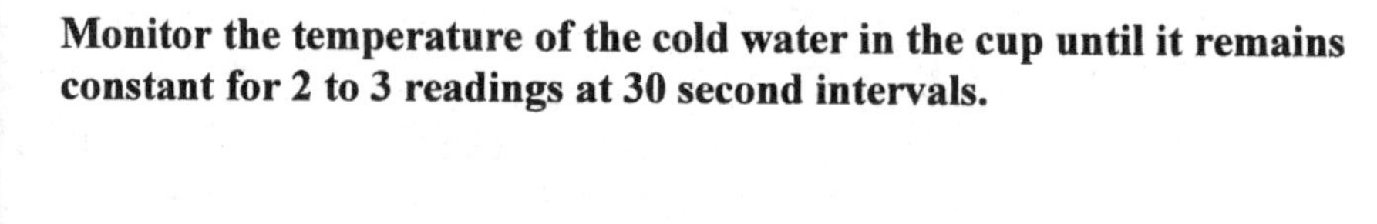

Monitor the temperature of the cold water in the cup until it remains constant for 2 to 3 readings at 30 second intervals.

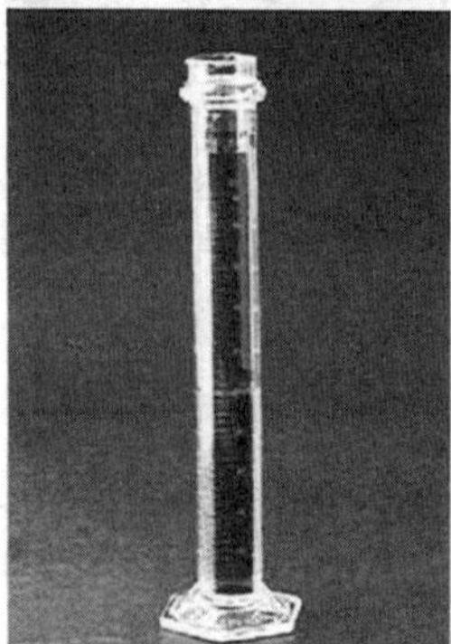

Measure 50 mL of your hot water into a graduated cylinder.

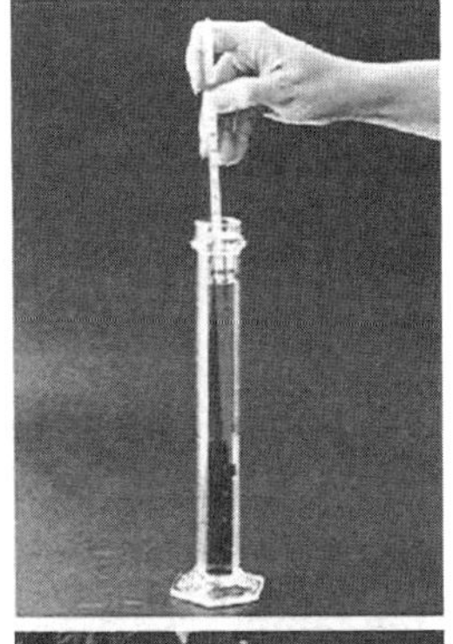

Using the thermometer record the temperature of the hot water.

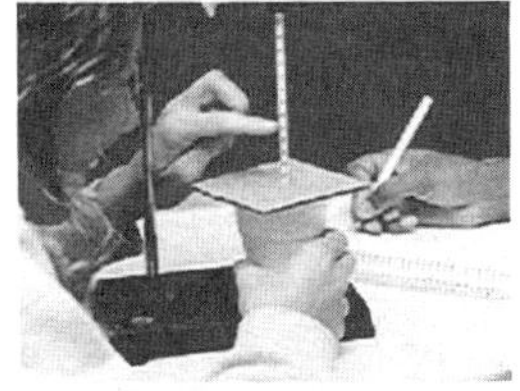

Return the thermometer to the calorimeter cup and record the temperature for 3 readings at 30 second intervals.

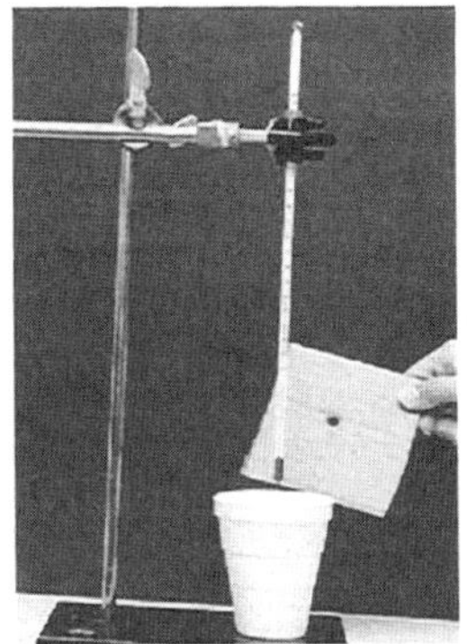

Lift the lid of the calorimeter and pour the 50 mL of hot water in, mixing continuously.

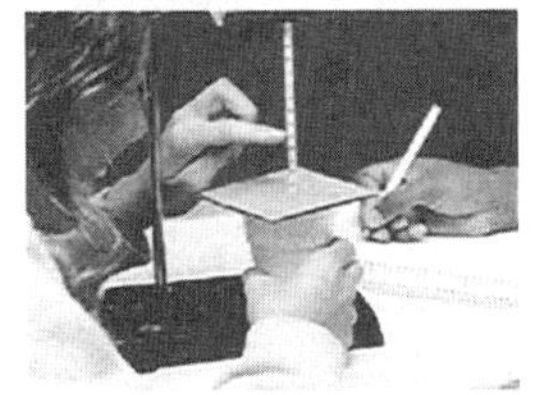

Record the temperature every 30 seconds until 10 minutes have elapsed.

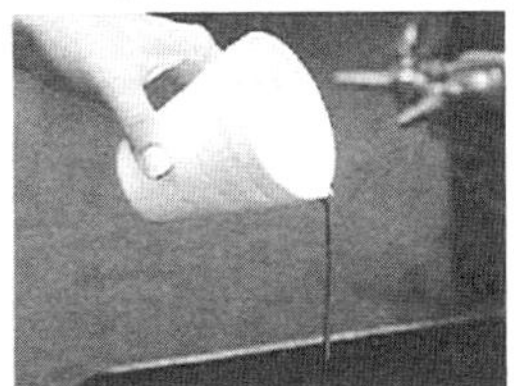

Dispose of the water in the calorimeter down the sink and repeat Part A of the experiment.

Experiment B: Determination of the Heat of Reaction of Mg(s) in HCl

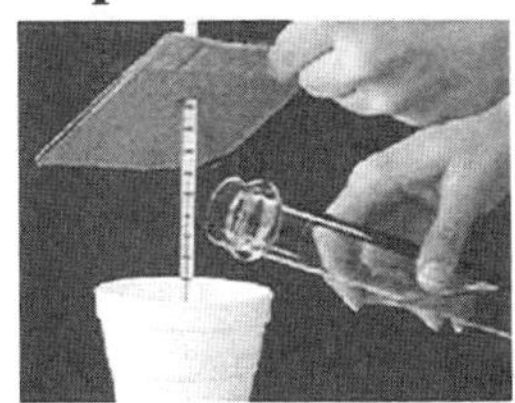

Make sure your calorimeter cup is clean and dry and then add 100 mL of 1.0 M HCl to the cup.

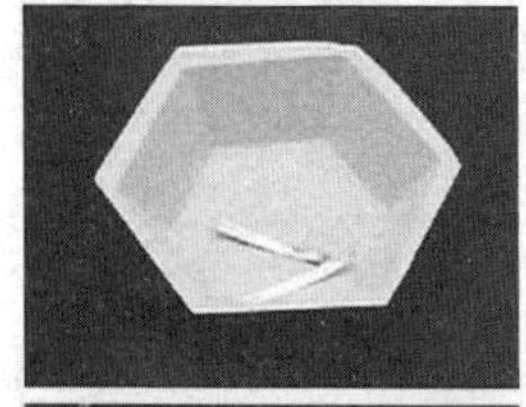

Weigh out four strips of magnesium metal to the closest 0.001g. (You can choose to use any weigh boat, watch glass, etc. to contain the magnesium for weighing.)

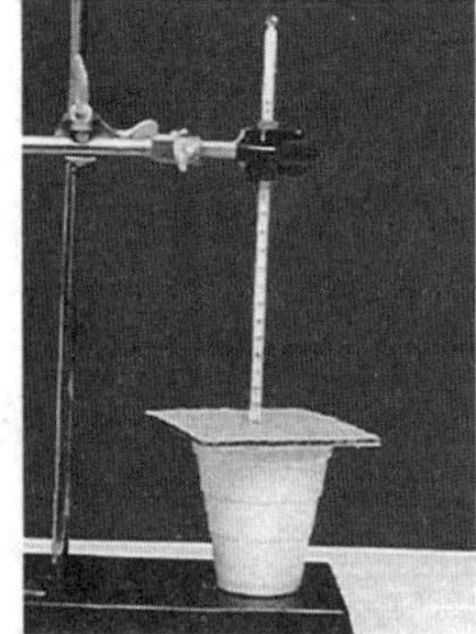

Place the thermometer in the calorimeter cup and record the temperature for 3 readings at 30 second intervals.

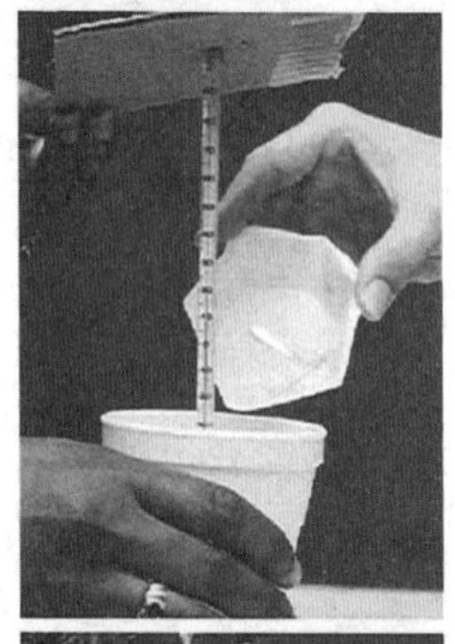

Lift the lid of the calorimeter and drop the pieces of magnesium in, mixing continuously.

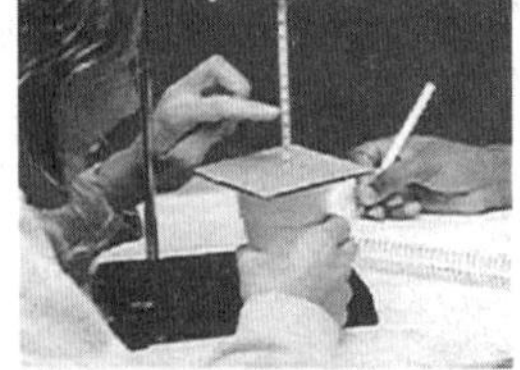

Record the temperature every 30 seconds until 10 minutes have elapsed.

Dispose of the solution in the waste jar in the hood and rinse and dry the calorimeter cup thoroughly.

Experiment C: Determination of the Heat of Reaction of MgO(s) in HCl

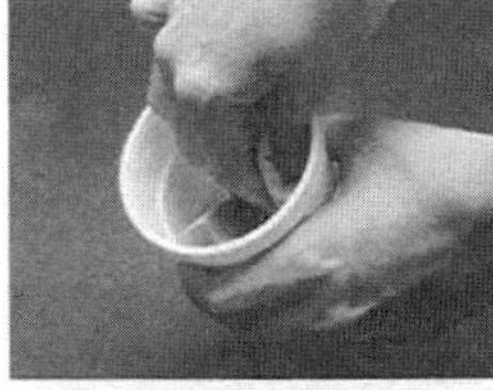

Add 100 mL of 1.0 M HCl to your calorimeter cup. (Make sure the cup was thoroughly rinsed and dried from the previous experiment.)

Weigh out ~0.50 g of magnesium oxide to the closest 0.001 g. (You can choose to use any weigh boat, watch glass, etc. to contain the magnesium for weighing.) Note: MgO is a powder that is vulnerable to being "blown" everywhere so be careful when transferring it to and from containers.

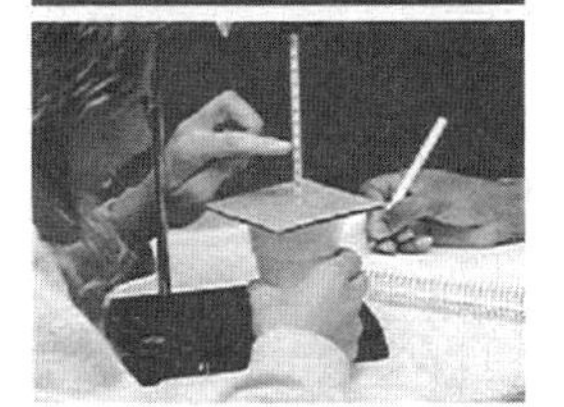

Place the thermometer in the calorimeter cup and record the temperature for 3 readings at 30 second intervals.

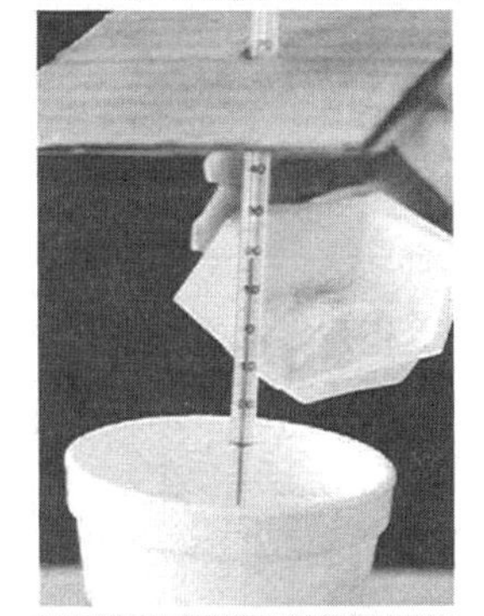

Lift the lid of the calorimeter and drop the magnesium oxide in, mixing continuously. (IMPORTANT: If you do not stir at this point the MgO will drop to the bottom of the cup and the reaction will not be complete.**)**

Record the temperature every 30 seconds until 10 minutes have elapsed.

Dispose of the solution in the waste jar in the hood and rinse and dry the calorimeter cup thoroughly.

Experiment 9

Pre-Laboratory Assignment

Name: ______________________ **Date:** ______________

Instructor: ______________________ **Sec. #:** __________

Show all work for full credit.

1) Use the graph of Temperature (°C) versus Time (s) provided to answer the following questions:

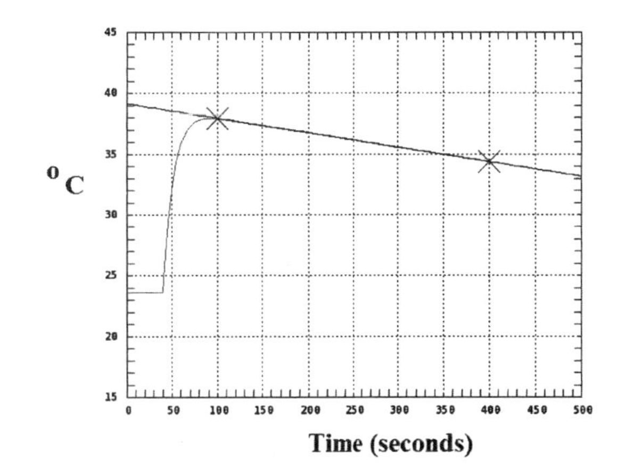

What is the initial temperature?

What is the final temperature?

The Cp for the calorimeter (including the HCl solution and the Mg) was calculated to be 0.226 kJ/°C. For the temperature curve shown above, 0.175 g Mg was reacted with HCl.

What is the ΔH of the reaction?

2) To calibrate your calorimeter cup, you first put 53 mL of cold water in the cup, and measure its temperature to be 21.1 °C. You then pour 41 mL of hot water, temperature = 53.7 °C, into the cup and measure the temperature every 30 seconds over a 10 minute period. You extrapolate this "cooling curve" back to the time of addition and find that the "final temperature" after mixing is 34.4 °C.

What is the heat change of the hot water, q_{HW}?
(Assume the density of the water is 1.00 g/mL, and remember that the specific heat of water is 4.184 J/g-K or J/g-°C.)

What is the heat change of the cold water, q_{CW}?

What is the heat change of the calorimeter cup, q_{cup}?

What is the heat heat capacity of the calorimeter cup, C_{cup}?

3) You would like to measure the heat of neutralization of an acid with a base. You mix 225 mL each of 0.77 M HCl and 0.77 M NaOH, both at a temperature of 18.7 °C in a calorimeter cup equilibrated to that same temperature. After following the temperature change for 10 minutes and extrapolating it back to the time of addition, you find that the "final temperature" after mixing was 23.7 °C. Previously you measured the heat capacity of the calorimeter cup (C_{cup}) to be 39 J/°C.

Assuming that the density of the two solutions is 1.00 g/mL, and that their specific heats are the same as water, 4.184 J/g-°C, what is the heat absorbed by the two solutions, q_{soln}?

What is the heat absorbed by the calorimeter cup, q_{cup}?

What is the heat change for the neutralization reaction, q_{rxn}?

What is the ΔH for the neutralization?

Write the net ionic equation for this neutralization reaction, including the states of matter (i.e., $H^+(aq)$) for the reactants and products.

4) Given the following data:

$SO_2(g) \rightarrow S(s) + O_2(g)$	ΔH = +296.8 kJ
$CO_2(g) \rightarrow C(s) + O_2(g)$	ΔH = +393.5 kJ
$C(s) + 2S(s) \rightarrow CS_2(l)$	ΔH = +89.4 kJ

Find the ΔH of the following reaction:

$$CO_2(g) + 2SO_2(g) \rightarrow CS_2(l) + 3O_2(g)$$

Experiment 9

Laboratory Report

Name: ______________________ **Date:** ______________

Instructor: ______________________ **Sec. #:** __________

PURPOSE: (*The purpose should be several well-constructed sentences describing what your experiment was designed to accomplish and the criteria used to determine success. These sentences should include both concepts and techniques.)*

PROCEDURE: (*The procedure section should reference the lab manual and include any changes made to the procedure during the lab.)*

DATA: (*The data section should include your own personal data and observations. The data section for this lab is very large. A fill-in-th- blank worksheet has been provided for you).*

Observations:

CALORIMETRY DATA TABLE:

PART A	TRIAL 1	TRIAL 2
Exact volume of cold water (to the nearest 0.1 mL)	______	______
Temperature of cold water (in cup)	______	______
Exact volume of hot water (to the nearest 0.1 mL)	______	______
Temperature of hot water (in cylinder)	______	______

PART A CALCULATIONS:

Mass of cold water (assume density = 1.00 g/mL)	______	______
T_f from graph by extrapolation	______	______
ΔT_{HW} for hot water	______	______
ΔT_{CW} for cold water	______	______
q_{HW} for hot water (use $q_{HW} = m\,c\Delta T_{HW}$)	______	______
q_{CW} for cold water (use $q_{CW} = m\,c\,\Delta T_{CW}$)	______	______
q_{Cal} for the cup (use $\lvert q_{HW}\rvert = \lvert q_{CW}\rvert + q_{Cal}$, solve for q_{Cal})	______	______
C_{cup} for the cup (use $q_{Cal} = C_{Cup}\,\Delta T$)	______	______

Which ΔT should be used?

PART B

Description of sample:

Exact volume of HCl (to the nearest 0.1 mL)	______
Initial temperature of HCl (in cup)	______
Exact mass of Mg (to the nearest 0.001g)	______

PART B CALCULATIONS:

T_f from graph by extrapolation	______
Mass of HCl solution (use the density of water for HCl = 1.00 g/mL)	______
ΔT_{HCl} for HCl	______
q_{HCl} for HCl solution (use $q_{HCl} = m\,c\,\Delta DT_{HCl}$)	______
q_{Cal} for the cup (use $q_{Cal} = C_{cup}\Delta T_{HCl}$)	______

q_{RXN} (use $q_{Rxn} = (q_{HCl} + q_{Cal})$) __________

ΔH_{RXN} for Mg (use $\Delta H_{Rxn} = q_{Rxn}/n_{Mg}$) __________

Write the net reaction which took place in the cup:

PART C

Description of sample:

Exact volume of 100 mL HCl (to the nearest 0.1 mL) __________

Temperature of HCl (in cup) __________

Exact mass of MgO (to the nearest 0.001g) __________

PART C CALCULATIONS:

T_f from graph by extrapolation __________

ΔT_{CW} for HCl __________

q_{HCl} for HCl solution (use $q_{CW} = m\ c\ \Delta T_{HCl}$) __________

q_{cal} for the cup (use $q_{cal} = C_{Cup}\ \Delta T_{HCl}$) __________

q_{RXN} (use $q_{RXN} = (q_{HCl} + q_{cal})$) __________

ΔH_{RXN} for MgO (use $\Delta H_{Rxn} = q_{Rxn}/n_{Mgo}$) __________

Write the net reaction that took place in the cup:

TIME VERSUS TEMPERATURE DATA:

Part A: Trial 1		Part A: Trial 2		Part B		Part C	
Time (s)	Temperature (°C)	Time (s)	Temperature (°C)	Time (s)	Temperature (°C)	Time (s)	Temperature (°C)

CALCULATIONS: (*The calculation section should include temperature vs. time graphs for each part. All graphing guidelines apply here. Sample calculations should be included for the following: the heat capacity of the calorimeter; heat of reaction for both parts B and C; and ΔH for both parts B and C.*)

Graph of the Part A, Trial 1 Temperature versus Time Data. All graphing guidelines apply here.

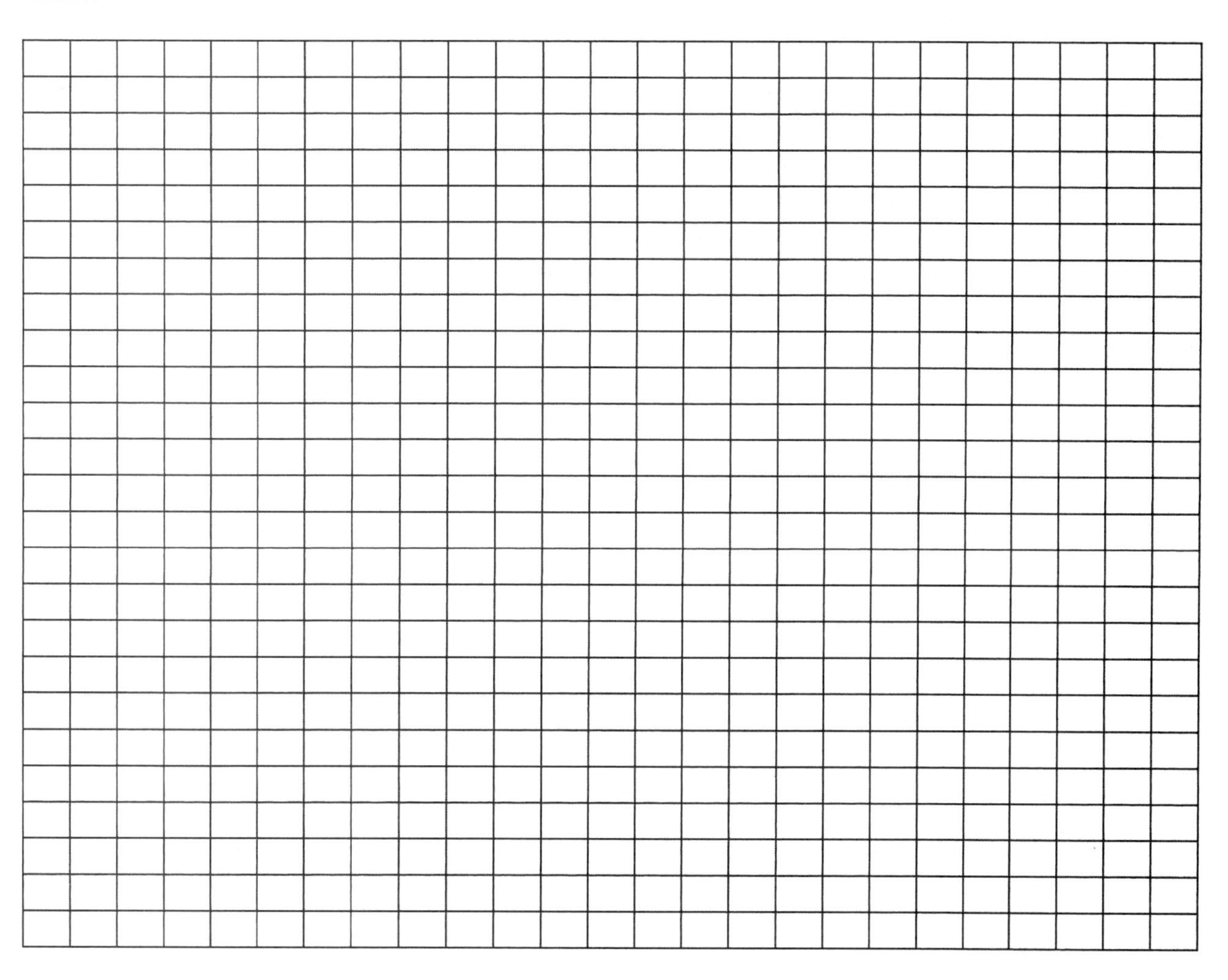

Graph of the Part A, Trial 2 Temperature versus Time Data. All graphing guidelines apply here.

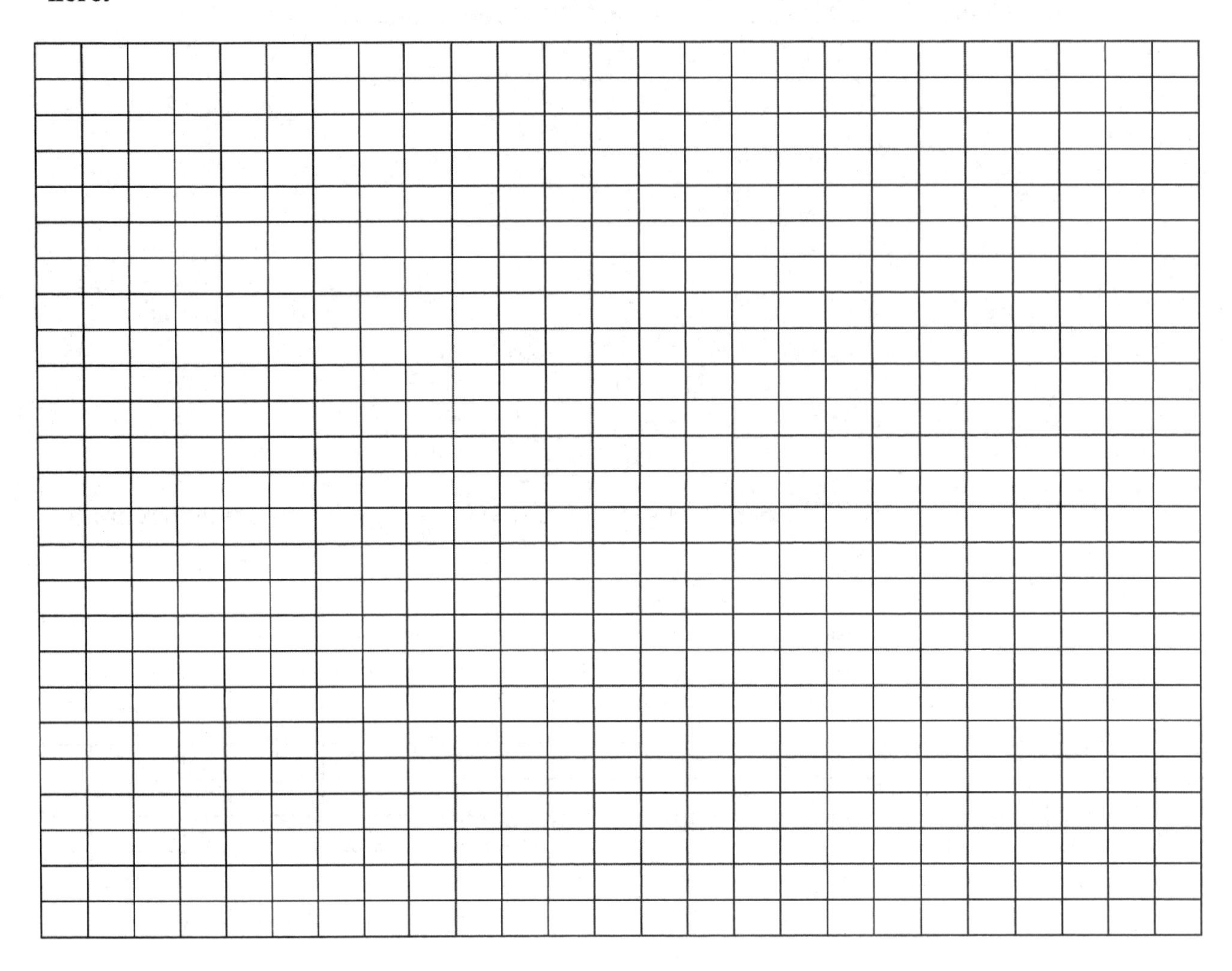

Graph of the Part B Temperature versus Time Data. All graphing guidelines apply here.

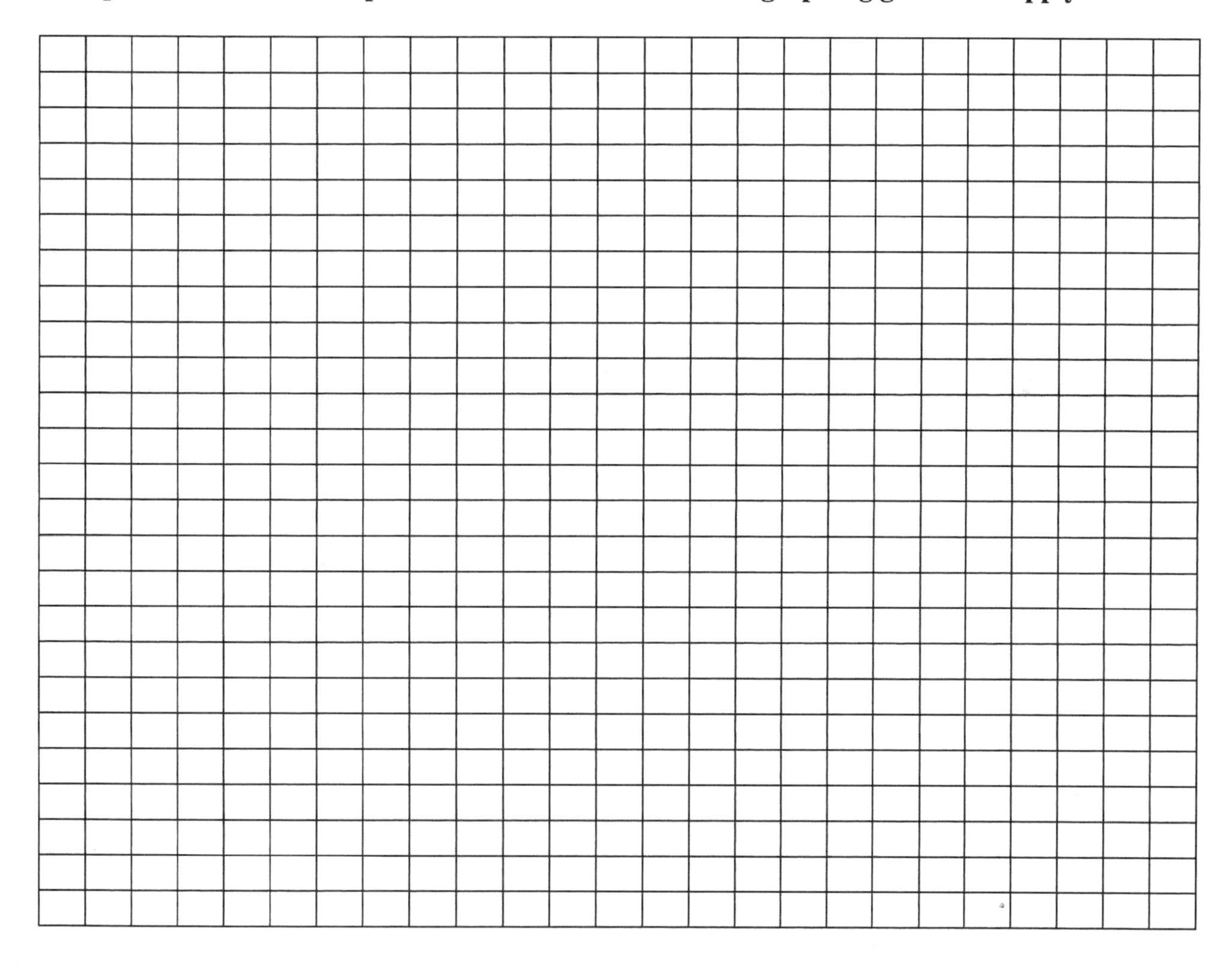

Graph of the Part C Temperature versus Time Data. All graphing guidelines apply here.

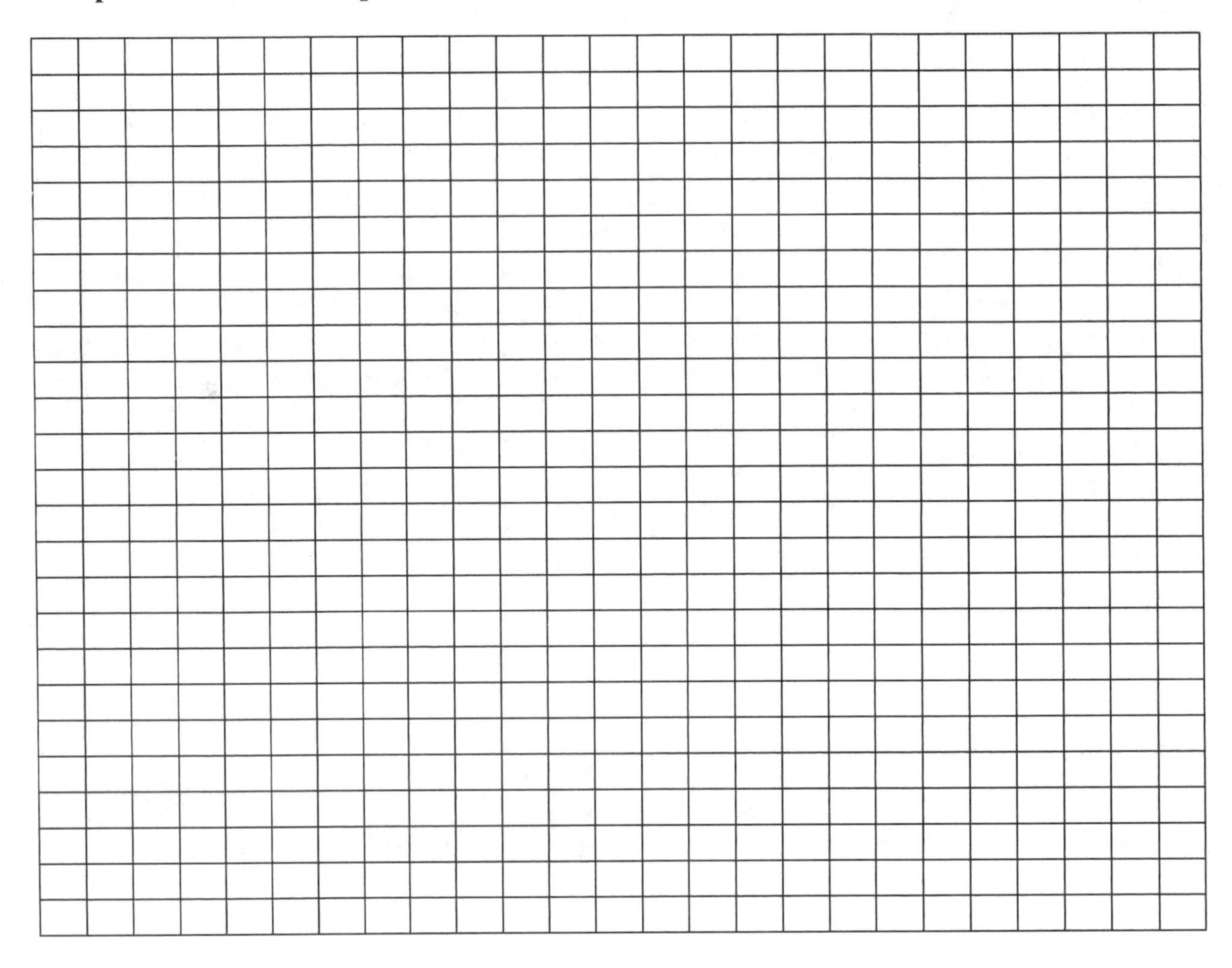

Using your experimental values of ΔH_{Rxn} from parts B and C and Hess's Law, calculate the ΔH_{Rxn} for

$$Mg(s) + \frac{1}{2}O_2(g) \rightarrow MgO(g) \quad \Delta H_{Rxn} = \text{Experimental Goal}$$

Be sure to show the equations and setup used.

Compare your value with the literature value of ΔH_{Rxn} and determine the percent error in your experiment:

CONCLUSION: *(The conclusion section for this experiment should be several well-developed paragraphs. These paragraphs should cover the following information: A discussion on how the various data was arrived upon, this should include but is not limited to: 1) the calculation of the maximum temperature of the reaction in Experiment A, 2) the heat of reaction for hot and cold water, 3) the heat of reaction of the calorimeter cup and 4) the heat capacity of the cup. For Parts B and C: 1) the maximum temperature achieved, 2) the heat of reaction of the solution, cup and reaction, and 3) ΔH_{Rxn}. Be sure to provide some type of discussion of these values. Also report the overall reaction being studied and the ΔH_{Rxn} values found using both experiment and literature values. Be sure to discuss differences in these values as well as the percent error in the experiment. Finally discuss any errors in the experiment.)*

Experiment 9

Post-Laboratory Assignment

Name: ______________________________ **Date:** __________________

Instructor: ______________________________ **Sec. #:** ____________

Show all work for full credit.

1. What is the difference between the heat (q) of a reaction and the reaction enthalpy (ΔH_{RXN}) ?

2. Why is T_{final} taken from extrapolating back to the beginning of the experimental run rather than from the last temperature reading taken?

Experiment 10

Gas Laws and Air Bags

Introduction

Throughout this semester we have provided experiments that illustrate some of the more complex concepts of chemistry as you were learning about them in lecture. We have also presented you with evidence that each of the concepts or techniques covered was either valuable to your future as a chemist or relevant to your everyday life. This experiment on "Gas Laws and Airbags" is both. Gas laws and the general concepts of gas volumes and pressures are vital for anyone deciding on a career in chemistry and even more important to those focusing on medicine. For example, oxygen saturation in the body is controlled by both internal and external pressures. Why is it harder to breathe at higher altitudes? What is hypoxia? What happens to a patient who receives too much oxygen? Not enough? And how do you as the doctor or nurse control the volume delivered? None of these questions can be answered without a firm understanding of the gas laws.

The experiment you will conduct today will not require you to administer oxygen to a patient (hopefully) but rather focuses on the automobile airbag, another very useful device that depends on the use of gas laws. Airbag deployment is chemically created because mechanical deployment would be too slow. An explosive production of nitrogen gas from sodium azide creates the filled airbag in a fraction of a second. Gas laws allow the construction of a completely full but not overfull bag of gas that keeps the driver and passenger(s) of a car from serious injury during a collision.

In order to study how gas laws can provide answers to the construction of airbags, you are going to design and build one. In addition to constructing the airbag, we are also going to revisit the use of a barometer to measure the atmospheric pressure. Some barometers use a Vernier scale, so make sure you remember how to read this type of scale.

Background

Though based on assumptions that gas molecules do not interact with each other and occupy no volume—assumptions you will later learn to be over-simplified—the ideal gas law is still incredibly useful in characterizing the properties of most gases. Mathematically, the ideal gas law is expressed by the equation $PV - nRT$, where P is the gas pressure, V is the volume in liters, n is the number of moles, T is the Kelvin temperature, and R is a constant.

The constant R is called the gas constant and takes on a number of different values depending on the units used for pressure. Note that although the units for pressure and volume can vary, the unit for temperature must always be in Kelvins.

Common Values of *R*
$8.314472\ J \cdot K^{-1} \cdot mol^{-1}$
$0.0820574587\ L \cdot atm \cdot K^{-1} \cdot mol^{-1}$
$8.20574587 \times 10^{-5}\ m^3 \cdot atm \cdot K^{-1} \cdot mol^{-1}$
$8.314472\ L \cdot kPa \cdot K^{-1} \cdot mol^{-1}$
$62.3637\ L \cdot mmHg \cdot K^{-1} \cdot mol^{-1}$
$62.3637\ L \cdot Torr \cdot K^{-1} \cdot mol^{-1}$
$1.987\ cal \cdot K^{-1} \cdot mol^{-1}$

In this experiment, you will use the ideal gas law to predict the volume of gas produced from a simple chemical reaction. This reaction will be used to create a mock automobile airbag.

Airbags

Airbags are safety devices found in most cars being produced today. The reaction used in commercial airbags depends on sodium azide (NaN_3), a fairly toxic chemical. When activated, the airbag's sodium azide rapidly undergoes a decomposition reaction generating sodium metal (Na) and nitrogen gas (N_2).

$$2NaN_3(s) \rightarrow 2Na(s) + 3N_2(g)$$

The nitrogen gas inflates the airbag to provide a safety cushion for the passengers in the car. Airbags deploy in 0.015 sec for high-speed crashes and 0.025 sec for low speed crashes.

The Reaction

Since your laboratory is not equipped to work with chemicals as toxic as sodium azide, we will use sodium bicarbonate ($NaHCO_3$) to produce the gas. When sodium bicarbonate reacts with acetic acid (CH_3COOH), carbon dioxide (CO_2) and two other products are formed as shown below:

$$NaHCO_3(s) + CH_3COOH(aq) \rightarrow CH_3COONa(aq) + 3H_2O(l) + CO_2(g)$$

The stoichiometry of this reaction is quite simple since all reactants and products are in a 1:1 molar ratio. Because the overall goal of this experiment is to design an "airbag" that inflates rapidly and fully, without wasting materials, you will need to convert molar amounts of reagents into gram and volume amounts in order to produce the precise volume of CO_2 that will fill your "airbag."

Stoichiometry and Manipulating the Ideal Gas Law

To be able to predict the volume of the gas being produced, you will use the ideal gas law equation. The variables in the ideal gas law are volume, temperature, pressure, and the number of moles of gas. We know that our unknown in the equation is going to be the amount of gas produced. Therefore, we can rearrange the ideal gas law to solve for n:

$$n = \frac{PV}{RT}$$

Since we know R is a constant, all we have to measure is the volume of our bag, the atmospheric pressure, and the temperature of our surroundings.

For this experiment we will use room temperature and pressure. A thermometer should be used to determine the room temperature and a barometer to measure the atmospheric pressure. With the temperature and pressure taken care of, all that remains is to determine the volume of your "airbag." This can be done by filling the bag with water and then measuring the volume of water. The water also serves the purpose of checking the bag for leaks which would ruin your experimental results.

By measuring the volume of gas needed and substituting it, along with room temperature and pressure, into the ideal gas law, you can predict the moles of gas required to fill the "airbag." You can then use simple stoichiometry to calculate the exact amounts of reactants ($NaHCO_3$ and CH_3COOH) to produce the correct amount of CO_2 needed to fill the bag.

Example Problem

To further clarify the principles and calculations in this experiment, let's look at an example incorporating real data from an airbag with sodium azide. We need to find out how many grams of NaN_3 are needed to completely fill a 75.0-L airbag to a pressure of 1.30-atm at 25.0 °C with N_2.

To find the mass of sodium azide required, we need to calculate the number of moles of N_2 needed.

$$n = \frac{PV}{RT} = \frac{(1.30 atm)(75.0 L)}{\left(0.08206 \frac{L \cdot atm}{mol \cdot K}\right)(298.15 K)} = 3.99\ mol\ N_2$$

Substituting these values into our equation as shown, we calculate that we need 3.99-mol of nitrogen gas. Knowing the number of moles of N_2 required, we can use the balanced equation given at the beginning of this discussion in order to determine the number of grams of NaN_3 needed.

$$3.00 \text{ mol } N_2 \left(\frac{2 \text{ mol } NaN_2}{3 \text{ mol } N_2}\right)\left(\frac{65.02 \text{ g } NaN_3}{1 \text{ mol } NaN_3}\right) = 173 \text{ g } NaN_3$$

Therefore, we need ~173 g of sodium azide to produce 75.0 L of N_2.. The calculations just completed are almost identical to those you will perform in this experiment.

Procedure

SAFETY NOTES: Acids can cause burns if they come in contact with your skin. Wash your hands immediately if you feel a burning or itching sensation. Use all necessary precautions when handling the acid container.

Part I: Calculating the Volume of the "Airbag"

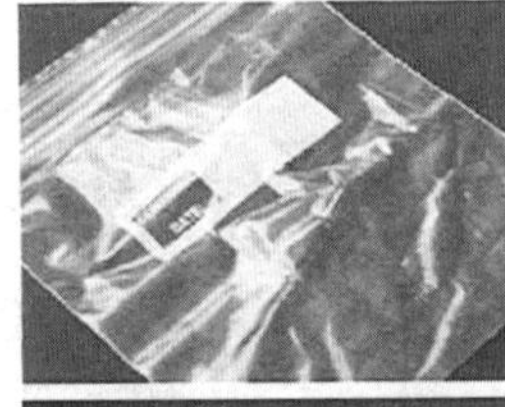

Obtain 1 plastic empty "airbag" from the front counter and determine its volume.

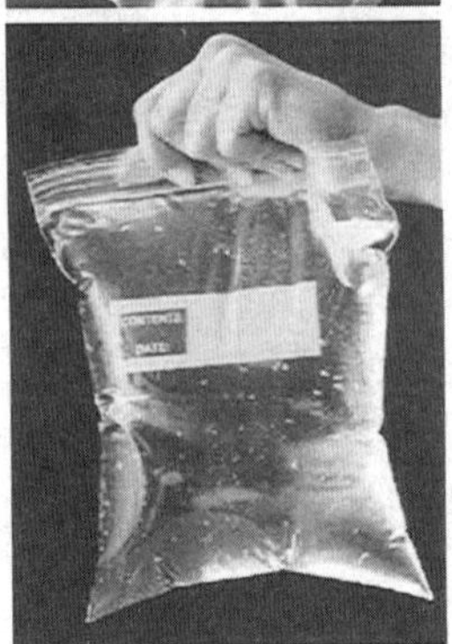

Fill your airbag as full as possible with water and use a graduated cylinder to calculate the volume of gas needed by determining the volume of water that filled the bag.

Part II: Testing Your "Airbag"

Use paper towels or Kim wipes to completely dry the inside of the bag.

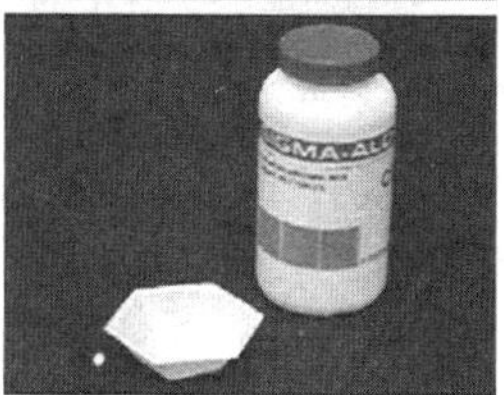

Weigh out your calculated amount of sodium bicarbonate and add it to your "airbag." Be sure to record the exact gram amount you add.

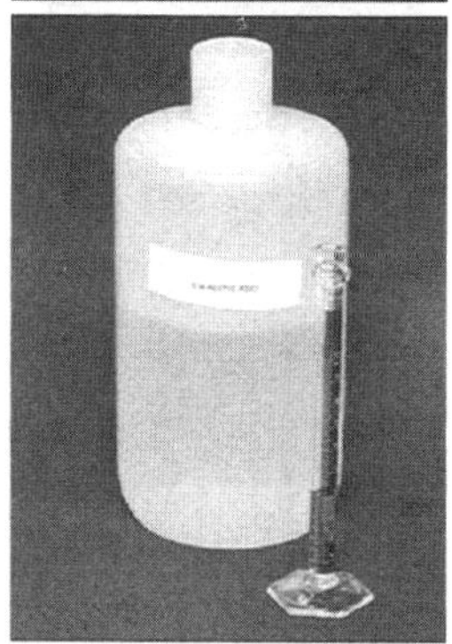

Measure out your calculated volume of 6.0 M acetic acid using a graduated cylinder. Record the volume to the nearest 0.1 mL. NOTE: Be very careful when handling the acid and wash your hands immediately if you get some on your skin. Acetic acid is not a strong acid but it will still cause irritation if left in contact with your skin.

Test your airbag by quickly but carefully pouring your acetic acid into the "airbag" and then seal it. Mix the ingredients completely by shaking and squishing the bag.

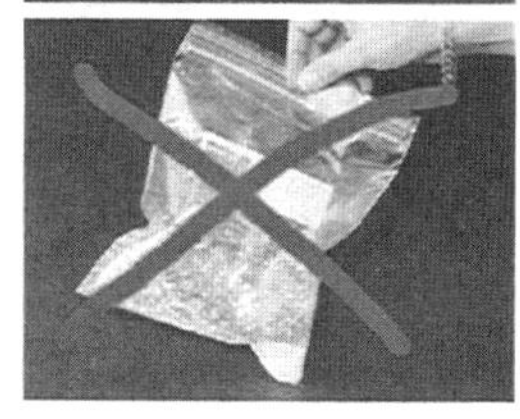

Once the reaction is complete, be sure to make observations of the fullness of the airbag and whether or not the reagents seem to be completely used up. A squishy bag with a lot of residue indicates failure.

Repeat steps 1 thru 6 making adjustments to your calculated amounts of reagents if necessary. Dump the remaining fluid from the bag in the waste jar into the hood and rinse and dry your bag thoroughly. Be sure to show a successful "airbag" to your TA before leaving.

Experiment 10

Pre-Laboratory Assignment

Name: ______________________________ **Date:** ________________

Instructor: ______________________________ **Sec. #:** ____________

Show all work for full credit.

1) Airbags in cars are devices that inflate very rapidly (in about 30 milliseconds) when there is a collision, creating a cushion that deflates slowly and cushions the impact of the collision on the driver or the passenger. The bag is inflated by a gas which is generated by the rapid heating of sodium azide (NaN_3) to about 300 °C, causing it to decompose into sodium metal and nitrogen gas.

   Complete and balance the equation for this reaction, indicating the states of the reactant ($NaN_3(s)$) and products ($Na(s)$ and $N_2(g)$).

   Enough nitrogen must be generated in the bag to create a total pressure of 3.50 atm, which then drops as the bag slowly deflates. Assuming the volume of the bag is 65.0 L and the temperature in the car is 28 °C, calculate the molar quantity of N_2 that must be generated.

   What molar quantity of sodium azide is required to generate this much nitrogen?

   What mass of sodium azide is required?

   The sodium metal produced in the reaction is highly reactive and potentially explosive, so it is removed in a reaction with excess $KNO_3(s)$ that produces $K_2O(s)$ and $Na_2O(s)$ and additional nitrogen. Write a balanced equation for this process, showing the states of the reactants and products.

2) In this laboratory experiment you are to generate sufficient carbon dioxide to inflate a plastic Ziploc® bag. You first must carry out some calculations to determine the quantity of reagents needed.

   Carbon dioxide will be generated by reacting solid sodium bicarbonate ($NaHCO_3$) with a solution of acetic acid (CH_3COOH) as follows:

$$NaHCO_3(s) + CH_3COOH(aq) \rightarrow H_2CO_3(aq) + CH_3COO^-(aq) + Na^+(aq)$$
$$H_2CO_3(aq) \rightarrow H_2O(l) + CO_2(g)$$

   You measure the plastic ziploc bag dimensions to be 6.90 inches by 7.15 inches. Assuming the average thickness of the bag when inflated will be 1.45 inches, what do you calculate the volume of the bag to be?

   Of course, this volume is only approximate as the width of the inflated bag is not uniform. To get a better value for the volume, you fill the bag with water, then measure the volume of the water to be 1174.5 mL. Now you measure room temperature to be 24 °C and the barometer on the wall gives a reading of 753 torr for the atmospheric pressure. Using the ideal gas law, calculate the number of moles of CO_2 that will be required to fill the bag at this temperature and pressure.

   Now inspect the reactions above generating CO_2 and calculate the molar quantity of $NaHCO_3$ and CH_3COOH needed.

   Quantity of $NaHCO_3$:

   Quantity of CH_3COOH:

   What mass of $NaHCO_3$ will be needed?

   You have a solution of 5.00 M acetic acid. What volume of this solution will be needed?

Experiment 10

Laboratory Report

Name: ______________________________ **Date:** ___________________

Instructor: ______________________________ **Sec. #:** ____________

PURPOSE: (*The purpose should be several well-constructed sentences describing what your experiment was designed to accomplish and the criteria used to determine success. These sentences should include both concepts and techniques.)*

____

____

____

____

____

____

____

____

PROCEDURE: (*The procedure section should reference the lab manual and include any changes made to the procedure during the lab.)*

____

____

____

____

____

____

DATA: (*The data section should include your own personal data and observations.)*

Write the balanced chemical equation for the reaction being studied:

Bag Dimensions	Bag Volume (L)	Moles CO_2 Needed	Moles $NaHCO_3$ Needed	Moles CH_3COOH Needed	Grams $NaHCO_3$ Needed	Volume of 6.0 M CH_3COOH Needed	Success or Failure

Room Temperature: ______________ Room Pressure:________________

Observations regarding the success or failure of the airbag:

____
____
____
____
____
____
____

CALCULATIONS: (*The calculation section should include sample calculations for each column in the data table. Be sure to include conversions of units.)*

CONCLUSION: (*The conclusion section should be in paragraph format and discuss the success or failure of each airbag design. This should include the volume of each air bag and the amount of materials used. Along with the components of the airbag you should include a discussion of the reasons for the success or failure of each bag. Also cite any problems or errors that may have occurred in the experiment and any second trials that may have taken place.*)

Experiment 10

Post-Laboratory Assignment

Name: ______________________________ **Date:** ____________________

Instructor: ______________________________ **Sec. #:** ____________

Show all work for full credit.

1) Is the gas produced in this experiment really an "ideal" gas? Explain.

2) The gas production in this experiment is fairly slow, but the production in a car airbag is quite rapid. What is the reason for the difference in the two reactions?

Experiment 11

Intermolecular Forces and the Triple Point of CO_2

Introduction

The ability of matter to change from one state to another is vital in geological, chemical, and even physical realms. As we all know, water can naturally be found as a solid, liquid, and gas. What you may not know is just how matter changes from one form to another and the factors affecting such a process. This experiment is designed to further elucidate the concept of phase transitions using a rather elementary scientific approach. After completing this lab—with only a few chemicals, a thermometer, and a "homemade" barometer—you will better understand phase changes and the effects of temperature, pressure, and intermolecular forces on those changes.

Whether you are aware of it or not, every one of us has witnessed a change of phase, whether it be through the evaporation of sweat from our bodies while working out, or something as simple as making a tray of ice cubes. Phase changes are important geologically, as there is growing concern for the melting of the polar ice caps caused by global warming. However, before we save the world from melting polar ice caps we have to understand phase transitions.

We all know that solid water can easily be converted into its liquid form simply by adding heat, but the processes occurring at the molecular level are not as obvious. Recall that phase changes are physical, not chemical, changes. This means that instead of breaking intramolecular bonds—the bonds between atoms that cause those atoms to form a specific molecule—intermolecular bonds are being broken. When the intermolecular attractions between water molecules in a drop of water are weakened and ultimately broken, water molecules escape into the gas phase. To emphasize the distinction between these intermolecular bonds and chemical bonds, we will instead refer to intermolecular attractions. To investigate this notion, we are going to observe the change in temperature that occurs in the surroundings when five different liquids are evaporated at room temperature. Then we will use this information to draw conclusions about the strength of each liquid's intermolecular forces.

The second part of this experiment will investigate the effect of temperature and pressure upon phase changes through the use of a barometer. A barometer can be any instrument that is used to measure atmospheric pressure. Temperature and pressure play an intricate role in the transition of matter from one phase to another, as shown generally by a phase diagram. However, graphs sometimes do not necessarily show how "explosive" such relationships can be, such as the one you will be investigating in this portion of the experiment.

The first part of this experiment will advance your knowledge of phases, the changes they may undergo, the underlying mechanisms responsible, and the effect of intermolecular forces upon them, while the second portion will show you the relationship between temperature and pressure in determining the triple point of CO_2.

Background

All matter occurs naturally in one of three physical states: solid, liquid or gas. A fourth state, plasma, will not be considered here (It requires significant magnetic fields for containment). The term ***state*** of matter is used interchangeably with the term ***phase***. A phase is defined as any physically distinct, homogeneous part of a system. The phase or state of a substance is determined by both the energy of the system and by the intermolecular attractions the substance undergoes. According to the Kinetic Molecular Theory, gases are composed of high energy molecules that do not experience intermolecular forces due to both the speed at which they travel and the space that exists between their molecules. On the other hand, the molecules of a liquid are spaced much closer together and move much more slowly. Therefore liquids undergo a number of intermolecular forces, ranging from the weak London dispersion forces to the much stronger networks of hydrogen bonds. Solids are composed of low energy molecules that are packed very closely together. The molecules in solids are said to vibrate, not move. Most ionic compounds are solids at room temperature because the ionic interactions they experience are extremely strong. The energy required to melt these ionic solids is normally very high and the energy required to promote them into the gas phase even greater.

Phase Changes

So how do phase changes take place? For a solid to change into a liquid (melt) or a liquid to change into a gas (evaporate) the intermolecular forces between the particles must be overcome. This requires the input of energy. The most common way to put energy into a system is by adding heat. You have experience with solid ice melting into liquid water and liquid water boiling to form gaseous steam, both phase changes resulting from the input of heat energy.

In this experiment, you will investigate the strength of the intermolecular forces of five liquids. Their identities are methanol, isopropanol, n-hexane, acetone and water. The liquids will be provided to you as unknowns in bottles marked A–E. It is your job to identify each of the unknown liquids by determining the relative strength of their intermolecular forces and by making any other scientific observations you can. By comparing your observations to the known properties you gathered in your pre-laboratory exercises, you should be able to identify the compounds.

Each of the five liquids will evaporate at room temperature. Those with stronger intermolecular forces will require more energy input to evaporate than those with weaker intermolecular forces because they have a higher enthalpy of vaporization (ΔH_{Vap}). The enthalpy of vaporization is a physical quantity that measures the amount of energy required to vaporize the liquid, i.e., cause the liquid to pass through a phase change from liquid to gas phase. Evaporation is an endothermic process that results in a drop in temperature of the surroundings. The compound requiring the least energy for vaporization (smallest ΔH_{Vap}) and thus the weakest intermolecular forces, will exhibit the largest temperature decrease. This occurs because as the molecules evaporate they carry heat away from the thermometer which registers each loss as a drop in temperature. Molecules with stronger intermolecular forces require a larger amount of energy to accumulate before they can evaporate so their removal of heat from the thermometer is a much slower process. The length of time between evaporation events allows the thermometer to recover heat from the atmosphere and thus we do not observe as great a drop in temperature for these molecules. If we monitor this temperature change, we can then relate the magnitude of the temperature drop to the relative strength of the intermolecular forces of each substance. Relating this information to their known ΔH_{Vap} values allows us to determine the identity of each of the unknowns.

Phase Diagrams and the Triple Point of CO_2

The second part of this experiment investigates the triple point. Phase diagrams illustrate the temperatures and pressures at which matter changes from one phase to another. The triple point of a substance is the temperature and pressure at which all three states of matter exist simultaneously. For this experiment we will determine the triple point of carbon dioxide. In its solid state, CO_2 is commonly referred to as "dry ice." If you look at the phase diagram of CO_2 you can see that at room temperature and pressure CO_2 is a gas. What is also clear is that if dry ice (solid CO_2) is present at room temperature and pressure, it will not melt, but instead will sublime, that is, it will transition directly from the solid to gaseous state. In order to observe the liquid state of CO_2 the pressure must be at or above the pressure of the triple point (5.1 atm). In this experiment, the pressure of the CO_2 is raised by placing dry ice in a sealed pipet and allowing it to sublime at room temperature. As more and more gas forms, the pressure inside the pipet increases, hopefully to the triple point. You will use a Boyle's law-type micro gauge to determine the pressure at the observed triple point (the point at which you observed the dry ice liquefy).

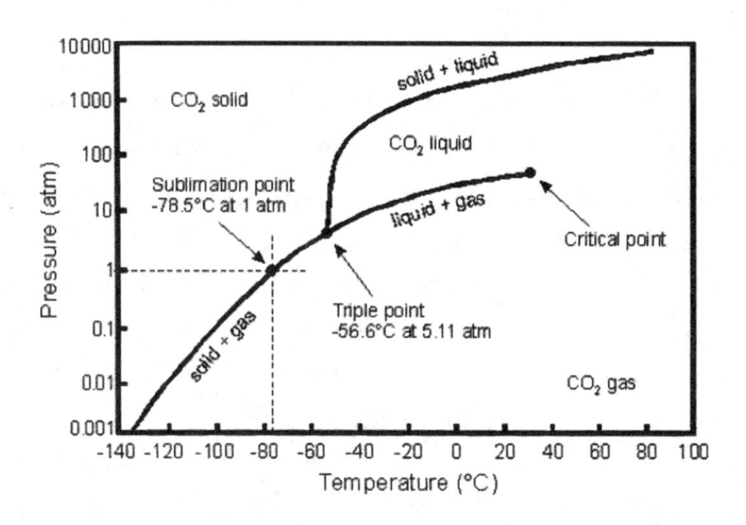

Pressure-Temperature phase diagram for CO_2.

Boyle's Law

Boyle's Law: $P_1V_1 = P_2V_2$

Where P_1 is read from the barometer in the lab.
V_1 is the total volume of your micro gauge
V_2 is the final volume read from your micro gauge at the triple point
P_2 is the experimentally determined pressure at the observed triple point.

Boyle's law relates the volume of a system to its pressure. The micro gauge we will construct will be made of a thin-nose plastic pipet sealed at one end. Normally, we would need to calculate the true volume of the pipet by measuring the radius of the tube and use the equation for the volume of a cylinder ($V_{cyl} = \pi r^2 h$) to determine its volume. However, the pipet's diameter will remain constant throughout the experiment and only the length portion (h) of the volume equation will change so we can omit the volume calculation on the principle that if a value is equal on two sides of an equation that value will cancel itself out. If we then complete the experiment and substitute the values into Boyle's law in the manner described above, we should be able to measure the approximate triple point of CO_2. I believe you will be surprised at how accurate your crude micro gauge can be.

Procedure

SAFETY NOTES:

Part 1. All of the organic solvents used in this part of the experiment are flammable. No flames should be used in lab today. The fumes of some of the solvents can be irritating. Please use your hand to waft the solutions towards you to smell them if necessary. Do not rinse any of the solvents down the sink. There are waste containers provided in the hood. Avoid direct contact with the solvents as they may cause irritation. If you should get any of the solvents on your skin wash the exposed skin with soap and water as soon as possible and notify your instructor.

Part 2. Dry ice can cause frostbite if held in the hand too long. Minimize direct contact with the dry ice as much as possible. Make sure the pipet bulb is submerged in the water while the dry ice is subliming. There is always the possibility that the bulb will burst under the pressure being exerted. The water in the cup will absorb the concussion and the debris. (Note: You might get wet from the splash).

Part I: Evaporation and Intermolecular Forces

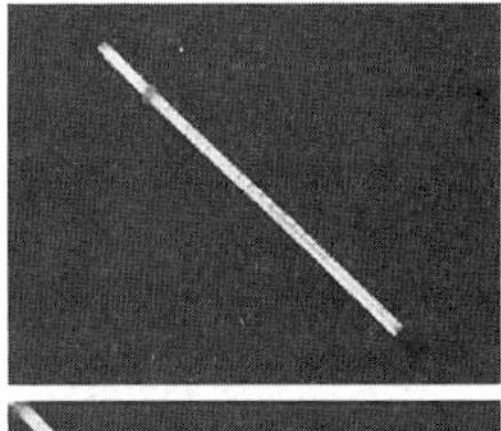

Collect a thermometer from the front counter.

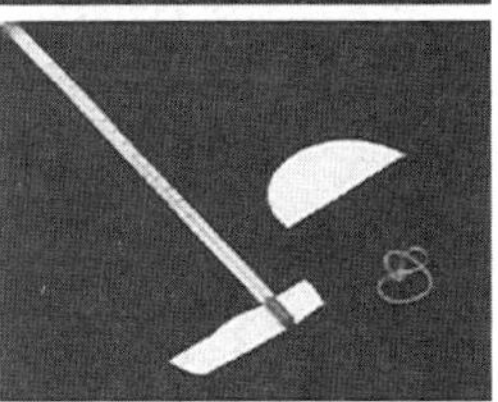

Prepare the thermometer by wrapping small cut pieces of filter paper secured by small rubber bands around the tip. Roll the filter paper around the tip in the shape of a cylinder and then slip the rubber band(s) around the filter paper to secure it in place. Make sure that the filter paper is even with the end of the thermometer.

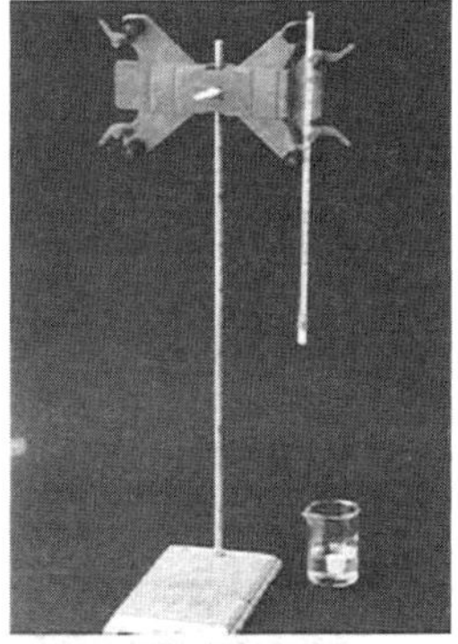

Collect a stand and clamp from the front counter. Place a clamp adaptor around the thermometer and clamp it in place. Make sure you can read the thermometer from ~30 °C and below.

Place ~10 mL of one of the unknown liquids A–E in a 100 mL beaker.

Lift the beaker of unknown liquid under the thermometer and submerge the tip in the liquid and hold it there for at least 30 seconds.

Begin collecting temperature data while the tip is still submerged. Take several readings for at least 1 minute, the average of these readings will be your T_I.

Lower the liquid away from the thermometer and observe the temperature change that occurs. Continue to take temperature readings until the temperatures pass through a minimum, this will be your T_F.
When done, clean the probe with DI water and repeat the above process for each of the unknown liquids.

Part II: The Triple Point of CO_2

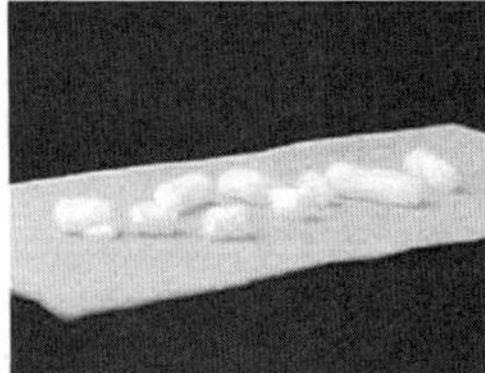

Collect some dry ice from the front of the lab (you only need a few small pieces).

Place the dry ice on a paper towel on the lab bench and watch what happens. Make notes.

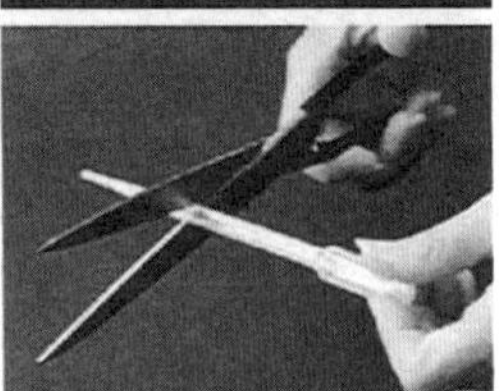

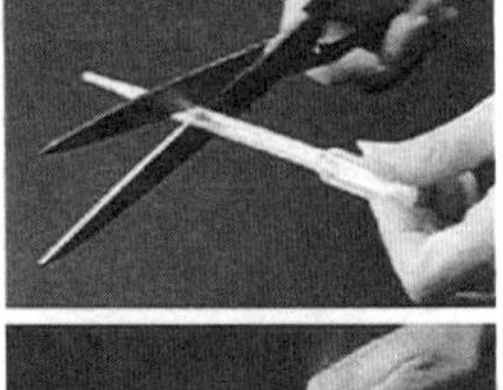

Take a plastic pipet and cut off the graduated tip of the pipet.

Crush some dry ice inside of a folded paper towel.

Place enough dry ice "crumbs" into the pipet to fill the bulb up about a third-full.

Fill a clear plastic cup to a depth of about 7–8 cm with water.

Fold the tip of the pipet bulb shut and clamp tightly using the pliers provided. Make sure the seal is very tight so that no gas can escape.

Using the pliers immediately lower the bulb portion of the pipet into the water.

Observe from the side of the cup what happens. Make notes. You can also release and tighten the pliers. Make observations on what occurs with the CO_2 each time you do.

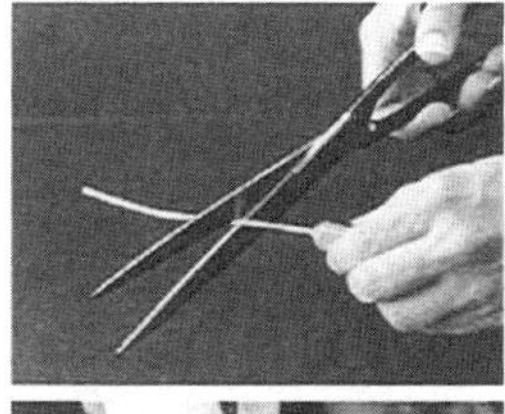

Collect a plastic thin stem pipet from the front of the lab. Using the technique demonstrated by your lab TA, cut the "nose." off of the pipet.

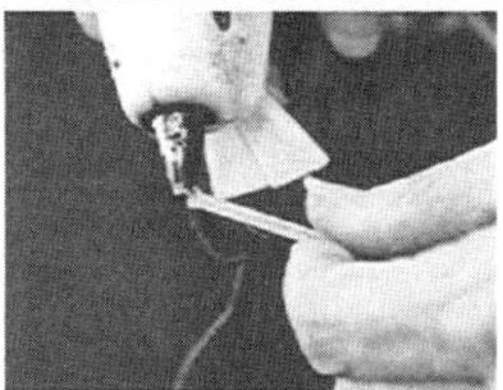

Collect a piece of string from the front counter and use the hot glue gun provided to attach the string to the pipet "nose". Be sure that the glue completely seals one end of the pipet "nose."

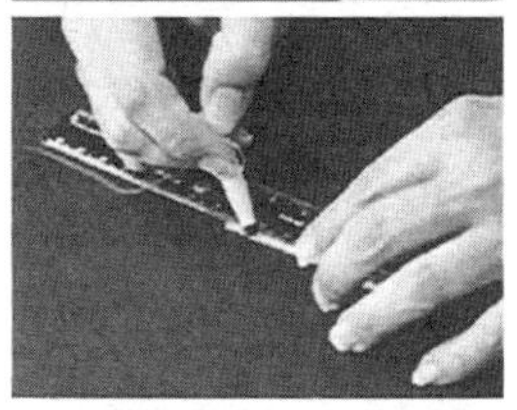

Using a wax pencil and a ruler, mark the newly created pipet micro gauge every centimeter from the sealed end as far up as you can. "0" is maked at the point where the glue seals the tube.

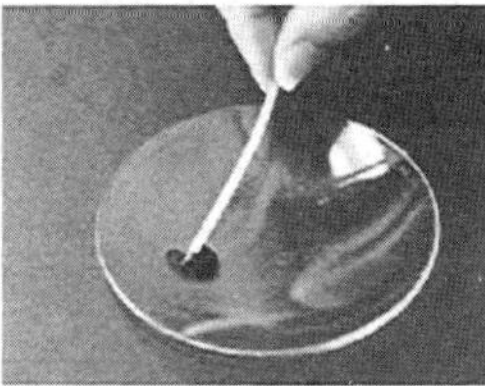

Place a drop of food coloring on a watch glass and add a few drops of water to dilute it. (Everyone at a bench can share.)

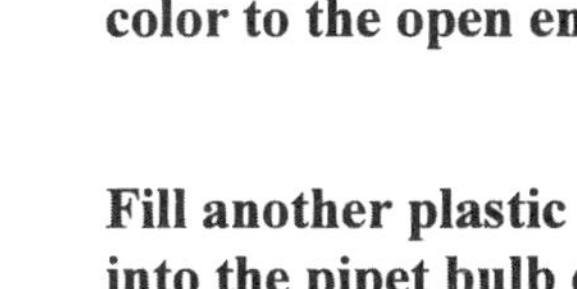

Cut the end of the micro gauge at an angle and add a drop of food color to the open end of the micro gauge.

Fill another plastic pipet with dry ice as before. Insert the micro gauge into the pipet bulb open end down but suspended by the string such that the micro gauge is just above the CO_2.

Clamp the pipet shut making sure you don't crush the micro gauge just the string.

Place the Pipet with the microgauge into the cup of water as earlier . Pay close attention to the micro gauge and record the change in volume at the observed triple point (the point at which you first observe liquid CO_2) by determining the distance the food color has traveled at that point. Read the microgauge from the bottom of the food color mark.

Experiment # 11

Pre-Laboratory Assignment

Name: ______________________________ **Date:** ________________

Instructor: ______________________________ **Sec. #:** ___________

Show all work for full credit.

Copy the information gathered in part 1 and 2 of this pre-lab into your lab notebook for later use.

1. For the following liquids: methanol, isopropanol, n-hexane, acetone and water, (a) write the molecular formula; (b) draw a Lewis structure for each compound; and (c) calculate the molecular weight of each compound.

2. The relative strength of the intermolecular forces depends on the size (i.e., MW), and structure of the molecules, and includes in rough order of strength: ion-ion, ion-dipole, hydrogen bonding, dipole-dipole (polar), dipole-induced dipole, and induced dipole-induced dipole (dispersion) forces. The latter increases with molecular weight and can sometimes become the most predominant force. There are no ions in these liquids, and all the molecules are alike, leaving primarily hydrogen bonding, dipole-dipole, and disperson forces to consider.

What are the strongest intermolecular forces in Acetone?

What are the strongest intermolecular forces in Methanol?

What are the strongest intermolecular forces in 2-Propanol?

What are the strongest intermolecular forces in Water?

What are the strongest intermolecular forces in Hexane?

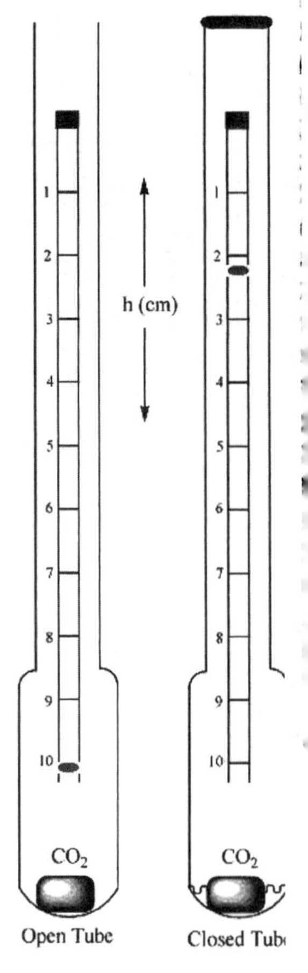

3) The figure at the right shows your experimental set-up for estimating the triple point of CO_2. Carefully mark off your capillary tube in ten 1-cm segments, and note the position of the drop of dye solution when the capillary is inserted in the open plastic tube. When the top of the plastic tube is sealed, the pressure inside will increase as the CO_2 sublimes, and the drop of dye will be forced up the capillary, compressing the air inside it. Note the position of the drop of dye when liquid CO_2 starts to form. The pressure inside the capillary should then correspond to the pressure at the triple point of CO_2.

You can calculate the pressure from a modification of Boyle's law, $\mathbf{P_1V_1=P_2V_2}$. The volume of the capillary $\mathbf{V} = \mathbf{hA}$ where h is the height and A is the cross-sectional area, which is constant. Boyle's law therefore reduces to $\mathbf{P_1h_1=P_2h_2}$.

where P_1 is the pressure in the open tube, h_1 is the distance of the drop from the top of the capillary in the open tube, h_2 is the distance of the drop from the top of the capillary in the closed tube, and P_2 is the pressure in the closed tube, the pressure at the triple point.

What are the heights of the air columns in the two capillaries on the previous page? (Measure from the top of the dye droplet).

Height of column in the open tube (h_1):
Height of column in the closed tube (h_2):

Calculate the pressure in the closed tube, assuming the pressure in the open tube is 0.96 atm.

Pressure in the closed tube =

Experiment 11

Laboratory Report

Name: ______________________________ **Date:** ___________________

Instructor: ______________________________ **Sec. #:** ____________

PURPOSE: (*The purpose should be several well-constructed sentences describing what your experiment was designed to accomplish and the criteria used to determine success. These sentences should include both concepts and techniques.*)

PROCEDURE: (*The procedure section should reference the lab manual and include any changes made to the procedure during the lab.*)

DATA: (*The data section should include your own personal data and observations.*)

Part 1: ΔT values for the Unknown Liquids

Unknown Letter	**Starting Temperature (°C)**	**Ending Temperature (°C)**	**ΔT**

Based on those ΔT values rank the intermolecular forces for unknown liquids A–E:

Using the rankings and what you know about the molecular structure of the liquids, propose identities of the unknown liquids:

Part 2: Observations of the CO_2:

____

____

____

____

____

____

P_1 read from the barometer:

V_1 total volume of your micro gauge:

V_2 final volume read from your micro gauge at the triple point:

P_2 pressure at the observed triple point:

CALCULATIONS: *(For Part I, show the calculation of ΔT. For part II determine the experimental value for the triple point of CO_2.)*

The triple point of CO_2 can be found using Boyle's Law, a variation of the ideal gas law, and solving for P_2.

Boyle's Law =

NOTE: In this equation P_1 is the pressure of the room, measured using a barometer. V_1 is the total volume (length) of the micro gauge and V_2 is the final volume read from the micro gauge at the triple point (the length of the gauge to the top of the food coloring). Both V_1 and V_2 in this experiment can be used in the Boyle's Law equation as distances (see Background for explanation), therefore V_1 is the total length of the micro gauge and V_2 is the distance the food coloring traveled. P_2 is the unknown variable, the pressure at the observed triple point. In order to solve for P_2 you must rearrange Boyle's Law in terms of P_2.

P_2 =

Once the value of P2 is known the percent error of your micro gauge can be determined using the following:

$$\%\ error = \left| \frac{experimental\ value - accepted\ value}{accepted\ value} \right| x\, 100\%$$

The experimental value is P_2 and the accepted value is the triple point found on the phase diagram of CO_2. Make sure the units for both triple points are the same before calculating percent error.

% error =

CONCLUSION: (*The conclusion section should be several paragraphs addressing the following: For Part I, the identity of the five unknown liquids with an explanation supporting the identification made; a discussion of the similarities and differences of your rankings compared to the heat of vaporization ranking done in the pre-lab. For Part II, summarize the observations of CO_2 made; state and compare the actual and experimental values for the triple point of CO_2 including the percent error. Finally, discuss any errors in the experiment.*)

Experiment 11

Post-Laboratory Assignment

Name: ______________________________ **Date:** ________________

Instructor: ______________________________ **Sec. #:** ____________

Show all work for full credit.

1. How is the change in temperature we observe as the various solvents evaporate related to their ΔH_{vap}? Explain this relationship.

2. In our use of the microgauge to determine the triple point pressure of CO_2, we only measure the height change (distance the dye moves) in the gauge, not the actual volume. Why does this approximation not ruin our results?

Experiment 12
The Purification of Water

Introduction

Clean drinking water is of great importance to people worldwide. The purification of drinking water is a laborious and complicated process. In this laboratory you are introduced to the various processes by which water quality is assessed and treated. As you are guided through these processes you will also be introduced to a number of valuable new chemical techniques.

The first step in any water processing is the collection of samples. As with any physical sample collection, the location and specific physical parameters (temperature, pH, sediment content, etc.) will affect the overall appearance and chemistry of your sample. The procedure provided for the proper collection of your water sample is designed to establish good collection techniques for your future as a scientist. Why is this so important? Suppose you collect your water sample in an old bleach bottle. If the bottle still contains remnants of the bleach it will contaminate your water sample. Another example would be if you collected your water sample just after a rainstorm, but did not note this in your report. It is likely that the levels of contamination in your samples would be considerably lower than in samples taken prior to the rain, as the rain would most likely dilute the contamination.

The second step in this experiment is the analysis of the possible contamination in your collected water sample. This establishes a baseline from which you can determine the effectiveness of your purification process. Most municipalities have entire departments dedicated to the monitoring of water quality in their lakes, ponds and drinking water. The process we will use to determine the phosphate content of water samples is identical to the one used across the country by many of these labs. In order to determine the phosphate concentration we will use an analytical technique called absorption spectrophotometry. Details of the specific type of spectrophotometry to be used in this experiment are outlined in the background section of this lab. This experiment introduces you to several other new techniques that are invaluable to a chemist: serial dilution, filtration and pH.

Spectrophotometry is the study of the interaction of electromagnetic radiation with matter. Through the many spectroscopic techniques available, the shape, composition and the way a compound or molecule reacts can be determined. It is a widely used technique with several variations, including but not limited to Raman, UV-visible, infrared and nuclear magnetic resonance. Each of these types of spectrophotometry observes the interaction between the sample and a different type of radiation. The actual instrumentation and the process for preparation of samples for each of these techniques are unique, but the general concept of spectrophotometry remains the same. A basic understanding of how to use a spectrometer to determine the concentration of a sample is yet another skill you should acquire as a chemist.

Serial dilution is a process that is vital to the conservation of chemical resources in a lab. For example, if you need about 10 mL each of a series of solutions from 1M to 5M spaced at half molar intervals (e.g., 1M, 1.5M, 2M, 2.5M…) you could simply use a 5M solution as stock and take enough stock solution to make each of the more dilute solutions. Making all these solutions would take a very long time and use a large amount of stock solution. Serial dilution would allow you to make these mixtures using a relatively small amount of stock solution. In a true serial dilution, each solution becomes the stock solution for the next solution needed, allowing you to work very efficiently.

In this lab we will create colored solutions of known concentration using serial dilutions and then determine their absorption value using a spectrophotometer. This will allow us to produce a calibration curve by which we can determine the concentration of the water samples collected in the field. The use of graphical analysis to determine an unknown value is another recurring theme in the lab and a handy technique you should become competent in performing.

Other techniques are explored during the purification process. Precipitation followed by filtration is a process by which some contaminants can be separated from a solution. Vacuum filtration uses suction to increase the speed at which you can filter your water. Precipitation uses chemical reactions to remove unwanted ions from your water. Referencing the solubility tables in your course textbook, you should find that the addition of CaO and $Al_2(SO_4)_3$ produces precipitates that remove various ions from your water samples. Filtrations (both gravity and vacuum) are then used to separate those precipitates from your samples.

Another technique we will introduce is the use of a pH meter to determine the acidity or basicity of the water sample. pH determination can indicate whether or not our water sample is contaminated and if it is, provide clues as to the nature of the contaminant. In order to use a pH meter it must first be calibrated to make sure that it is reading accurately. Your instructor will guide you through the calibration of the pH meters used in your lab.

As a conclusion to the lab, you are asked to repeat the phosphate determination, this time using your “purified” water sample. This allows assessment of the success of the process you used to purify your samples.

Background

Nothing is more vital to the survival of living organisms than water. In fact ~60% of the human body is water; the brain is composed of ~70% water and the lungs nearly ~90% water. About ~83% of our blood is water, which helps digest our food, transport waste and control body temperature. It is estimated that each day humans must have about 2.5 liters of drinking water to maintain a reasonable quality of life. As Earth's population continues to grow, an adequate supply of clean water becomes a challenge. Water may be everywhere one looks and cover approximately 2/3 of the world, but only a small amount of that water is considered "drinkable."

Water can become unfit for drinking from biological or chemical impurities. Bacteria, algae, viruses, fungi, minerals, and other chemicals can contaminate water sources such as rivers, lakes, and oceans. Some of these simply make water unappealing in smell, taste, or appearance; others are dangerous. Most of these impurities must be removed for the water to be considered drinkable.

The water molecule itself is extremely important. Water is a polar molecule: it has an uneven distribution of electron density. It has a partial negative charge near the oxygen atom and a partial positive charge near the hydrogen atoms. The polarity of water allows for ions and polar molecules to dissolve in it causing contamination.

Water Contamination

Chemical impurities found in water include many dissolved salts. Ions such as potassium, calcium, magnesium, iron, sulfate, and carbonate become dissolved in water as rain and groundwater pass over rocks, which slowly dissolve into the water. Chemical impurities can also come from human sources. Two common chemical impurities are nitrite (NO_2^-) and phosphate (PO_4^{3-}). Nitrite contamination is caused by fertilizer run-off from farms and agricultural lands, while phosphate contamination is caused by fertilizers, wastewater treatment plants, soaps, detergents and industrial processes.

Being able to test for the presence of both biological and chemical impurities in water is important. This experiment is designed to give you experience in some of the testing and purification processes commonly performed in a water quality lab or water treatment plant. BEFORE your lab period you MUST collect ~ 1L of water from a natural water source (for example, a lake, river, or runoff) in the area. Once in lab the following techniques will be used to test for the concentration of phosphate in the water sample and purify the water: a test of pH, spectroscopic determination of concentration of phosphate, serial dilutions to prepare standards, and filtration to remove sediment and other solids.

pH

pH is a good indicator of water contamination. The pH of a neutral solution is 7, while an acidic solution is less than 7 and a basic solution is greater than 7. The pH of natural source water is going to vary because of several factors. The pH may be more than 7 because of photosynthesis. Photosynthesis uses dissolved CO_2 which acts like carbonic acid in water. When water plants use the dissolved CO_2 the acidity is reduced. The presence of large quantities of bacteria or other biological contaminants in water can produce ammonia that will raise the pH of the water. Acidic water can result from acid rain or from the presence of many industrial wastes, including sulfur and nitrogen oxides, which become sulfuric and nitric acids when delivered by rain or runoff into lakes or streams.

The pH is normally recorded in the field at the water source but as long as the water sample is refrigerated to keep any biological entities from changing the pH, it should be reasonably the same when taken in the lab. For this experiment, the pH will be recorded using a pH meter. The pH will be measured twice, before and after the purification process.

Determining Phosphate Concentration

The phosphate concentration will be measured using spectrophotometry. In order to understand how spectrophotometry works one must first understand the basic principles of light and energy. When an electron absorbs energy, i.e., light, the electron moves from one energy level to the next, usually from the ground state to an excited state. An absorption spectrum is produced when continuous electromagnetic radiation (every wavelength, all colors) passes through a substance and certain wavelengths are absorbed. In other words, an observed absorption spectrum is made up of the wavelengths of light that are not being absorbed by the substance. A molecule can absorb light only if it can accommodate the additional energy by promoting electrons to higher energy levels. The energy of the light being absorbed must match the energy required to promote an electron. Therefore, not all wavelengths of light are absorbed equally by a sample.

In this experiment we will be using a UV-Vis spectrometer to determine the concentration of phosphate in your water sample. A color-treated water sample is placed into a cuvette, and then placed into the instrument. Inside the spectrometer, light is passed through the sample cell at the appropriate wavelength selected by a diffraction grating. The light is then focused through an aperture to pass through the sample cell. The light can scatter, be absorbed by the molecule of interest and re-emitted in any direction, or pass through the sample cell without interacting with the sample at all. We are interested in measuring the amount of light absorbed and emitted because it is concentration dependent. The amount of light that passes through the sample is quantified by the detector.

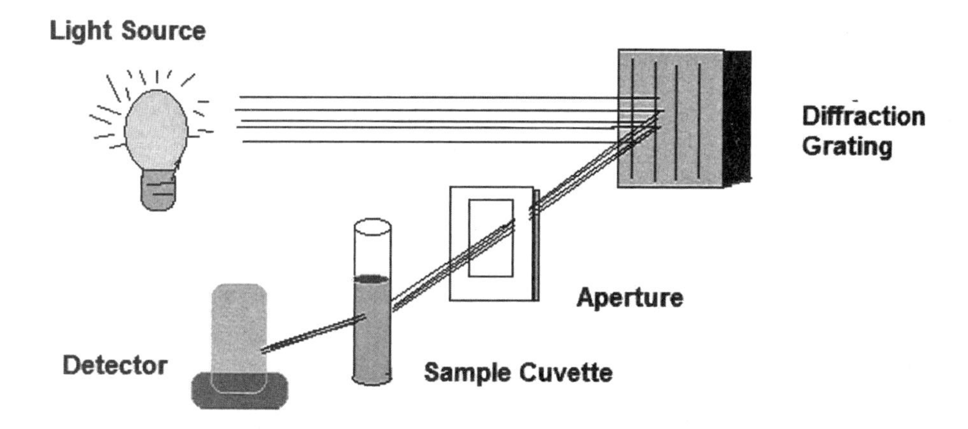

Spectrometers like the one above can be set to read in two different units, either absorbance (A) or percent transmission (%T). In this experiment we will set the spectrometer to read absorbance (A) because the Beer-Lambert Law expresses the direct correlation between the absorbance of a sample, its concentration, and its thickness. The law can be written as:

$$A = \varepsilon bc$$

where ε is the molar absorptivity (this is a constant which depends on the nature of the absorbing system and the wavelength passing through), b is the path length (the width of the sample cell or cuvette, usually 1 cm), c is the concentration of the sample and A is the absorbance of the sample.

If the molar absorptivity and the path length are held constant, the relationship between absorbance and concentration can be easily studied. If there is a direct relationship, a plot of different concentrations of the solution will generate a straight line and the system is said to obey the Beer-Lambert Law. When a curved plot is obtained, the system deviates from the law and the correlation between concentration and absorbance cannot be easily established. (This can happen at high concentrations.)

Colorimetric Methods

Since the phosphate ion in water is colorless and therefore not easily detected using spectrophotometric techniques (that determine concentration from color intensity), the phosphate must be reacted with a color-producing agent. In this experiment we will react the phosphate with ammonium molybdate $(NH_4)_6Mo_7O_{24}{\bullet}4H_2O$ and stannous chloride $(SnCl_2{\bullet}2H_2O)$ in an acidic solution. When combined with phosphate, these reagents produce a blue compound whose absorption can be observed at 650 nm. Measuring the absorbance of a set of standards with known concentrations of phosphate will allow us to develop a linear graph that can be used to determine concentrations of unknowns based on their absorbance values. The absorbance value of your water sample can then be used to determine the concentration of phosphate in your water.

Serial Dilution

Serial dilutions are a way to make solutions of varying concentrations very easily in succession. In this experiment you will be provided a stock solution of 10 ppm phosphate. From that solution you will make phosphate solutions of 8 ppm, 6 ppm, 4 ppm and 2 ppm. It is extremely important to understand how to make serial dilutions before coming to lab.

Example:
Using a 15 ppm stock solution, create solutions of 10 ppm, 7 ppm and 3 ppm using serial dilution.

Sample cuvettes only need about 2–3 mL for a measurement, so starting with 10.0 mL of 15 ppm stock and using the equation:

$$\mathbf{M_1V_1 = M_2V_2}$$

where M_1 is concentration of solution one, V_1 is the volume of solution one, M_2 is concentration of solution two, and V_2 is volume of solution two, you can calculate the amount of stock solution you will need to make 10.0 mL of 10 ppm solution:

$$(10.0 \text{ mL})(10 \text{ ppm}) = X(15 \text{ ppm})$$
$$X = 6.7 \text{ mL}$$

The results indicate that 6.7 mL of 15 ppm solution contains the appropriate amount of phosphate to make a 10 ppm solution when diluted to 10.0 mL. The 10 ppm solution is therefore made by measuring out 6.7 mL of the 15 ppm stock phosphate solution and then adding 3.3 mL of water to make a 10.0 mL total volume.

This process is repeated in order to make the 7 ppm solution. But the stock solution used this time is the newly made 10 ppm solution. Repeating the calculations above:

$$(10.0 \text{ mL})(7 \text{ ppm}) = X(10 \text{ ppm})$$
$$X = 7.0 \text{ mL}$$

The 7 ppm solution is therefore made by measuring out 7.0 mL of the 10 ppm phosphate solution and then adding 3.0 mL of water to make a 10.0 mL total volume. The process continues using each newly made solution as the stock for the next lower concentration until all of the concentrations needed are produced. The results are a series of solutions that are similar in appearance to the ones below:

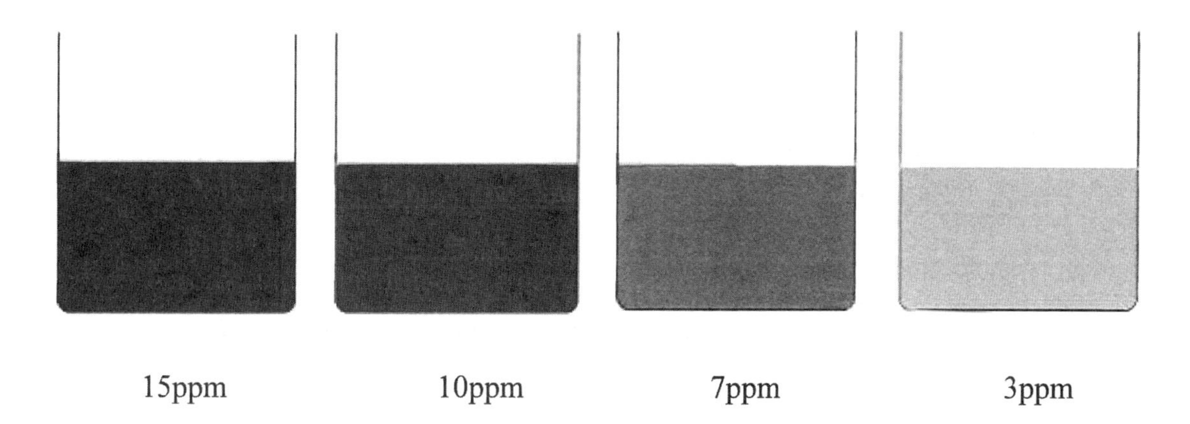

15ppm 10ppm 7ppm 3ppm

Phosphate Concentration

After the solutions are prepared by serial dilution, the absorbance of each sample must be determined using the spectrometer. Looking at the Beer-Lambert Law, you can see that if you plot absorbance versus concentration, the slope of the line is the path length multiplied by the molar absorptivity (εb).

$$A = \varepsilon bc$$
$$y = mx$$

Since the value of εb is a constant, the concentration of phosphate in the water sample you collected can be determined from the absorbance of the sample. The absorbance value is measured in the same manner absorbances were obtained for the standards. The Beer-Lambert Law is then rearranged and solved for the sample's concentration (c). For example, if the absorbance is 0.253 and the value of εb is 1.45 x 10^4 M, you can solve for the concentration of the sample:

$$A = \varepsilon bc \Rightarrow$$
$$0.253 = \left(1.45x10^4 \,{}^{L}\!/_{mol\cdot cm}\right)(1cm)(c) \Rightarrow$$
$$c = 1.70x10^{-5}\,M$$

Water Purification

In order to make the natural water sample you collected "drinkable," you must remove the chemical and biological impurities. The first step is to filter the water through filter paper to remove large particles from your sample. The next few steps of the process remove the remaining ionic impurities through the addition of chemical reagents and several more filtrations. Once the filtration process is complete, the pH and absorbance measurements are repeated to determine if the natural water source can be considered "drinkable."

Procedure

Part I: Collection of Samples

You will be collecting your own water sample for use in this laboratory. ***The collection should take place no more than 24 hours before lab.*** **This is because analysis must take place within 48 hours to be valid. The sample should come from a fresh water source because there are specialized tests required to analyze salt water samples that we are not equipped to run.**

Obtain a clean and dry ~1L glass or plastic bottle. You may use a 20 oz + soda bottle if you clean it thoroughly with hot water and rinse it with vinegar. Do not use soap as this might add phosphate.

Find a source of fresh water. Make copious notes regarding the location and condition of the source.

Collect the sample by inserting the bottle upside-down into the water. Rotate the open end toward the direction of the flow, and allow the bottle to fill under the surface. Sample about 12 inches under the surface. Avoid surface scum and bottom sediment.

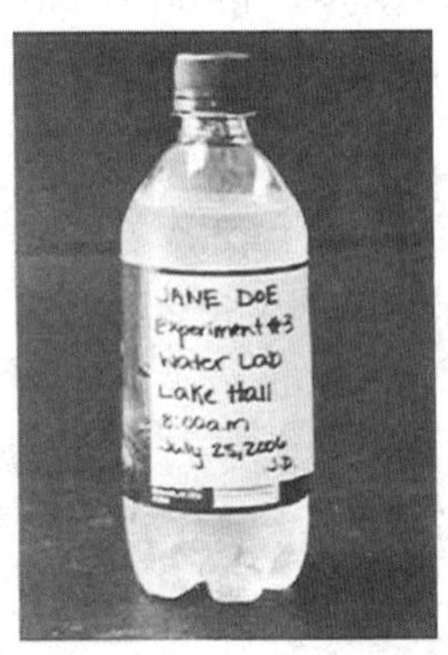

Label the bottle with the sample name, location and time collected and the initials of who collected the sample.

Store in a refrigerator until time for lab. (You can bring samples by the lab early in the day or keep them on ice in a small cooler until lab.)

Part II: Analysis of pH

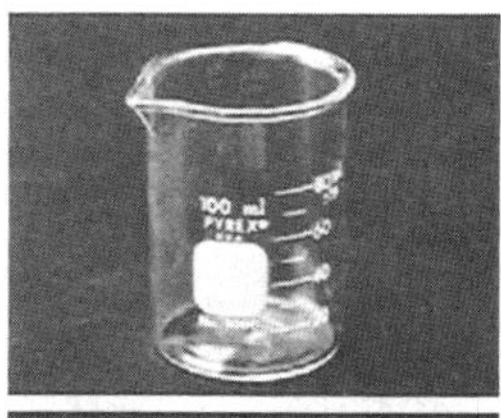

Take 50 mL of your water sample and place it in 100 mL beaker that has been cleaned and rinsed with the acid wash and distilled water provided.

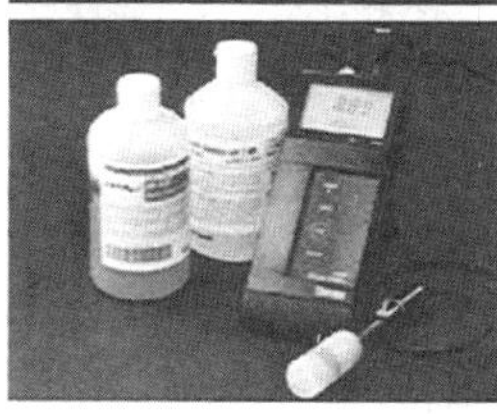

Using a pH meter, take the pH of the sample and record it in your notebook.

Part III: Analysis of PO_4^{3-}

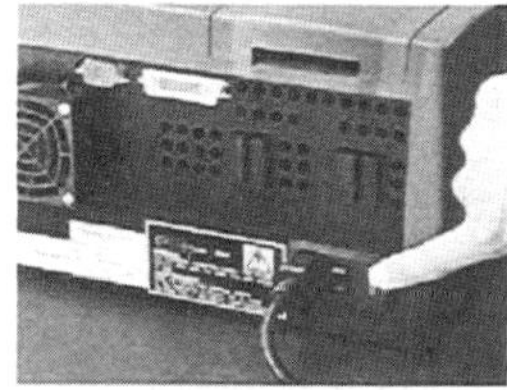

Turn on your spectrometer.

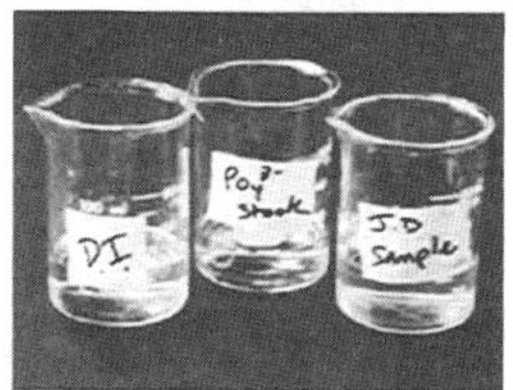

Take 25.0 mL of your water sample, 25.0 mL of stock phosphate solution and 25.0 mL of distilled water and place each of them in separate 100 mL beakers that have been cleaned and rinsed with the acid wash provided.

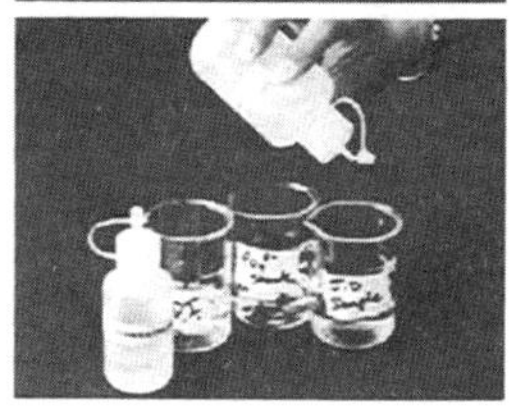

To the beakers containing the unknown water samples (one from each student at the bench) and to the phosphate standard beaker (only one which is shared) add 1 mL of the Ammonium Molybdate color reagent and 2 drops of stannous chloride.

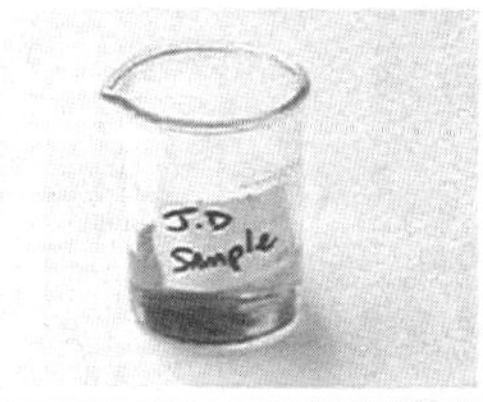

Mix the solutions for 5 minutes, allowing color to develop.

Using the calculations from your prelab, create the four serial dilutions of your stock solution. These will be used to create the standard calibration graph to determine the concentration of your unknown so be meticulous in your technique when making the dilutions.

You are now ready to measure the absorbance of your samples.

Make sure your spectrometer has been on for a minimum of 10 minutes.

Obtain a cuvette from the TA. Be responsible for it and return it at the end of the class.

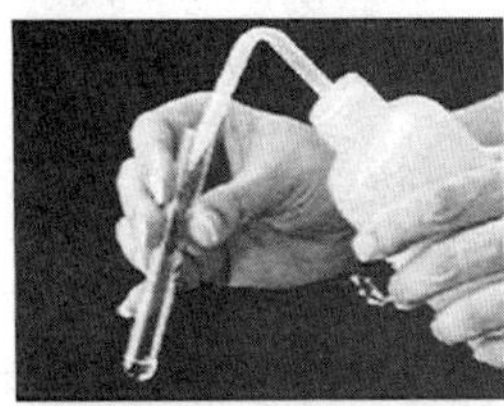

Wash the cuvette carefully and rinse with distilled water. Do not use brushes to wash the cuvettes, as they scratch very easily.

Fill one of the cuvettes to about two-thirds (2/3) full with your distilled water blank.

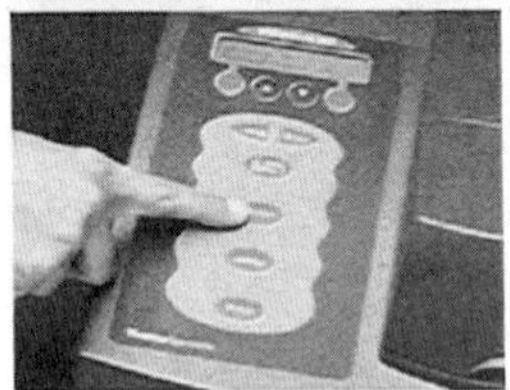

Press the A/T/C button on the spectrometer to select the absorbance mode. The current mode appears on the display.

Using the nm (up/down) controls, set the wavelength at 543 nm.

Insert the blank cuvette into the sample compartment and close the compartment's lid.

Observe the white reference mark on the cuvette. Align this mark with the one on the sample compartment every time a measurement is taken.

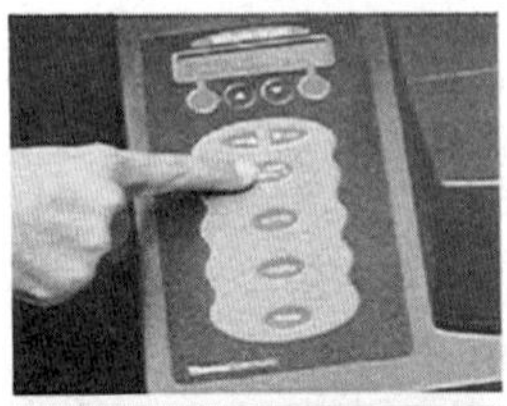

Press 0 ABS/100%T to set the blank to 0 absorbance. Remove the "blank" cuvette and discard the blank water sample.

Rinse the sample cuvette with a small portion of the standard stock solution (2.50g/L) and discard the rinse. Fill the cuvette to about 2/3 with the stock solution and insert into the sample compartment. Record the absorbance.

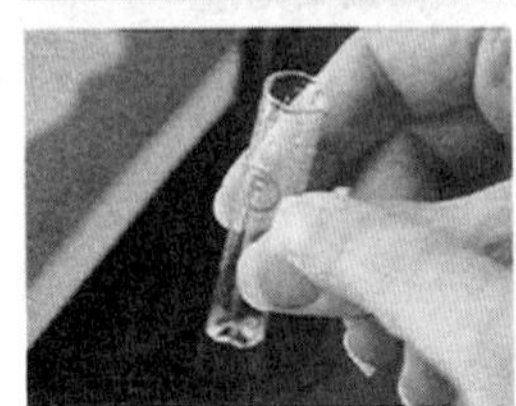

Remove the sample cuvette, discard the stock solution in the waste jar and rinse the cuvette well with DI water.

Rinse the sample cuvette again with a small portion of your next standard solution and discard the rinse. Fill the cuvette to about 2/3 with the standard solution and insert into the sample compartment.

Record the absorbance.

Repeat until all the standards and all of the student's unknown samples have been measured.

Part IV: Water Purification (Each student should purify their own water sample.)

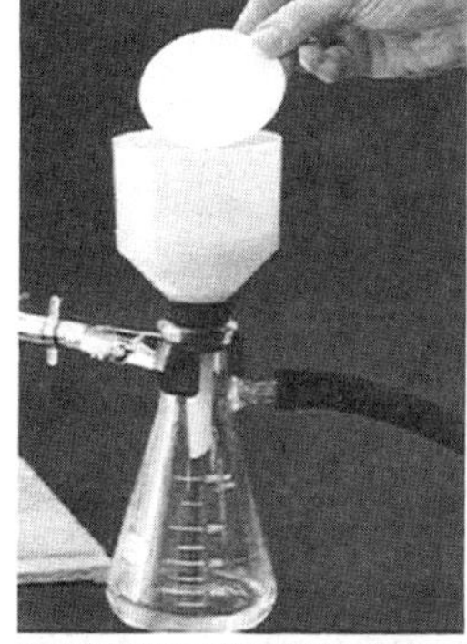

Using the 100 mL of your water sample, *slowly* filter the sample using a suction filtration set-up to remove large particulates. Use 0.1 mm filter paper.

After a liquid sample has been passed through a filtration system it should now be called a "filtrate."

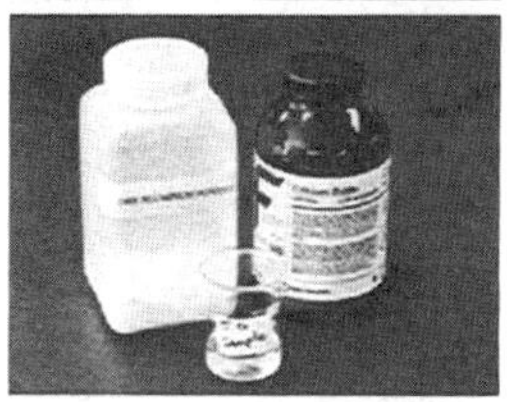

Add 1 spatula of dry lime (CaO) powder and 1 spatula of dry alum (Al_2SO_4) powder to the filtrate and stir with a clean, glass stirring rod. Allow the powders to react for ~10 minutes. Make observations.

Using your filter funnel again, place a filter paper in the funnel and then add enough sand to cover the filter paper with ~.5cm of sand.

Pass the filtrate through the sand slowly.

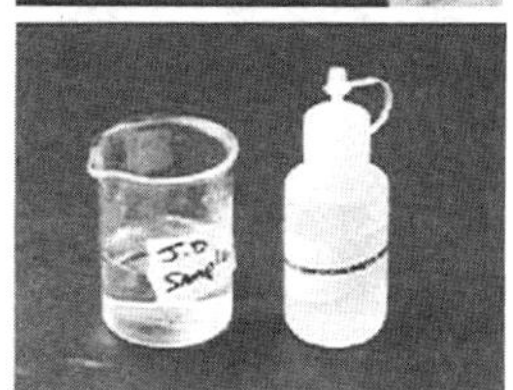

Add 1 mL of sodium hypochlorite reagent to the filtrate. Stir to mix completely.

Set up the charcoal filtration funnel. Place a filter paper in the Buchner funnel. Add a 1 cm layer of activated charcoal on the filter and then cover with a second filter. Attach a hose to the side nozzle of the water faucet in the sink and the side nozzle of a filtration flask. Use an adaptor and fit the funnel into the flask.

Rinse the charcoal with distilled water until the water in the flask is clear. Make sure you discard the charcoal rinse water and dry the filtration flask before filtering your water sample.

Pass your water sample through the fully rinsed charcoal filter.

Make observations regarding color and clarity of the "purified" water.

Part V: Water Quality After Purification

Repeat the pH test on your purified water sample. If the pH is > 8, add a drop of 1.0M HCl to neutralize it.

Repeat the phosphate test for your "purified" water sample only. You don't need to repeat the measurement for the standards.

Experiment 12

Pre-Laboratory Assignment

Name: ______________________________ **Date:** ________________

Instructor: ______________________________ **Sec. #:** ___________

Show all work for full credit.

1) Calculate the $[PO_4^{3-}]$ concentrations corresponding to the following ppm values:

Concentration = 0.41 ppm; $[PO_4^{3-}]$ =
Concentration = 275 ppm; $[PO_4^{3-}]$ =
Concentration = 21 ppm; $[PO_4^{3-}]$ =

Calculate the ppm values corresponding to the following $[PO_4^{3-}]$ concentrations:

$[PO_4^{3-}] = 4.56\times10^{-6}$ M; Concentration =
$[PO_4^{3-}] = 3.88\times10^{-3}$ M; Concentration =
$[PO_4^{3-}] = 7.66\times10^{-5}$ M; Concentration =

2) Use the following table of standard concentrations and absorbances for the phosphate ion to create a calibration graph. What are the slope and intercept of the calibration curve?

Phosphate Analysis	
Concentration (ppm)	**Absorbance (AU)**
0	0.109
2	0.288
4	0.476
6	0.651
8	0.847
10	0.997

What is the concentration of phosphate in a solution that has an absorbance of 0.708 AU?

One of your samples shows a reading of 1.207 AU, which is beyond the range of the calibration curve. The lab is over and it is too late to redo the calibration over a broader range, or to dilute the sample and determine it a second time. Assume that Beer's Law still holds up to the value of the sample, and calculate the concentration using the equation for the calibration line.
(Path length = 1.0 cm)

3) Phosphate standards used to generate the calibration curve are usually made by first preparing a stock solution of known concentration, then diluting it to the required concentration.

First prepare a stock solution that is 1.00×10^{-2} M in phosphate ion. How many g of Na_3PO_4 would you add to a 250 mL volumetric flask to prepare this solution?

Mass Na_3PO_4:

What is this concentration in ppm of phosphate? (That is, as PO_4^{3-}, not Na_3PO_4).

PO_4^{3-} concentration:

What volume of this ppm solution would you dilute with water into a 1.000 L volumetric flask to make a solution of 10.0 ppm?

This much of the work will be done for you by the laboratory assistant. You will be given a stock solution of 10.0 ppm phosphate to use in calibrating your phosphate assay. To preserve the ammonium molybdate reagent, you will create the calibrating standards by doing a serial dilution, using an aliquot of each diluted solution to create the next dilution.

After adding 1 mL of ammonium molybdate reagent and 2 drops of stannous chloride to 25 mL of the 10.0 ppm solution to develop the color, use a 20 mL aliquot to dilute with 5 mL of water to make 25 mL of a solution of 8.0 ppm.
Using the dilution equation: $V_1 \times Conc_1 = V_2 \times Conc_2$ and rearranging:

$$20mL = 25mL \frac{8\,ppm}{10\,ppm}$$

To create 25 mL of a 6.0 ppm solution, what volume of the 8.0 ppm solution and what volume of water should be mixed?

To create 25 mL of a 4.0 ppm solution, what volume of the 6.0 ppm solution and what volume of water should be mixed?

To create 25 mL of a 2.0 ppm solution, what volume of the 4.0 ppm solution and what volume of water should be mixed?

4) What is the molar concentration of each of your calibration solutions?

Concentration = 10 ppm; $[PO_4^{3-}]$ =
Concentration = 8 ppm; $[PO_4^{3-}]$ =
Concentration = 6 ppm; $[PO_4^{3-}]$ =
Concentration = 4 ppm; $[PO_4^{3-}]$ =
Concentration = 2 ppm; $[PO_4^{3-}]$ =

Experiment 12

Laboratory Report

Name: ____________________________ **Date:** _________________

Instructor: ____________________________ **Sec. #:** ___________

PURPOSE: (*The purpose should be several well constructed-sentences describing what your experiment was designed to accomplish and the criteria used to determine success. These sentences should include both concepts and techniques.*)

PROCEDURE: (*The procedure section should reference the lab manual and note any changes made to the experiment. For this experiment the procedure must include the location, date, and time of sample collection. Also be sure to note any changes made to the procedure and the names of your team members.*)

DATA: (*The data section should include your own personal data and observations.*)

Phosphate Concentrations and Absorbance Values of the 5 Standard Solutions:

Standard in PPM	$[PO_4^{3-}]$	Absorbance

Personal Sample	PPM	$[PO_4^{3-}]$	Absorbance
Before Purification			
After Purification			

Observations of your water sample made before, during and after purification:

CALCULATIONS: (*The calculation section should include sample calculations for each column in the data table. Be sure to include conversions of units.)*

Calibration graph of concentration vs. absorbance for PO_4^{-3} (*All graphing guidelines apply here):*

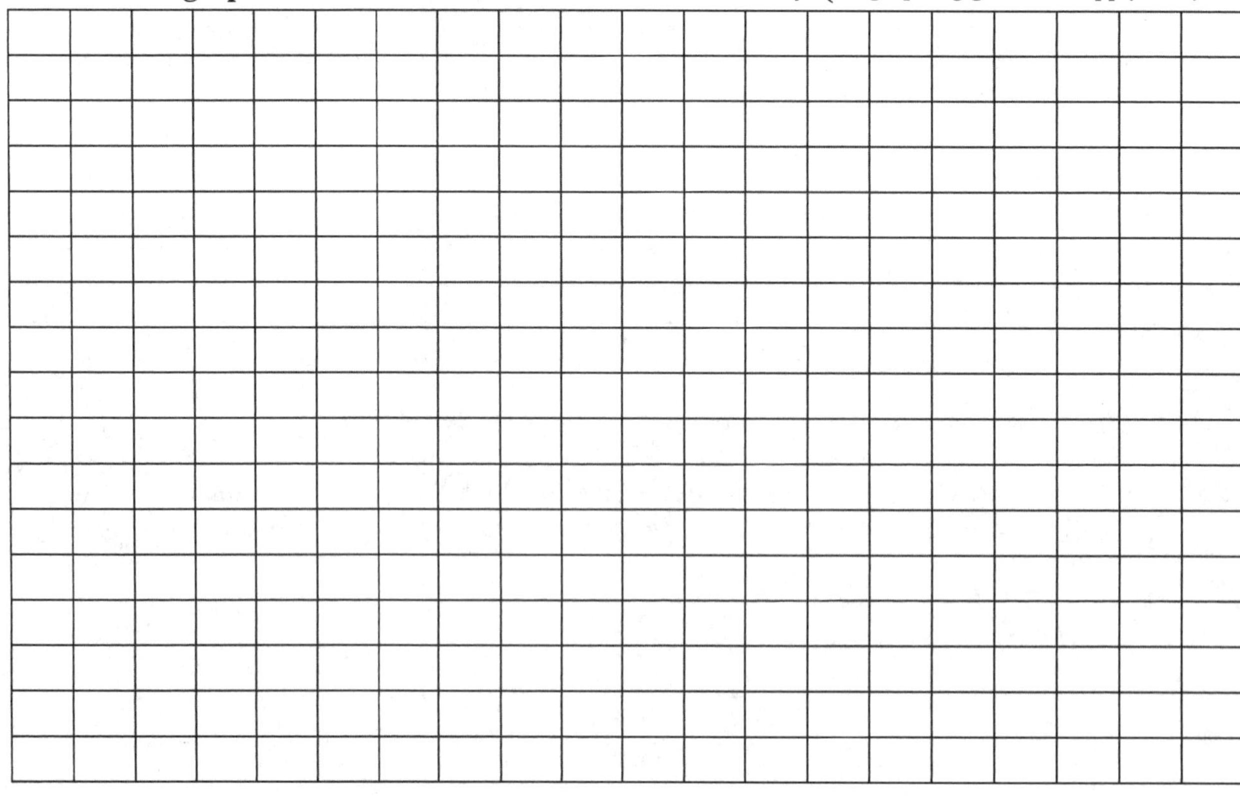

Sample calculation of the concentration of the phosphate concentration of your water sample before and after purification using Beer's Law ($A=\varepsilon bc$, where A is the absorbance, ε is the molar absorptivity which is equal to the slope of the line, b is 1.00 cm and c is the concentration:

Sample calculation of the conversion from ppm to molarity should also be included. Finally, you must include an example calculation of the purity of the water sample. This can be determined using the following formula:

$$\%\text{Purity} = \frac{\text{ppm } PO_4^{3-} \text{ before purification} - \text{ppm } PO_4^{3-} \text{ after purification}}{\text{ppm } PO_4^{3-} \text{ before purification}} \times 100\%$$

CONCLUSION: (*The conclusion should be several paragraphs addressing the following: The location where the water sample was obtained; the phosphate concentration and pH of the water sample before and after purification; a discussion of the possible reasons for the phosphate concentration level and for the pH value of the water sample before and after; in other words what might be causing these values. A discussion on the "purity" of your water sample before treatment. As always, discuss any possible errors in the experiment.*)

Experiment 12

Post-Laboratory Assignment

Name: ____________________________ **Date:** _________________

Instructor: _____________________________ **Sec. #:** ___________

Show all work for full credit.

1. Why should we care about the amount of PO_4^{3-} found in a natural water source?

2. If you really wanted to be sure your water was drinkable, what further purification step(s) would you take?

3. Blue light has a wavelength between 450–495 nm. Why don't we measure the absorbance of a blue solution within this wavelength range?

Experiment 13

Kinetics: The Iodine Clock

Introduction

Chemical kinetics is a topic that is conceptually difficult for a number of students. The rate laws that are created and studied are varied and are often even more complicated than the chemical reactions they represent. In the laboratory, we will explore the concept of reaction kinetics by actually measuring the rate at which a reaction is completed.

One of the questions that often remains unanswered in our desire to teach a student all of the complex mathematics that go along with kinetic calculations is why kinetics is an important topic. The most readily recognized use of kinetics is in manufacturing. In the production of various chemicals it is vital to know the rate at which a product is both produced and also broken down.

For example, in the production of steel the proportions of iron and carbon are varied based on the intended use of the steel. The cooling of the molten steel is controlled to allow a desired crystal structure to develop; these crystal structures have specific names and properties. When bainite (a type of iron) forms in austenite (a mixture of iron and carbon), the carbon content of the austenite increases as the transformation occurs. If the austempering stage is too short, martensite may form in the residual austenite. If it is too long, carbide precipitation occurs. Both phenomena are detrimental to the mechanical properties of the steel that is being produced. It is therefore necessary to know how long the austempering treatment should be.

Another more commonplace example is that of expiration dates or shelf lives of foods and drugs. Knowledge of the kinetics of the chemical reactions that break down certain chemicals present in a food or drug can be used to determine how potent or unsafe a food or drug may be after a certain period of time. The same knowledge of reaction rates for specific chemicals can also be used to determine the dosage of a drug. For example, if the drug is slow to react within the body, the dosage can be larger because it is naturally "time-released" and thus won't hit the system all at once. If the reaction rate is very fast, too high a dose might kill the patient. Determination of reaction rates is of vital importance to a number of industries.

Our purpose in this laboratory is to determine the rate of reaction for the "Iodine Clock." The "iodine clock" reaction is a good example to use because the rate law can be determined by observing the completion rate of the reaction based on colorimetric evidence. Although you will not be learning any new techniques in this lab, you will get the opportunity to practice what you have learned previously regarding the proper way to make solutions. Since this is a quantitative laboratory, good quantitative transfer technique is vital.

Background

Chemical kinetics is the study of reaction rates and reaction mechanisms. In studying the chemical kinetics of a system we are interested in what the reaction rate means, how to determine the reaction rate experimentally, how temperature affects the reaction rate, how concentration affects the reaction rate and the detailed pathway taken by the atoms as the reaction proceeds.

Measuring Rates Experimentally

The fundamental observation in reaction kinetics is of the reaction occurring in a given time interval. Concentration data may then be used to calculate the rate of reaction for a given experiment according to the following equations.

$$\text{Rate of Reaction} = \frac{\text{Amount of Reactant Consumed}}{\text{Time of Observation}} = \frac{-d[\text{Reactant}]}{dt}$$

or

$$\text{Rate of Reaction} = \frac{\text{Amount of Product Created}}{\text{Time of Observation}} = \frac{d[\text{Product}]}{dt}$$

In order to measure a reaction rate, there must be some method of determining the concentration of reactants or products at the beginning and at the end of a time period. We will vary the initial concentration of a reactant, and measure the time until that reactant is completely consumed. For most reactions, the rate decreases continuously with the progress of the reaction. This is due to decreasing concentrations of kinetically significant reactants. For example, assume a reaction of the general type:

$$A + B \Leftrightarrow 2C$$

The rate law would be expressed as follows:

$$rate = \frac{-d[A]}{dt} = \frac{-d[B]}{dt} = \frac{1}{2}\frac{d[C]}{dt} = k[A]^m[B]^n$$

Assuming the superscripts m and n are positive integers, the rate of the reaction will decrease as the number of moles of A and B diminish as they form C. The exponents m and n in the above equation are called reaction orders and their sum is the overall reaction order. For example, if m = n = 1, the reaction order in A is 1, the reaction order in B is 1; the reaction is first order in A and B; and the overall reaction order is 1 + 1 = 2 or second order. Both m and n are determined experimentally and cannot be deduced from the stoichiometry of the reaction.

The Iodine Clock Reaction

In this experiment we will investigate the kinetics of the following reaction, known as the "iodine clock". The iodine clock reaction (or Landolt reaction) is a classic chemistry reaction in which two colorless solutions are mixed and their chemical reaction results in a colored solution. This reaction demonstrates that reaction rates depend on the concentrations of the reagents involved in the overall reaction. In the iodine clock reaction, you are measuring the rate of the reaction of iodide ion (I^-) with peroxydisulfate ion ($S_2O_8^{2-}$):

$$S_2O_8^{2-} + 2I^- \rightarrow 2SO_4^{2-} + I_2 \quad \text{Reaction 1}$$

The rate of the reaction is determined by measuring the amount of I_2 formed over a particular amount time, or $\Delta [I_2]/\Delta t$. The I_2 is measured by "titrating" it with a fixed amount of thiosulfate ion ($S_2O_3^{2-}$) which rapidly converts the I_2 formed in Reaction 1 back to iodide ion:

$$I_2 + 2S_2O_3^{2-} \rightarrow 2I^- + S_4O_6^{2-} \quad \text{Reaction 2}$$

While Reaction 2 is said to be "fast" compared to Reaction 1, it can actually proceed no faster than the rate at which I_2 is produced, so its rate is controlled by the rate of Reaction 1. When the $S_2O_3^{2-}$ is completely consumed, then I_2 begins to accumulate. It combines with a starch indicator to form a dark blue complex, so the "end-point" of the "titration" is determined by the time required for the solution to turn blue. The time it takes for the color to appear is called the period of the clock, or Δt, the time it takes for Reaction 2 to go to completion. The rate of $S_2O_3^{2-}$ disappearance is therefore given by its initial concentration divided by the clock period, or $[S_2O_3^{2-}]/\Delta t$. Because the rate of disappearance of I_2 in Reaction 2 is controlled by its rate of appearance in Reaction 1, the rate of both reactions can be determined from the value of $1/2 \times [S_2O_3^{2-}]/\Delta t$. (Two $S_2O_3^{2-}$ ions are consumed for every I_2 consumed.) Actually, the I_2 in solution combines with iodide ion to form triiodide ion, I_3^-, which is pale yellow. Starch is added to form a blue color that is much easier to observe. The reaction between the triiodide ion and starch is not shown to keep the equations simple, but adding it would not change the conclusions.

The Rate Law

The rate law for the reaction with respect to the concentrations of each reactant is expressed as:

$$\text{rate} = \frac{-d[I^-]}{dt} = \frac{-d[S_2O_8^{-2}]}{dt} = k[I^-]^m[S_2O_8^{-2}]^n$$

In order to determine the value of the coefficients m and n in the rate law, we must first determine the rate of the reaction with respect to each reactant. In other words, what effect does the concentration of each reactant have on the overall reaction rate? The computation of the rate is complicated by the need to allow for continuous change in the rate. However, for the "clock reaction" it is possible to choose concentrations of reactants such that the rate remains constant until one reactant, the limiting reagent, is entirely consumed. Under these conditions the calculation of the rate becomes the following:

$$\text{Rate of Reaction} = \frac{\text{Initial Amount of Limiting Reactant}}{\text{Time of Observation}}$$

If the number of moles of limiting reagent put into the reaction solution at the start of the reaction is known, as well as the total volume of the reaction solution, the initial concentration can be calculated. The time (in seconds) required for the complete reaction (at constant rate), is then recorded in lab and the rate of the reaction can be determined.

Typically, reactions occur in more than one step. The slow step determines the rate. The "clock reaction" consists of a slow step followed by a fast step:

Slow $S_2O_8^{2-} + 2I^- \rightarrow 2SO_4^{2-} + I_2$ Reaction 1

Fast $I_2 + 2S_2O_3^{2-} \rightarrow 2I^- + S_4O_6^{2-}$ Reaction 2

The overall reaction will be observed in lab by using the reaction of iodine and starch, which forms a blue complex. The appearance of the blue complex is an indication that the reaction is complete. As mentioned above, the thiosulfate ion ($S_2O_3^{2-}$) is the limiting reactant in the clock reaction. The I^- (or I_3^-) that is needed to react with the starch, indicating the end point of the reaction, does not begin to form until all of the $S_2O_3^{2-}$ is gone. The rate of the reaction can therefore be determined by measuring the amount of time required for the blue color to appear.

In order to determine the rate law, the rate order with respect to each reactant in the slow step will be determined. If the peroxydisulfate ion ($S_2O_8^{-2}$) is used in large excess so its concentration remains essentially constant throughout the reaction and only the KI concentration is varied, we can determine how the rate of the reaction varies with respect to iodide ion. If next we hold the concentration of KI in excess and vary the peroxydisulfate ion concentration, we can determine its contribution to the rate as well.

Colorimetric Determination

When has the limiting reagent been consumed? Iodine is formed in the slow step, but is removed rapidly by the reaction with the limiting reagent $S_2O_3^{2-}$. When all of the $S_2O_3^{2-}$ is used up, yellowish I_2 begins to accumulate. (The other reactants and products are colorless.) However, the iodine color is faint and difficult to observe. Sensitivity can be improved by using starch as an indicator. In the presence of a suitably prepared starch solution, I_2 forms a deep blue starch-tri-iodide complex. In effect, the starch indicator amplifies the color of the iodine. At exactly the time when all of the thiosulfate ($S_2O_3^{2-}$) has been consumed the solution will change from colorless to blue. We are measuring the rate of the slow step because the reaction shown in the fast step is too fast.

Determining the Rate Law

Once we have determined the rate of the reaction we want to determine the rate law for the reaction. A rate law of a reaction is a mathematical expression relating the rate of a reaction to the concentration of either reactants or products. The rate law may be theoretically determined from the rate determining step (slow step) of the reaction mechanism. Many chemical reactions actually require a number of steps in order to break bonds and form new ones. The rate law of a reaction must be proven experimentally by looking at either the appearance of products or the disappearance of reactants.

The following examples show how the rate law can be determined for a reaction similar to that of the "Iodine Clock."

Example Problem

Consider the following overall chemical reaction:

$$2NO(g) + O_2(g) \Leftrightarrow 2NO_2(g)$$

Here is an example of a rate law for the above reaction:

$$k = [NO]^2[O_2]$$

In this rate law, the rate of disappearance of NO and O_2 is proportional to the concentrations of NO and O_2, where k is a rate constant (a proportionality constant). The exponent associated with each concentration term is referred to as the order. The sum of the individual orders gives the overall order of the reaction. In the above rate law, the order for NO is 2 and is often referred to as 2nd order. The order for O_2 is 1 and is often referred to as 1st order. The overall order is 3 (3rd order). It is important to note that the coefficients found in the balanced equation are not necessarily related to the exponents found in the rate law.

Let's look at a set of experimental data from which the rate law may be deduced. In this case, we will examine the rate in terms of the disappearance of reactants. Experimentally one compiles the reaction rates relative to initial concentrations of reactants:

	Rate M/s	**[NO] M**	**[O_2] M**
Experiment 1	**1.20×10^{-8}**	**0.10**	**0.10**
Experiment 2	**2.40×10^{-8}**	**0.10**	**0.20**
Experiment 3	**1.08×10^{-7}**	**0.30**	**0.10**

Where the rate is determined by:

$$\text{rate} = \frac{d[\text{Reactant}]_i}{dt}$$

where $[\text{Reactant}]_i$ is the initial concentration of the limiting reactant and dt is the time it takes for the reaction to run to completion.

The rate law can be determined using two methods. The first method is by examining the experimental data. Pick two trials to compare in which the concentration of one reactant is changed while the concentration of the other reactant is held constant. We will determine the order for the reactant whose concentration has changed. By keeping the concentration of the other reactant constant, we ensure that the rate will not be affected by that reactant.

Let's begin by choosing the results for Trial 1 and Trial 2. In these two trials the concentration of NO is held constant while the concentration of O_2 has been doubled. Also, note that doubling the concentration of O_2 doubles the rate of the reaction. This may be mathematically expressed as

$$\frac{[O_2]_f^x}{[O_2]_i^x} = \frac{0.20}{0.10} = 2 \text{ and } \frac{\text{rate final}}{\text{rate initial}} = \frac{2.40x10^{-8}}{1.20x10^{-8}} = 2$$

or

$$[O_2]^x = rate$$

Mathematically for the rate to be directly proportional to the change in concentration x (the reaction order) must equal 1. This makes the order for O_2 first order.

The same procedure is used to determine the order of NO. Pick two trials in which the concentration of O_2 is held constant while the concentration of NO is varied. These criteria are met with Trial 1 and Trial 3. The concentration of O_2 is held constant while the concentration of NO has tripled. Under these conditions, the rate of the reaction is increased nine fold. This can be expressed mathematically as:

$$\frac{[NO]_f^x}{[NO]_i^x} = \frac{0.30}{0.10} = 3 \text{ and } \frac{\text{rate final}}{\text{rate initial}} = \frac{1.08x10^{-7}}{1.20x10^{-8}} = 9$$

or

$$[NO]^x = rate$$

For the rate to have increased by a factor of nine when the concentration was tripled, x must be 2. This makes the order for NO second order. We are now able to write the rate law: Rate = $k[NO]^2[O_2]$, where k is the rate constant.

To solve for the rate constant, choose any experimental trial and substitute the values for the rate and concentrations of NO and O_2. Trial 1 data will be arbitrarily chosen in this case:

$$k = \frac{Rate}{[NO]^2[O_2]} = \frac{1.2x10^{-8}\ M/s}{(0.10M)^2(0.10M)} = 1.2x10^{-5} M^{-2}s^{-1}$$

Thus the rate law becomes: Rate = $1.2 \times 10^{-5}\ M^{-2}\ s^{-1}[NO]^2[O]$

The rate law for the above reaction can also be determined graphically by expressing the rate law in a straight-line form, which is what you will be doing in this experiment.

$$\mathbf{Rate = k[NO]^x[O_2]}$$

By rearranging the rate law, taking the log of each side, and assuming one concentration is a constant:

$$k = \frac{\text{Rate}}{[\text{NO}]^x[\text{O}_2]} \quad \text{Multiply both sides by } [\text{O}_2]$$

$$k[\text{O}_2] = \frac{\text{Rate}}{[\text{NO}]^x} \quad \text{Take the log of both sides}$$

$$\log k[\text{O}_2] = \log \text{Rate} - x\log [\text{NO}] \quad \text{Rearrange}$$

$$\log \text{Rate} = x\log [\text{NO}] + \log k[\text{O}_2]$$

Graphing the log of the rate (y-axis) versus the log of the [NO] (x-axis) should result in a straight-line graph with a slope equal to x, the reaction order for NO. This same process can be done for both reactants simply by holding the concentration of one reactant constant each time.

Once the rate orders have been determined, the rate constant k can be determined by substituting the data for each run and solving for k. If the data is good the k values should all be reasonably close and the average k for all the runs can be reported.

Temperature Effects on Rate

In this experiment we will also collect data at several temperatures, which will allow us to calculate the reaction's activation energy. The mathematical relationship between temperature and activation energy is given by this variation of the Arrhenius equation:

$$\ln k = \frac{-E_a}{R}\left(\frac{1}{T}\right) + \text{Ln } A \quad \text{lower case l for Ln here}$$

In this equation, A (a constant called the frequency factor) is related to the frequency of collisions and the probability that the collisions are favorably oriented; R is the universal gas constant, expressed in J/mol.K; E_a is the activation energy; T is the temperature in Kelvin; and k is the rate constant. Since this version of the Arrhenius equation is in the form of a straight line equation, a plot of ln (k) on the y-axis vs. 1/T on the x-axis will give a straight line. The slope of the line will equal $-E_a/R$ and the y-intercept equals ln A. Therefore we can use the Arrhenius equation to determine the activation energy and frequency factor for the "Iodine Clock" Reaction.

The chemical kinetics of the "Iodine Clock" can be studied by simply measuring the rate of the reaction with varying concentrations of reactants and at several temperatures. Once the rate of the reaction is determined one can mathematically determine the rate constant for the reaction at the given conditions and the amount of energy it takes to activate the reaction.

Procedure

SAFETY NOTES: Use good chemical safety techniques as usual. Make sure all waste goes into the containers provided and not down the sink. Wear your goggles!

GENERAL INSTRUCTIONS: You will work with a partner for this lab. Before starting determine the room temperature. Careful labeling of reagent s and the graduated cylinders used to measure them is essential to the success of this experiment.

Part I: Kinetics at Room Temperature

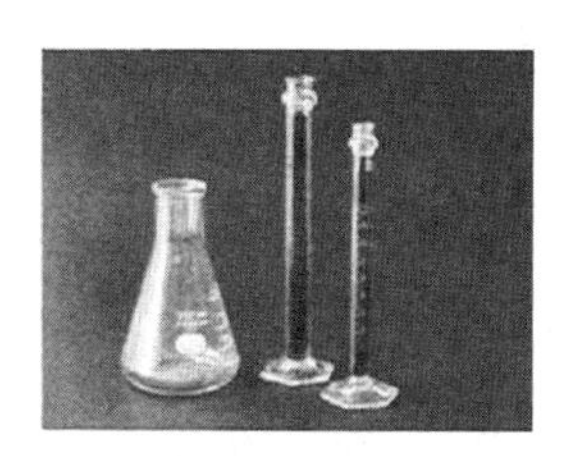

Use a clean 250mL Erlinmeyer flask to combine the reagents for each run (1-5, 8) listed in the table in the pre-lab. Use a different graduated cylinder to measure each reagent. The reagents should be added one at a time in the order they are listed from left to right.

Be ready to start timing the reaction when you add the final reagent, $(NH_4)_2S_2O_8$

Stop timing each reaction when the first blue color appears. Record the elapsed time in your notebook.

Part II: Kinetics at Elevated Temperature

For run 6, collect a hotplate from the front counter. Turn it on and set it to 10–15 °C above room temperature. Place a 400 mL beaker with 200 mL of water on the hotplate.

Put all of the reagents except for the $(NH_4)_2S_2O_8$, in an Erlinmeyer flask and place it in the water bath to equilibrate.

Place the $(NH_4)_2S_2O_8$ in a large test tube and place it in the water bath as well.

Leave the reagents in the bath until they both stabilize to the temperature of the bath water. Then add them together and record the time as before.

Part III: Kinetics at Lowered Temperature

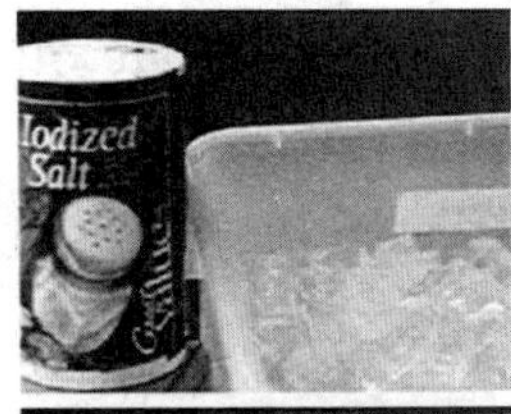

For run 7, collect a plastic tub and ice from the front counter. Add water and salt to make an ice bath.

Place the reagents in the ice bath in the same way you placed them in the hot water bath.

Wait for them to equilibrate, then combine them and record the reaction time as before.

Part I: Kinetics at Room Temperature

Initial Concentrations of Reactants					
Run #	KI 0.20 M	$Na_2S_2O_3$ 0.00175M	Water	Starch	$(NH_4)_2S_2O_8$ 0.050M
1	30 mL	20 mL	0 mL	3 Drops	30 mL
2	20 mL	20 mL	10 mL	3 Drops	30 mL
3	10 mL	20 mL	20 mL	3 Drops	30 mL
4	30 mL	20 mL	10 mL	3 Drops	20 mL
5	30 mL	20 mL	20 mL	3 Drops	10 mL
6	30 mL	20 mL	0 mL	3 Drops	30 mL
7	30 mL	20 mL	0 mL	3 Drops	30 mL
8	30 mL	10 mL	10 mL	3 Drops	30 mL

Experiment 13

Pre-Laboratory Assignment

Name: ______________________ **Date:** ____________

Instructor: ______________________ **Sec. #:** __________

Show all work for full credit.

1) For KI, $Na_2S_2O_3$ and $(NH_4)_2S_2O_8$ calculate the initial concentration (M) based on the table of solution concentrations and volumes given in the procedure section. ***Copy the completed table into your lab notebook for later use***. The shaded columns will be completed in lab.

Run Number	**[KI] (M)**	**[$Na_2S_2O_3$] (M)**	**[$(NH_4)_2S_2O_8$] (M)**	Time (sec)	Rate of Reaction (M/sec)	Temp. (K)	K (determine units)
1							
2							
3							
4							
5							
6							
7							
8							

2) A first order chemical reaction has a rate constant of $9.1\times10^{1}\ s^{-1}$ at 12 °C and a rate constant of $4.1\times10^{2}\ s^{-1}$ at 36 °C.

Calculate the activation energy, E_a, for the reaction.

3) In the iodine clock reaction, you are measuring the rate of the reaction of iodide ion (I^-) with peroxydisulfate ion ($S_2O_8^{2-}$):

$$S_2O_8^{2-} + 2I^- \rightarrow 2SO_4^{2-} + I_2 \qquad \text{(Reaction 1)}$$

The rate of the reaction is measured by measuring the amount of I_2 formed in a particular time, or $\Delta[I_2]/\Delta t$. The I_2 is measured by "titrating" it with a fixed amount of thiosulfate ion ($S_2O_3^{2-}$) which rapidly converts the I_2 formed in Reaction 1 back to iodide ion:

$$I_2 + 2S_2O_3^{2-} \rightarrow 2I^- + S_4O_6^{2-} \qquad \text{(Reaction 2)}$$

(While Reaction 2 is said to be "fast" with respect to Reaction 1, it actually can proceed no faster than the rate at which I_2 is produced, so its actual rate is controlled by the rate of Reaction 1).

When the $S_2O_3^{2-}$ is completely used up, then I_2 begins to accumulate. It combines with starch to form a dark blue complex, so the "end-point" of the "titration" occurs when the solution turns blue. The time it takes for the blue color to appear is called the period of the clock, or Δt, the time it takes for Reaction 2 to go to completion. The rate of $S_2O_3^{2-}$ disappearance is therefore given by its initial concentration divided by the clock period, or $[S_2O_3^{2-}]/\Delta t$. Because the rate of disappearance of I_2 in Reaction 2 is controlled by its rate of appearance in Reaction 1, the rate of both reactions can be determined from the value of $1/2\times[S_2O_3^{2-}]/\Delta t$. (Two $S_2O_3^{2-}$ ions are consumed for every I_2 consumed.)

(Actually, the I_2 in solution combines with iodide ion to form triiodide ion, I_3^-, which is the species that forms the blue color with starch. That step is not shown to keep the equations simple, but adding it would not change the conclusions.)

The following graphs illustrate what is happening to the concentration of each of the species involved in these two reactions.
Show you understand the reaction by identifying the species associated with each of the lines in the graphs.
Initial conditions:
$[S_2O_8^{2-}] = [I^-] = 0.055$ M;
$[S_2O_3^{2-}] = 0.0015$ M)

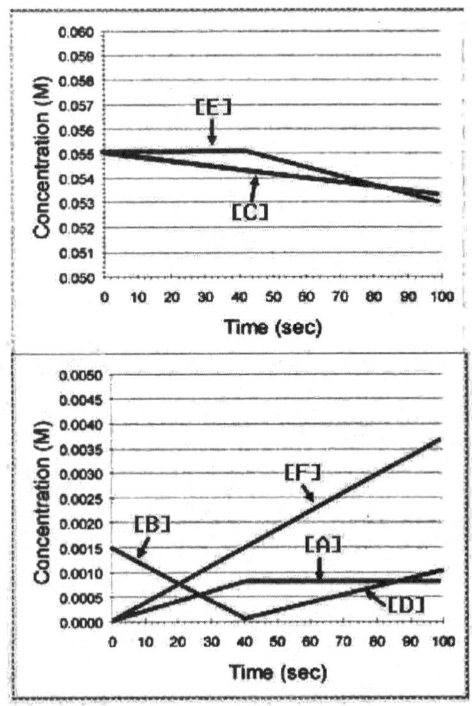

$[SO_4^{2-}] =$ __________

$[I^-] =$ __________

$[I_2] =$ __________

$[S_2O_3^{2-}] =$ __________

$[S_4O_6^{2-}] =$ __________

$[S_2O_8^{2-}] =$ __________

Experiment 13

Laboratory Report

Name: ______________________________ **Date:** ____________________

Instructor: ______________________________ **Sec. #:** ____________

PURPOSE: (*The purpose should be several well-constructed sentences describing what your experiment was designed to accomplish and the criteria used to determine success. These sentences should include both concepts and techniques.)*

PROCEDURE: (*The procedure section should reference the lab manual and include any changes made to the procedure during the lab.)*

DATA: (*The data section should include your own personal data and observations.)*

Run Number	[KI] (M)	$[Na_2S_2O_3]$ (M)	$[(NH_4)_2S_2O_8]$ (M)	Time (sec)	Rate of Reaction (M/sec)	Temp. (K)	*k* (determine units)
1							
2							
3							
4							
5							
6							
7							
8							

Run Number	log[*k*I] (M)	$\log[(NH_4)_2S_2O_8]$ (M)	log Rate of Reaction (M/sec)	k
1				
2				
3				
4				
5				
6				
7				
8				

m =

n =

Part II:

E_A =   A =

Observations:

CALCULATIONS: (*The calculation section should include sample calculations for each column in the data table. Be sure to include conversions of units.)*

Solution Concentrations: (*To determine the [KI], determine the number of moles in the volume of KI you are using in each trial and divide that amount by the total volume for each run. (Hint: 20 drops = 1 mL). Repeat these calculations to determine the [*$Na_2S_2O_3$*] and [*$(NH_4)_2S_2O_8$*].*)

Example calculation of the rate of the reaction: (*Divide the [*$Na_2S_2O_3$*] by the time*)

Graph of the log of the rate versus log of [KI] using runs 1–3; the slope of this graph will give the value of m in the following equation:

$$\log Rate = m\log k[I^-] + n\log[S_2O_8^{2-}]$$

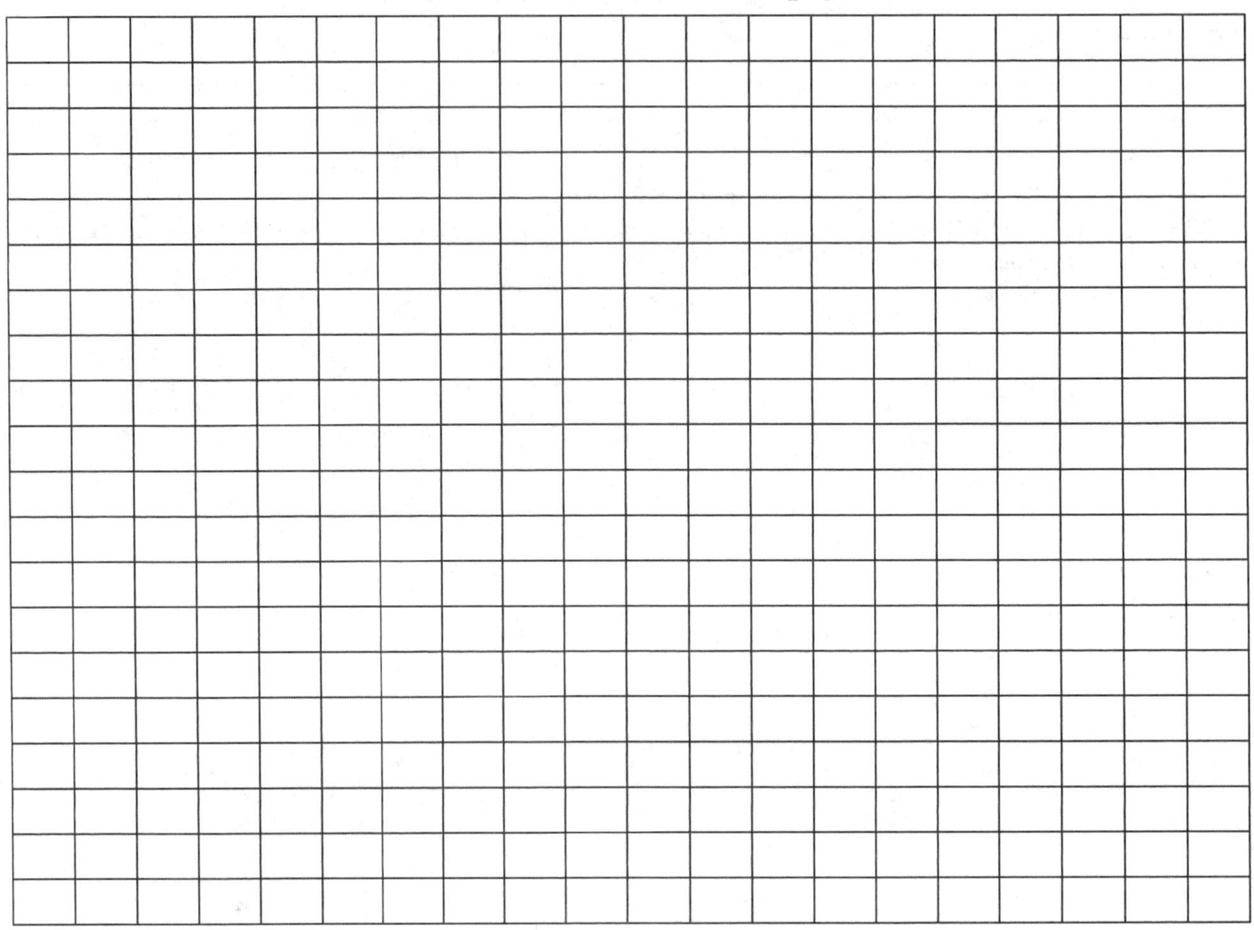

Graph of the log of the rate versus log of [$(NH_4)_2S_2O_8$] using runs 1, 4 & 5.

Using Rate = $k[I^-]^m[S_2O_8^{2-}]^n$, show an example calculation for k:

Graph of ln k vs. 1/T for trials 1, 6 and 7.

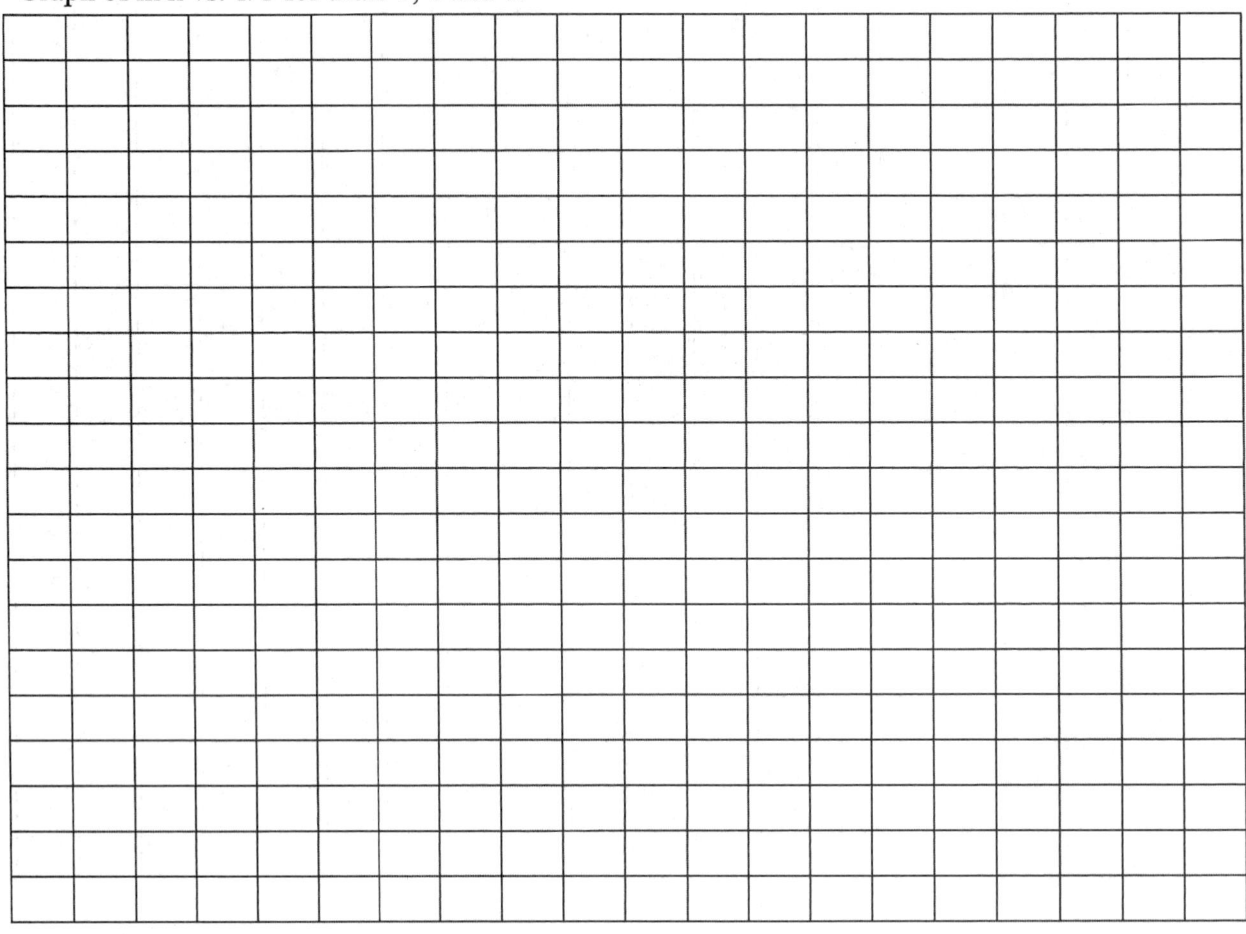

Example calculation of E_A (*slope of this graph is equal to* $-E_a/R$):

Example calculation of A (*y-intercept is ln A*):

CONCLUSION: *(The conclusion section should be several paragraphs containing all of the values obtained, including the rate law. This section should also include a comparison of the rate constants at the same temperature; a well-developed discussion of the rate constants for all runs, including both similarities and differences; and an explanation of the value for* E_a*. As always, discuss any possible errors in the experiment.)*

____

____

____

____

Experiment 13

Post-Laboratory Assignment

Name: ______________________________ **Date:** ____________________

Instructor: ______________________________ **Sec. #:** ____________

Show all work for full credit.

1) Which of the eight values of k should be the same (assuming no experimental error)?

2) Explain what factors affect the rate law constant.

3) Which reaction proceeds faster, one with a large E_a or a small E_a? Explain.

Experiment 14

Photometric Determination of an Equilibrium Constant

Introduction

In this experiment you will investigate the equilibrium of the formation of a complex ion $Fe(SCN)_x^{(3-x)}$. By determining the exact concentration of each species in equilibrium, the equilibrium constant, K, can be derived. There are several reasons why a scientist would want to determine the equilibrium constant. K can give us indications of solubility, acidity and even structural conformations. Of primary interest, however, is the determination of how to best produce a desired material. When the K value is very large, over 1, the reaction is said to lie to the right (of the arrow) indicating that the amount of product is greater than the amount of reactants; values less than 1 indicate a higher concentration of reactants than products, and values close to one indicate similar amounts of products and reactants. If a scientist can develop reaction conditions (temperatures, pressures, concentrations) that generate a $K > 1$, then more of a desired product will be formed. Obviously, making more product with the same amount of reactant(s) is a goal of any manufacturing process.

In order to determine the value of K experimentally, a method by which the concentrations of all species in the reaction may be determined must be used. Luckily, in the reaction we will be investigating, the complex ion product, $Fe(SCN)_x^{(3-x)}$, is colored. This will afford us the use of a technique called absorption spectroscopy.

Absorption spectroscopy has already been used in the "Solutions and Colloids" lab. Unlike the spectroscopic use in that lab, our purpose here is not only to create a calibration curve, but also to determine the concentration of $Fe(SCN)_x^{(3-x)}$ by use of Beer's Law, $A = \varepsilon bc$. Stoichiometric calculations will then be used to derive the equilibrium concentrations of the other species.

The use of a Mohr or graduated pipet is a technique introduced in this lab. It is very important that the volumes of the solutions used are accurate, and the Mohr pipet will allow us the best opportunity to create correct concentrations.

Finally, while not required, it will be advisable to refresh your use of Excel®. Not only is a graph needed in this lab, but you will also be making repetitive calculations that could be quite arduous if done by hand. Excel can simplify the calculations since minor changes in a single cell can be translated into every other cell.

Background

In its most abstract form, equilibrium is defined as a state in which a particular system at a given temperature has allowed its energy to be distributed in the statistically most probable manner. In simpler terms, equilibrium is reached when the forces, influences, and reactions in a system balance each other out so that there is no net change. For example, your body is said to be in thermal equilibrium when there is no net heat exchange between it and your surroundings.

In this experiment you will be introduced to chemical equilibrium. The generalized chemical equation for a reversible reaction where the capital letters (A, B, C, and D) represent the species involved in the reaction, and the lower case letter (a, b, c, and d) correlate to the numerical coefficients indicating how many moles of each reactant and product are involved is:

$$\mathbf{aA(aq) + bB(aq)\ldots \Leftrightarrow cC(aq) + dD(aq)\ldots}$$

The Equilibrium Constant

Perhaps the most important component of equilibrium is the equilibrium constant. This value, expressed as K_c when concentrations are used and K_p when partial pressures are used, reveals the position of equilibrium. K_c is generally written as a ratio of the concentration of the products, with their appropriate powers, to the concentration of the reactants, with their respective powers. For our example, the equilibrium constant is:

$$K_c = \frac{[C]^c[D]^d\ldots}{[A]^a[B]^b\ldots}$$

In the previous paragraph it was mentioned that the magnitude of the equilibrium constant reveals the position of equilibrium. Recalling that the constant is a ratio of products to reactants, it is rather intuitive that a small value of K indicates that the amount of product (C and D) is relatively small compared to the amount of reactant (A and B). In other words, when K is small, the reverse reaction is dominant. The equilibrium expression, as you will see in a later lab, can also indicate just how the equilibrium will shift when reactants or products are added to or removed from the system.

Example Problem

In order to further clarify the significance of the equilibrium constant and to familiarize you with some of the calculations it involves, look at the simple reaction shown below:

$$2NH_3(g) \Leftrightarrow N_2(g) + 3H_2(g)$$

Assume that at a particular temperature, 6.00-mol of NH_3 is introduced into a 3.00-L flask and allowed to reach equilibrium. At equilibrium, it was revealed that only 3.00-mol of NH_3 remained, and that we need to determine the equilibrium constant.

In order to begin this problem, an expression for the equilibrium constant must be written:

$$K = \frac{[N_2][H_2]^3}{[NH_3]^2}$$

Since the number of moles of NH_3 remaining at equilibrium and a volume are given, it is easiest to deal with concentrations as opposed to partial pressures. With that being said, take the 3.00-mol of NH_3 remaining at equilibrium and divide by the volume of the vessel to obtain the equilibrium concentration of ammonia.

$$[NH_3]_{Eq} = \frac{3.00mol}{3.00L} = 1.00M$$

The equilibrium expression now looks like this:

$$K_c = \frac{[N_2][H_2]^3}{[1.00]^2}$$

However, there are still two unknowns, $[N_2]$ and $[H_2]$, whose concentrations must be obtained to calculate the equilibrium constant. You can always use an ICE table to calculate the equilibrium concentrations based on the equilibrium concentration of NH_3. But, looking back at the equation for this reaction and observing the coefficients for each species, it is apparent that the concentration of NH_3 is twice that of N_2 and two-thirds of H_2. Mathematically, the respective concentrations can also be calculated as shown below:

$$[N_2]_{Eq} = \frac{1}{2}[NH_3] = \frac{1}{2}[1.00M] = 0.50M$$

$$[H_2]_{Eq} = \frac{3}{2}[NH_3] = \frac{3}{2}[1.00M] = 1.50M$$

Using the concentrations of all the species at equilibrium, the value of the equilibrium constant becomes:

$$K = \frac{[0.50][1.50]^3}{[1.00]^2} = 1.69$$

With an equilibrium constant of ~1.7, the reaction is slightly dominated by the forward reaction. In other words, for this particular reaction at the given temperature the concentration of the products is slightly more than that of the reactants.

The Experiment

In this experiment the equilibrium concentration of a complex ion is to be determined by use of absorption spectroscopy. The first part of the experiment, however, requires the determination of the extinction coefficient, ε. As in the Solutions and Colloids lab, the Beers-Lambert relationship between concentration and absorption is used to create a calibration graph. As long as one concentration in the reaction remains constant, the graph will then be (hopefully) linear with a slope equal to ε.

The absorbance of a series of solutions with different concentrations of the colored species is measured at a specific wavelength. A plot of Absorbance (y-axis) versus [KSCN] (x-axis) is created to generate a linear graph corresponding to the equation, $A = \varepsilon c$, where the extinction coefficient is equal to the slope. This extinction coefficient is then used in a later series of reactions to calculate actual concentrations of the colored complex.

Determining the Equilibrium Constant, K

As with the example problem above, the first step in determining the equilibrium constant is the writing of a balanced chemical equation.

$$\mathbf{Fe^{3+}(aq) + xSCN^{-}(aq) \Leftrightarrow Fe(SCN)_x^{(3-x)}(aq)}$$

The reaction above poses some interesting challenges because the exact chemical formula of the product is unknown. The subscripts and coefficients represented with an "x" shown in the reaction above could be 1, 2 or 3, yielding three different balanced reactions, respectively:

$$\mathbf{Fe^{3+}(aq) + SCN^{-}(aq) \Leftrightarrow Fe(SCN)^{2+}(aq) \quad when\ x = 1}$$

or

$$\mathbf{Fe^{3+}(aq) + 2SCN^{-}(aq) \Leftrightarrow Fe(SCN)_2^{+}(aq) \quad when\ x = 2}$$

or

$$Fe^{3+}(aq) + 3SCN^{-}(aq) \Leftrightarrow Fe(SCN)_3(aq) \quad \text{when } x = 3$$

In order to determine the equilibrium constant, the equilibrium concentrations of all the products and reactants must be determined several times starting with different initial concentrations of reactants. Although this seems complicated, all it really means is a little more work. Once the equilibrium concentration of $Fe(SCN)_x^{(3-x)}(aq)$ is experimentally determined based on the reactant combinations above, calculations of the equilibrium concentrations of the reactants are completed in the exact same manner as shown in the example problem. The calculations simply must be completed 3 times; once for each possible x value. The use of a spreadsheet greatly simplifies these calculations.

The only question that will remain at that point is which x value is the correct one. The answer to that will be readily apparent when you calculate the average and standard deviations of the equilibrium constant for each run. The value that produces the most consistent value of K (smallest standard deviation) will be the correct value to assume for the reaction.

Procedure

SAFETY NOTES: The solutions being used are acidified. Use all appropriate precautions. Wear goggles at all times and wash your hands immediately if your skin begins to itch or tingle.

Part I: Making Solutions to Determine the Extinction Coefficient

In clean, dry 100 mL beakers, collect 50 mL of 0.200M acidified $Fe(NO_3)_3$ and 0.001M KSCN.

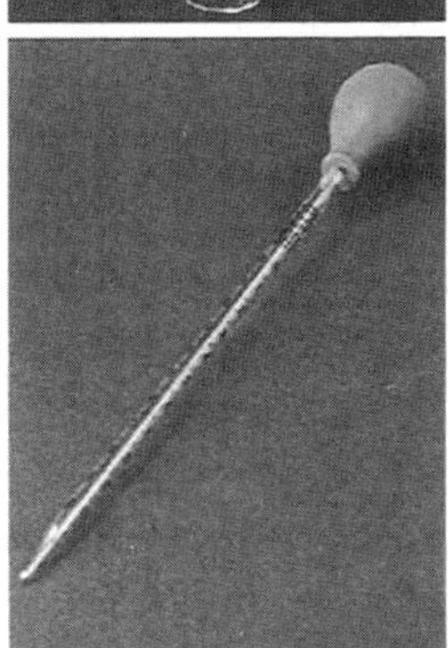

Using a Mohr pipet create the solutions A–E described in the background. Be very careful not to contaminate your stock solutions. If a rust color should form in the pipet it is contaminated. Clean it and remake the solution you were working on.

Note: Each solution should have the exact same total volume.

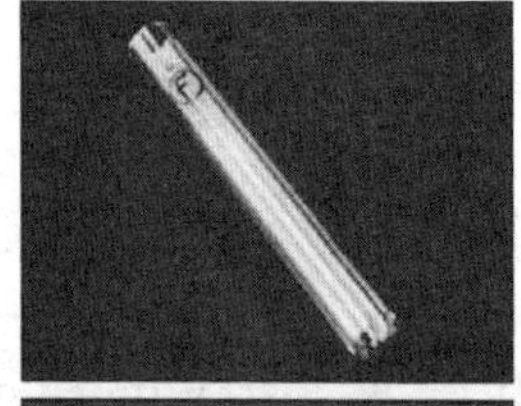

Collect a cuvette from the front counter. Make sure it is clean and dry.

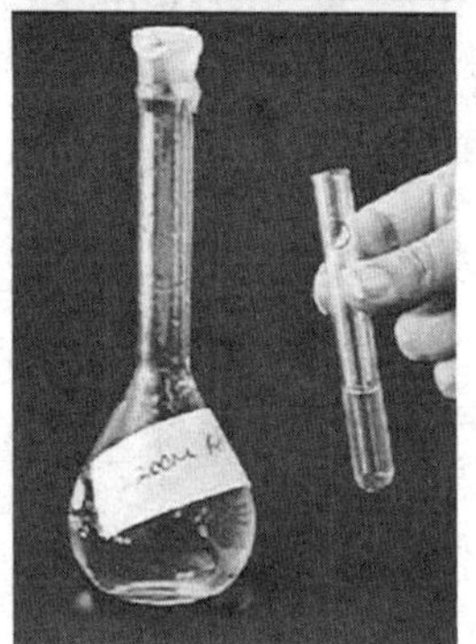

Use a blank of 0.200M $Fe(NO_3)_3$

Rinse the cuvette with the blank and then fill the cuvette with blank and zero the spectrometer.

Rinse the cuvette with the first solution and then fill it with the solution. Measure the absorbance and record it in your notebook.

Repeat the process until all the samples are measured.

Part II: Making Solutions to Determine the Equilibrium Constant

In clean dry 100 mL beakers, collect 50 mL of 0.002M acidified $Fe(NO_3)_3$ and 0.002M KSCN.

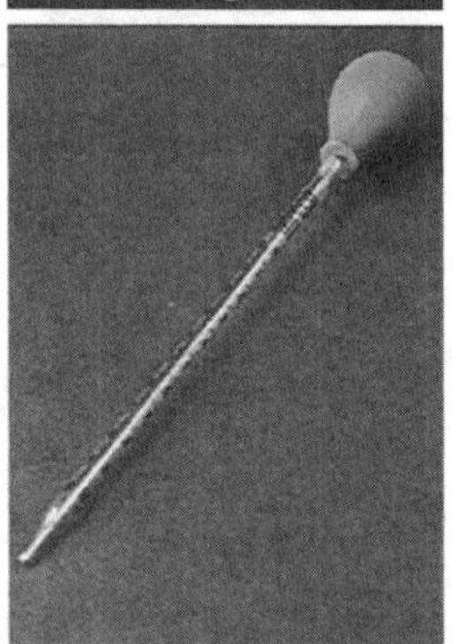

Using a Mohr pipet create the solutions F–J described in the background. Be very careful not to contaminate your stock solutions. If a rust color should form in the pipet it is contaminated. Clean it and remake the solution you were working on.

Note: Each solution should have the exact same total volume.

Part III: Measuring the Absorbance

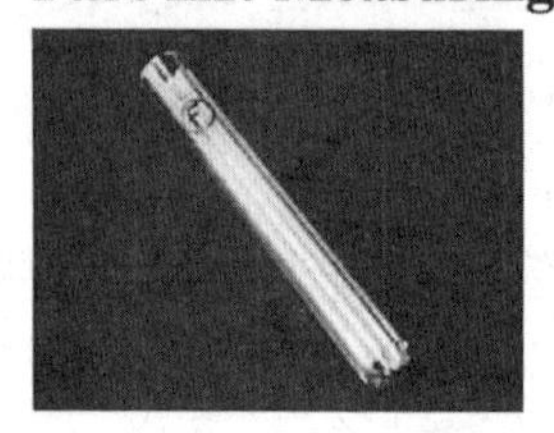

Collect a cuvette from the front counter. Make sure it is clean and dry.

Use a blank of 0.002M $Fe(NO_3)_3$

Rinse the cuvette with the blank and then fill the cuvette with blank and zero the spectrometer.

Rinse the cuvette with the first solution and then fill it with the solution. Measure the absorbance and record it in your notebook.

Repeat the process until all the samples are measured.

Experiment 14

Pre-Laboratory Assignment

Name: ______________________________ **Date:** __________________

Instructor: ______________________________ **Sec. #:** ____________

Show all work for full credit.

1) You would like to determine the equilibrium constant for the binding of a ligand to a metal ion: $M^{3+} + L^{-} <=> ML^{2+}$ where $K = [([ML^{2+}])/([M^{3+}][L^{-}])]$. The ML^{2+} complex has a unique absorbance in the visible region, so its concentration at equilibrium can be determined by measuring the absorbance of the solution at an appropriate wavelength. The concentrations of M^{3+} and L^{-} can be determined from their initial concentrations and the concentration change in going to equilibrium using an ICE table. The following experiment is carried out first in order to determine the extinction coefficient of the complex at the wavelength chosen.
Calculate the complex concentration in each solution, and then its extinction coefficient.

Solution	0.200 M Metal Ion (mL)	0.1 M HNO_3 (mL)	0.0011 M Ligand Ion (mL)	Solution Concentration (M)	Absorbance	Extinction Coefficient ($M^{-1}cm^{-1}$)
A	5.0	4.0	1.0		0.113	
B	5.0	3.0	2.0		0.227	
C	5.0	2.0	3.0		0.337	
D	5.0	1.0	4.0		0.431	
E	5.0	0.0	5.0		0.562	

Plot the absorbance versus the solution concentration and determine the slope and intercept of the resulting line.

2) Now you will determine the equilibrium constant, K, at several starting concentrations of M^{3+} and L^{-}. The following table summarizes your measurements.

Solution	0.0023 M Metal Ion (mL)	$[M^{3+}]_{init}$	0.0011 M Ligand Ion (mL)	$[L^{-}]_{init}$	Absorbance
F	3.0		7.0		0.286
G	4.0		6.0		0.312
H	5.0		5.0		0.308
I	6.0		4.0		0.276
J	7.0		3.0		0.221

Using an ICE table, calculate the equilibrium concentrations of ML^{2+}, M^{3+}, L^{-} and K for each of the five solutions.

Solution	$[ML^{2+}]$=x	$[M^{3+}]_{equil}$	$[L^{-}]_{equil}$	$[ML^{2+}]$=x	K
F					
G					
H					
I					
J					

What value of K do you get for Solution H?

What average value of K do you get?

Experiment 14

Laboratory Report

Name: ______________________ **Date:** ______________

Instructor: ______________________ **Sec. #:** __________

PURPOSE: (*The purpose should be several well-constructed sentences describing what your experiment was designed to accomplish and the criteria used to determine success. These sentences should include both concepts and techniques.*)

PROCEDURE: (*The procedure section should reference the lab manual and include any changes made to the procedure during the lab.*)

DATA: (*The data section should include your own personal data and observations.*)

Calibration Data:

Solution	0.200 M Metal Ion (mL)	0.1 M HNO_3 (mL)	KSCN Solution (mL) (Conc.)	$FeSCN^{2+}$ Concentration (M)	Absorbance	Extinction Coefficient $M^{-1}cm^{-1}$
A	5.0	4.0	1.0			
B	5.0	3.0	2.0			
C	5.0	2.0	3.0			
D	5.0	1.0	4.0			
E	5.0	0.0	5.0			
					Slope	
					Intercept	

Part I and II Absorbance Data:

Solution	mL 0.200M $Fe(NO_3)_3$	mL 0.002M KSCN	mL 0.1M HNO_3	Absorbance
A	5	1	14	
B	5	2	13	
C	5	3	12	
D	5	4	11	
E	5	5	10	

Solution	mL 0.002M $Fe(NO_3)_3$	mL 0.002M KSCN	Absorbance
F	3	7	
G	4	6	
H	5	5	
I	6	4	
J	7	3	

Complete the following table for x = 1, 2 and 3:

X=1

Assume: $Fe^{3+} + SCN^{-} = Fe(SCN)^{2+}$

Soln	Initial Volume of Fe^{3+} (mL)	Initial Volume of KSCN (mL)	Total Volume (mL)	Initial $[Fe^{3+}]$ (M)	Initial [SCN-] (M)	$[Fe(SCN)_x^{(3-x)}]$ (M)	Eq $[Fe^{3+}]$ (M)	Eq [SCN-] (M)	K
F									
G									
H									
I									
J									

X=2

Assume: $Fe^{3+} + 2\ SCN^{-} = Fe(SCN)_2^{+}$

Soln	Initial Volume of Fe^{3+} (mL)	Initial Volume of KSCN (mL)	Total Volume (mL)	Initial $[Fe^{3+}]$ (M)	Initial [SCN-] (M)	$[Fe(SCN)_x^{(3-x)}]$ (M)	Eq $[Fe^{3+}]$ (M)	Eq [SCN-] (M)	K
F									
G									
H									
I									
J									

X=3

Assume: Fe^{3+} + 3 SCN^- = $Fe(SCN)_3$

Soln	Initial Volume of Fe^{3+} (mL)	Initial Volume of KSCN (mL)	Total Volume (mL)	Initial $[Fe^{3+}]$ (M)	Initial [SCN-] (M)	$[Fe(SCN)_x^{(3-x)}]$ (M)	Eq $[Fe^{3+}]$ (M)	Eq [SCN-] (M)	K
F									
G									
H									
I									
J									

CALCULATIONS: (*The calculation section should include sample calculations for each column in the data table. Be sure to include conversions of units.)*

Graph of Absorbance versus [$FeSCN^{2+}$ Complex] *(The slope of this graph is the extinction coefficient used in Part II to determine the equilibrium concentration of [$Fe(SCN)_x^{(3-x)}$]$_{Eq.}$)*

Example calculations for the concentrations used in the calibration graph and each calculation done for the equilibrium data:

CONCLUSION: *(In the conclusion section include several paragraphs discussing the values determined for the extinction coefficient, the equilibrium constant K and the value for x. Be sure to discuss the data in support of these conclusions. Compare and contrast the values calculated when each of the three values for x are used. Do not forget to discuss any errors.)*

Experiment 14

Post-Laboratory Assignment

Name: ______________________________ **Date:** ____________________

Instructor: ______________________________ **Sec. #:** ____________

Show all work for full credit.

1) How constant were your K_c values at room temperature? Explain any variation.

2) See if you can find literature and/or Internet references for the equilibrium constant for this equilibrium. Please cite the reference(s). How does your result compare with the one(s) you cited? Perhaps you will even find a similar experiment, with sample results, posted out on the Web.

Experiment 15

Titration of 7-Up®

Introduction

Two important concepts in chemistry are the focus of this experiment: titration and acid-base reactions. Titration is the method of determining the concentration of a solution by allowing a carefully measured volume to react with a standard solution of another substance whose concentration is known. In today's experiment, this means that by adding a carefully measured volume of a base of known concentration to a carefully measured volume of acid of unknown concentration, or vice versa, and using your knowledge of stoichiometry and acid-base neutralization, you can determine the concentration of the unknown.

The key phrase above for successful completion of this experiment is "your knowledge of stoichiometry and acid-base neutralization." Understanding these concepts is necessary for you to determine the unknown concentrations of three chemicals: NaOH, phosphoric acid and citric acid in 7Up. The other concept mentioned above, titration, is a new technique that you will use to make these determinations. Titration is a much-used process in chemistry and is an invaluable skill to master. It is one of the simplest and least expensive methods by which you can determine the concentration of an unknown solution, only requiring a titrant, a buret or other volumetric delivery device and a color indicator.

The specific purpose of this experiment is to titrate a solution of a weak polyprotic acid (citric acid, $C_6H_8O_7$) and a stronger polyprotic acid (phosphoric acid, H_3PO_4) to determine their concentrations as well as their K_a values. The base (NaOH) used in these titrations will first be standardized using a commercially available standard HCl solution. The titration curves will be plotted and the concentrations and K_a values of each acid will be determined. By doing these titrations, you will have the opportunity to learn more about the chemical differences between weak and strong acids and also the meaning of polyprotic and how it affects the outcome of a titration.

Background

The first of a series of similar experiments using titration to determine concentration, this investigation focuses upon acid-base chemistry. In general terms, titrations utilize a known property of one solution to determine a similar property of an unknown solution. Specifically, these include acid-base titrations, potentiometric titrations (redox), complexometric titrations, and even titrations utilized to determine specific concentrations of bacteria or viruses. For example, the alkalinity and acidity of water in streams and rivers is an important topic to environmental chemists. In order to observe such characteristics, they use the same technique you will learn in this experiment— acid-base titration.

Titration Set-up

The generalized setup of a titration is shown below (Figure 1). The base is placed in the buret, so that a precise amount of solution can be added to the acid. The buret's precision is attributed to its graduations up and down the tube, making it one of the more expensive pieces of glassware you will use in the lab. The precision of the buret is dependent upon reading it correctly: volumes delivered by a buret are read to the hundredth of a milliliter. The knob on the buret is called a stopcock, and its sole purpose is to deliver the titrant to the solution below in a controlled manner.

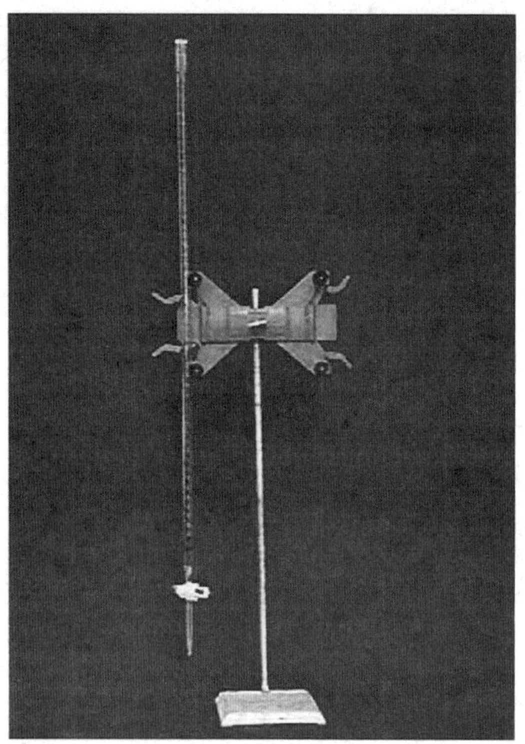

Figure 1: Titration Set-up

Acid-Base Titrations

When an acid solution is titrated with a strong base such as NaOH, the initial pH of the solution is low. As base is added to the acidic solution, the pH gradually rises until the volume added is near the equivalence point, the point during the titration when equal molar amounts of acid and base have been mixed. Immediately before the equivalence point, the pH increases very rapidly and then levels off again with the addition of excess base (Figure 2). In this experiment, a carefully measured volume of HCl of a known concentration is titrated with NaOH of an unknown concentration. Since the buret allows us to determine the precise amount of base needed for neutralization, the precise concentration of the base can be calculated using $M_AV_A = M_BV_B$.

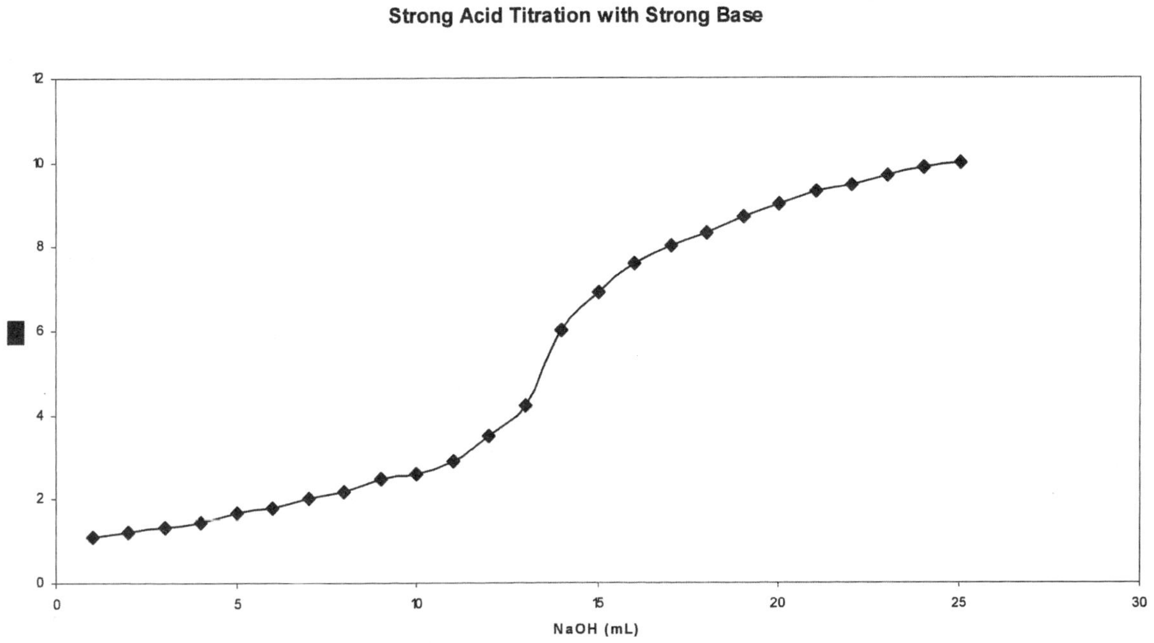

Strong Acid Titration with Strong Base

NaOH (mL)

Figure 2. Titration curve of strong acid by strong base

Visualizing the "end" of a particular titration, specifically referred to as the endpoint, is essential to a successful titration. The endpoint is the point in the titration where the indicator changes and the equivalence point is the point in the titration when the exact amount of titrant (equivalence volume) has been added such that the stoichiometric reaction has been completed. The indicator is generally chosen so that endpoint ≡ equivalence point. Indicators, often added in minute amounts to the solution of interest, are chemical compounds that undergo dramatic changes of color when a particular property of a solution is changed. Indicators are specific to the reaction being analyzed.

pH and pH Meters

The hydrogen ion concentration, expressed in terms of pH, is one of the most important properties of aqueous solutions as it can control the solubility of various species, the formation of complexes, and even the kinetics of an individual reaction. In order to obtain precise data of the particular hydronium concentrations of the solutions in this experiment, and to clearly observe the change in pH at the equivalence point, a pH meter is used. In general, a pH meter measures the differences in electromotive force between two electrodes. A pH meter contains an electrode sensitive to the concentration of the hydrogen ion as well as one used solely for a reference.

For accurate measurements, it is necessary to calibrate the instrument using a buffer solution of approximately the same pH as the sample to be used. This calibration takes care of temperature effects and minor variations in the potential due to changes in the membrane. Your instructor will provide details regarding the calibration of the pH meters used in your laboratory.

The Experiment

There are a total of four acid-base titrations to be completed in this experiment: 1) standardize the titrant, a solution of sodium hydroxide (NaOH), using a known volume and concentration of hydrochloric acid (HCl); 2) repeat the standardization to confirm the concentration; 3) with the standardized solution of NaOH, determine the concentration of the relatively strong polyprotic acid, phosphoric acid (H_3PO_4) and 4) determine the concentration of the much weaker polyprotic acid, citric acid ($H_3C_6H_5O_7$), found in a sample of the soft drink 7-Up.

Weak Acid Equilibria

A weak acid (HA) is one that does not fully dissociate in water. In other words, if the weak acid represented is allowed to ionize, as shown in the equation below, then a significant amount of HA will remain un-ionized.

$$HA(aq) + H_2O(l) \Leftrightarrow H_3O^+(aq) + A^-(aq)$$

At equilibrium, the dissociation of a weak acid is generally described by its acid-dissociation constant (K_a) and is mathematically represented as follows:

$$K_a = \frac{[H_3O^+][A^-]}{[HA]} \quad (1)$$

In this investigation you are asked to experimentally determine the acid-dissociation constants of both phosphoric acid and citric acid. With the knowledge that at equilibrium the concentration of the free hydronium ions (H_3O^+) is equal to the concentration of the conjugate base (A^-), if the concentration of either of these chemicals is determined experimentally, then stoichiometry can be used to determine the concentrations of the other components in the solution. Mathematically, the relationship for the reaction above is expressed as:

$$[HA]_{Eq} = [HA]_{Init} - [H_3O^+]_{Eq} = [HA]_{Init} - [A^-]_{Eq}$$

From this logic, combined with the fact that pH is equal to the negative log of the hydrogen ion concentration, we can arrive at an expression for K_A incorporating only the initial concentration of the weak acid, and the experimentally determined pH at the equivalence point.

$$K_a = \frac{[10^{-pH}]^2}{[HA]_{Init} - [10^{-pH}]}$$

The acid-dissociation constant of a weak acid can also be obtained by another method. This method involves the "half equivalence point," where just enough NaOH has been added to the weak acid to convert half of the acid to its salt. At this point, the concentration of the weak acid, [HA], is equal to the concentration of its conjugate base, [A-]. Utilizing this fact, our generalized equilibrium expression equation (1) can now be defined as shown below because [A-] and [HA] can be canceled out of the expression.

$$K_a = [H_3O^+]_{Eq} \qquad pK_a = pH$$

Overall, by performing all three of these titrations and plotting the pH versus volume of NaOH added, you can see how the pH of the solution changes as an acid, or base, is added. In fact, if you are precise enough, you will get an idea of just how the shape of a titration curve can be influenced by both the concentration and nature of the acid or base. Further, by titrating both strong and weak acids you will see different titration curves and how they change with the strength of the acid.

Polyprotic Acids

Both phosphoric acid and citric acid are triprotic acids, one form of polyprotic acid. This means that each of these acids has three ionizable hydrogens and thus three separate pK_a values; one for each dissociation. What is important to understand about a polyprotic acid dissociation is that at any point along the curve there is some percentage of each acid form present in the solution. This means that unlike a monoprotic dissociation that is "all or nothing," the pH of a polyprotic acid solution is dependent on several forms of the acid. This also means that you should observe more than one inflection point in the titration curves.

Procedure

SAFETY NOTES: You will be handling strong acids and bases in this lab. Wash your hands immediately if you get some acid or base on your skin. Notify your instructor immediately regarding any spills that occur. You must wear goggles at all times.

Part I: Standardization of NaOH

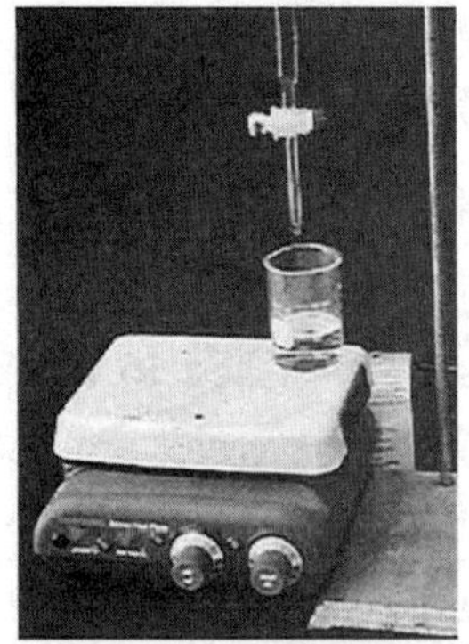

Obtain about 65–75 mL of NaOH and 30–35 mL of 0.01 *M* standard HCl.

Clean a 50 mL buret and rinse it with the NaOH solution as follows. Add a few mL of the titrant and tip the buret to allow the solution to run almost all the way to the open end. Then rotate the buret slowly, so that all surfaces contact the solution. Drain this solution into a waste beaker by opening the stopcock. Repeat this procedure two more times, using no more than 15–20 mL total. Fill the buret with the NaOH, and open and close the stopcock to force air bubbles out of the tip. Use the buret clamp to place your buret on the ring stand.

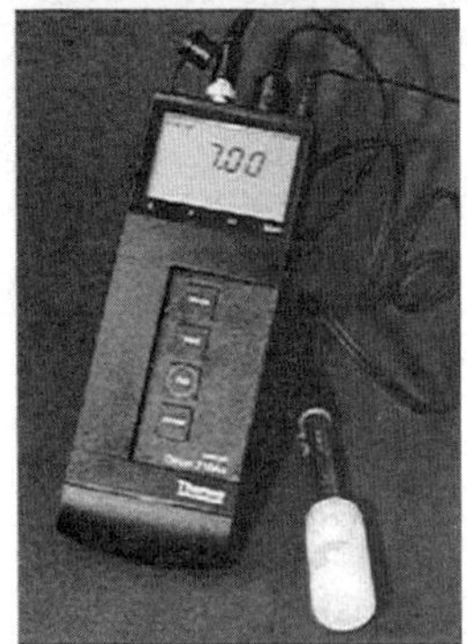

Use a utility clamp to suspend a pH electrode on the ring stand. Obtain a short stir bar, stir plate, and flask for titration.

Calibrate the pH meter for data collection based on your instructor's directions.

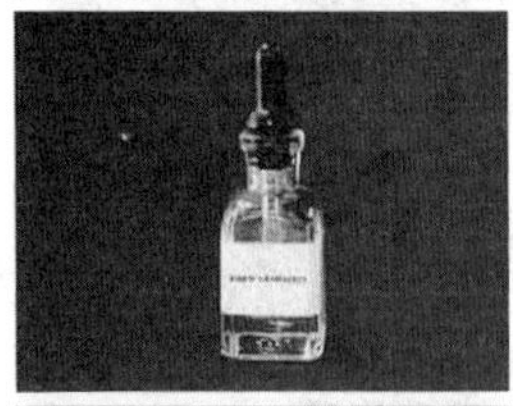

Pipet 25.00 mL of standard 0.01 M HCl into a 250 mL beaker. Add a few drops of phenolphthalein to the beaker.

You are now ready to begin the titration. Read your buret to two decimal places, being sure that your eye is level with the meniscus. Always record the buret reading at the beginning and end of a titration. Your goal in this first titration will be to identify the pH region that surrounds the equivalence point(s). You will do this by consistently adding 2 mL of NaOH and noting the change in pH after each addition. You will continue adding NaOH until you reach a point where the pH no longer changes (> pH 11).

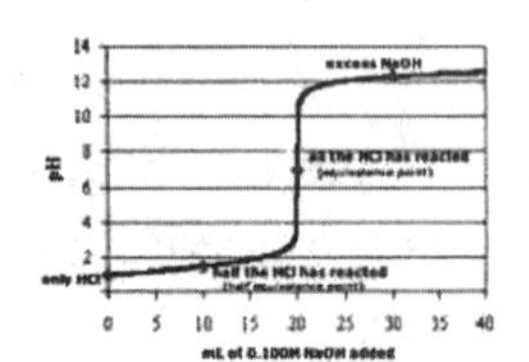

From the data collected above, you should be able to identify the regions of the titration where the pH changes very slowly and also the region of the equivalence point. You will now repeat the titration. The goal of this second titration is to generate enough data points to create an accurate and smooth curve.

Refill the buret if necessary, note the volume and begin the second titration as previously instructed.

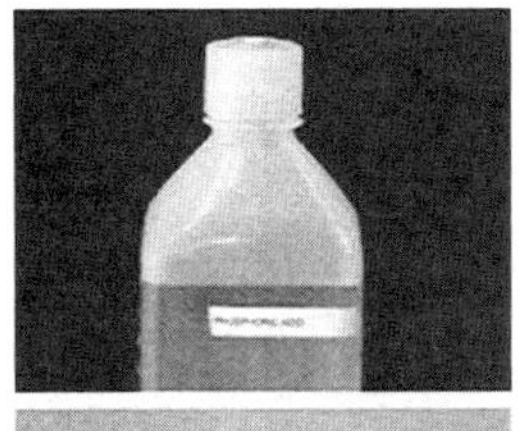

Repeat the Steps above of the procedure using a 10.00 mL sample of phosphoric acid instead of the HCl. Add increments of ~0.2 mL until you reach a point where the pH no longer changes.

Repeat the Steps above of the procedure using a 10.00 mL sample of 7-Up instead of the HCl. Add increments of ~0.2 mL until you reach a point where the pH no longer changes.

Experiment # 15

Pre-Laboratory Assignment

Name: ______________________ **Date:** ______________

Instructor: ______________________ **Sec. #:** __________

Show all work for full credit.

1) To standardize your NaOH solution, you titrate 25.0 mL of 0.0822 M HCl solution, measuring the pH after each small addition of NaOH. Following is the graph you obtain. The equivalence point occurs where there is a steep rise in pH.

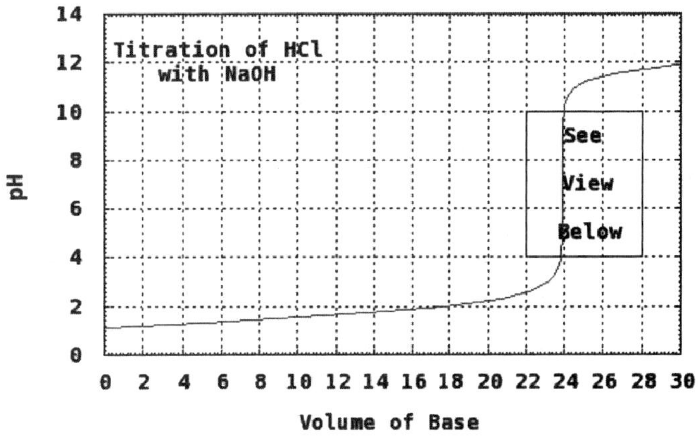

What molar quantity of HCl is being titrated?

What molar quantity of NaOH has been added at the equivalence point?

What volume of NaOH solution has been added at the equivalence point?

Expanded view of the equivalence point:

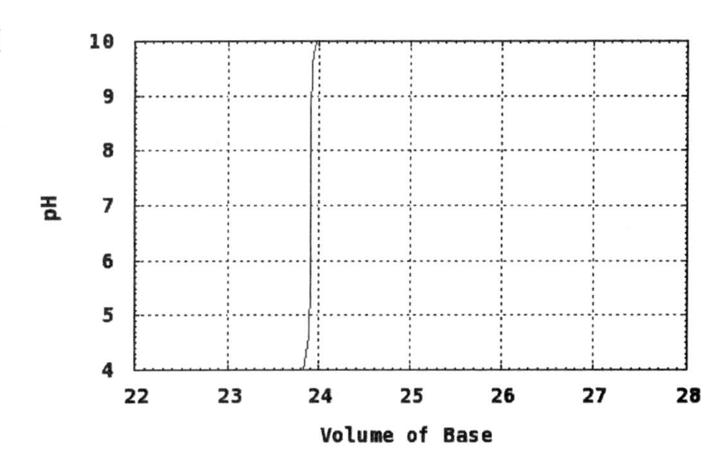

What is the concentration of the NaOH solution?

2) A triprotic acid requires three moles of base to neutralize it, and the protons are removed one at a time as follows:

$$\mathbf{H_3A + OH^- \leftrightarrow H_2A^- + H_2O}$$
$$\mathbf{H_2A^- + OH^- \leftrightarrow HA^{2-} + H_2O}$$
$$\mathbf{HA^{2-} + OH^- \leftrightarrow A^{3-} + H_2O}$$

A titration curve, which plots the pH of the acid solution versus quantity of base that is added, can reveal several things about the nature of the acid. First of all, some or all of the equivalence points can be observed in the graph as places where there is a maximum in the slope of the curve.

As an example, you have a 40.0 mL solution of a triprotic acid, H_3A, with a concentration of 0.0588 M. You titrate it with a 0.350 M solution of NaOH. How many millimoles of H_3A are you titrating? (Give units.)

How many millimoles of NaOH are required to complete the titration? (Give units)

What volume of NaOH will be needed to completely titrate the acid?

What volume of NaOH will be needed to reach the first equivalence point?

What volume of NaOH will be needed to reach the second equivalence point?

3) Citric acid is a weak organic acid found in citrus fruits and some soft drinks. It is an important intermediate in the citric acid cycle and therefore occurs in the metabolism of almost all living things. It is a **triprotic** acid, requiring three moles of OH^- to remove all the acidic protons. Titration of citric acid, as with all triprotic acids, occurs in a stepwise fashion in which one proton is removed at a time. The stepwise dissociation can therefore be represented by three equations, each with an acid dissociation equilibrium constant.

$$\mathbf{H_3C_6H_5O_7 \leftrightarrow H_2C_6H_5O_6^- + H^+ \qquad (K_{a1})}$$

$$\mathbf{H_2C_6H_5O_7^- \leftrightarrow HC_6H_5O_6^{2-} + H^+ \qquad (K_{a2})}$$

$$\mathbf{HC_6H_5O_7^{2-} \leftrightarrow C_6H_5O_6^{3-} + H^+ \qquad (K_{a3})}$$

The following graph shows how the concentration of each species changes when NaOH is added to a solution containing 100 moles of citric acid.

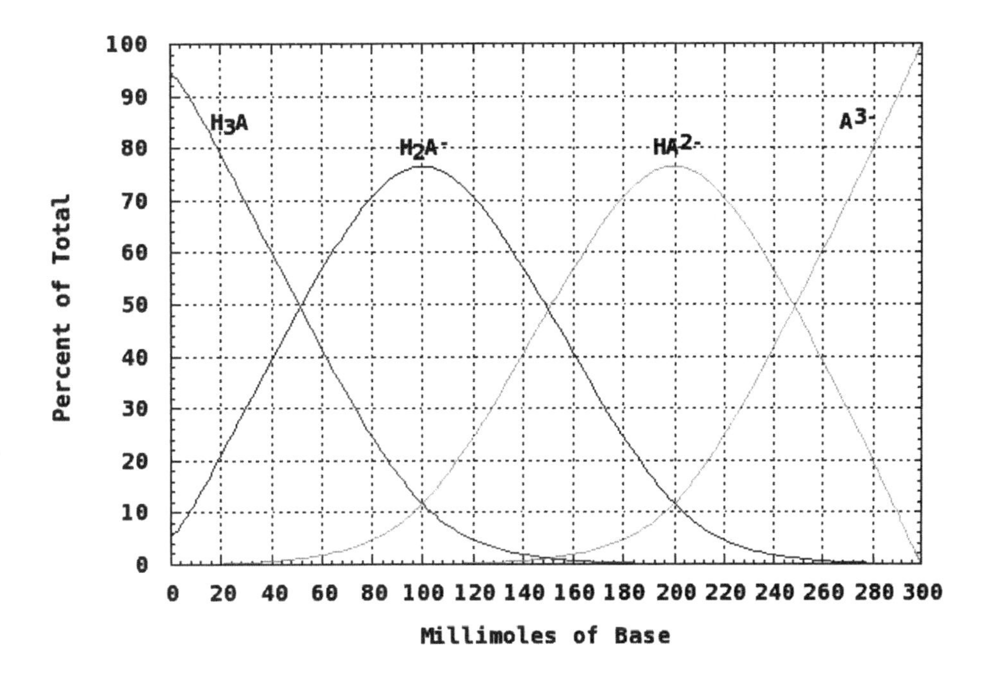

What is the molecular weight of citric acid?

What is the mass of citric acid used in this titration?

After addition how much base does $[H_2C_6H_5O_7^-] = [C_6H_5O_7^{3-}]$?
After addition how much base does $[H_2C_6H_5O_7^-] = [HC_6H_5O_7^{2-}]$?
After addition how much base does $[H_3C_6H_5O_7] = [HC_6H_5O_7^{2-}]$?

Experiment 15

Laboratory Report

Name: ____________________ **Date:** ______________

Instructor: ____________________ **Sec. #:** __________

PURPOSE: (*The purpose should be several well-constructed sentences describing what your experiment was designed to accomplish and the criteria used to determine success. These sentences should include both concepts and techniques.*)

PROCEDURE: (*The procedure section should reference the lab manual and include any changes made to the procedure during the lab.*)

DATA: (*The data section should include your own personal data and observations.*)

Part I: Balanced Equation of NaOH and HCl:

Titration 1:

mL NaOH	pH	mL NaOH	pH	mL NaOH	pH

Titration 2:

mL NaOH	pH	mL NaOH	pH	mL NaOH	pH

Part II: Balanced Equation of NaOH and H_3PO_4:

Titration Data:

mL NaOH	pH	mL NaOH	pH	mL NaOH	pH

Part III: Balanced Equation of NaOH and 7Up:

Titration Data:

mL NaOH	pH	mL NaOH	pH	mL NaOH	pH

Experimentally Determined Concentrations:

Substance	Concentration	K_a
NaOH		--------------
Phosphoric Acid		
7Up		

Observations and Notes:

CALCULATIONS: (*The calculation section should include four titration curves: two for HCl, one for phosphoric acid and one for citric acid. Each graph should be pH vs. the volume of titrant added. All graphing guidelines apply. DO NOT add a trend line to these graphs; if you would like to add a line, a moving average would be acceptable.)*

Example Calculations:

Determination of the molarity of NaOH:

Determination of the molarity of phosphoric acid:

Determination of the molarity of citric acid:

Determination of the molarity of K_a values:

Using the K_a values you calculate for each acid, calculate a percent error based on literature K_a values and cite your resource for these values:

Finally, show a sample calculation of the number of grams of citric acid in a can of 7-Up.
NOTE: There are 240 mL of 7-Up in a can.

CONCLUSION: *(The conclusion should include several paragraphs with the following: the concentrations of HCl, NaOH, phosphoric acid and citric acid; a discussion of the K_a values for the weak acids and the errors associated with them; a discussion of the differences in the titration curves for a weak acid and a strong acid; and a discussion of all possible errors.)*

Experiment 15

Post-Laboratory Assignment

Name: ______________________________ **Date:** ___________________
Instructor: ______________________________ **Sec. #:** ____________

Show all work for full credit.

1) Are you able to determine all of the K_a values for citric acid? Why or why not?

Experiment 16

Hydrogen Phosphate Buffer Systems

Introduction

By now you should be starting to discuss buffer systems in class and you might wonder if there is a purpose to the subject beyond what appears to be some very complicated math. Buffers are vital to the health of every living thing on Earth. Your body is full of buffer systems. We have learned previously that the H^+ concentration (or pH) of a solution can cause molecules to gain and lose protons and electrons (ionize) based on their pK_a values. Proteins are biological molecules that make up the muscles and enzymes in your body. If proteins are not kept at the proper pH, they will not function properly. Enzymes, which are the protein catalysts by which most of your physiological processes are regulated, are rendered inoperative by the changing of ionic states, creating nonfunctional protein shapes. pH is probably the most important single thing the body regulates since it affects so many other aspects of biochemistry. Proteins themselves often act as buffers; sodium bicarbonate and phosphate are others.

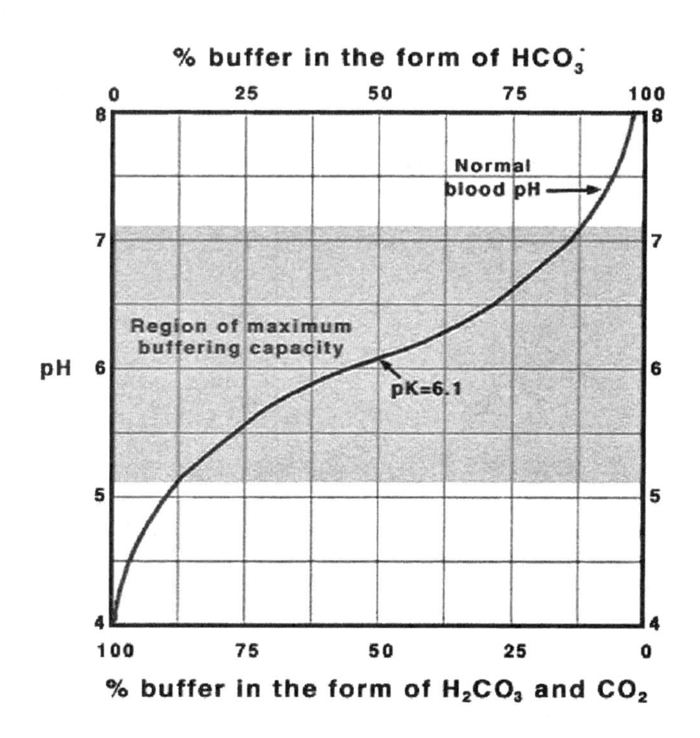

Your blood is buffered by the bicarbonate system. If disease overcomes the system, conditions known as acidosis or alkalosis (too much acid and too much base, respectively) occur. Mild symptoms of both are nausea, migraine, cramps, and fainting; severe forms can lead to coma and even death.

In order to study various cellular proteins in vitro, scientists have to maintain the pH conditions that are found in the human body. This means that as a chemist or biochemist, you should be able to construct a buffer solution at a particular pH as needed.

This experiment is designed to give you practical experience in the calculations needed to create a buffer in the lab and to allow you to observe the buffering capacity of a phosphate buffer system. The repetitive nature of the experiment is designed to give you lots of practice with that "complicated" math.

Background

The pH of an acidic or basic solution is dependent on the concentration of hydrogen ions, H^+, also called free protons or **hydronium** ions:

$$\mathbf{pH = -\log[H_3O^+]}$$

This concentration can be altered by adding substances that either react with or supply hydronium ions. Some solutions, however, exhibit the ability to resist pH changes—to a point—as additional acid or base is added. Such mixtures are said to be "buffered" and the solutions are known as buffer solutions.

A buffer solution consists of an equilibrium system containing a weak acid (or base) and its conjugate. A simple example would be the mixture of acetic acid and sodium acetate. The acetate ion (CH_3COO^-) is the conjugate base of acetic acid. When only acetic acid is present, the amount of acetate ion in the equilibrium mixture is very small and can be determined from the K_a value and the molarity of the acid. The amount of acetate ion is sometimes described by the term "% dissociation."

$$\mathbf{CH_3COOH + H_2O \Leftrightarrow CH_3COO^- + H_3O^+}$$
$$\mathbf{K_a = 1.8 \times 10^{-5}}$$

The small value of K_a tells us that this reaction has very little tendency to progress in the forward direction, i.e., acetic acid is a weak acid and does not dissociate completely in water. Only about 1% of the acetic acid molecules actually donate protons to water at any given time. Thus the equilibrium concentration of acetate ion is very small.

If additional acetate ion is added to acetic acid (in the form of sodium acetate or some other soluble acetate salt) H_3O^+ that have been formed by the small amount of dissociation will have a strong tendency to donate protons to the added acetate ions. The result of this "common ion effect" is that the H_3O^+ concentration decreases and the pH of the mixture rises a little. We would predict exactly this behavior by Le Châtelier's principle. Because the concentration of free H_3O^+ is low to begin with, adding acetate will most likely leave an excess of acetate ion beyond that consumed by the reverse reaction. It will also most likely exceed the concentration that would be present due to the normal dissociation of acetic acid. This sets up the conditions for "buffering": ready supplies of the weak acid (or base) and the conjugate. When these supplies are of equal concentration the buffer is said to be "centered." This means it is able to absorb equal moles of acid or base before a significant change in pH occurs.

The key to understanding buffer behavior is to keep in mind that the pH of the mixture is related to the amount of free H_3O^+. If excess acid (e.g., some HCl) is added to the acetic acid/acetate buffer mixture it readily donates protons to the acetate ion. As long as there are acetate ions available, added acid will be "absorbed" by this process and will not affect the concentration of H_3O^+ (or not much). Thus the pH remains close to its original value. Note that this depends on having a large enough concentration of buffer so that some acetate will still remain after the addition of the acid.

Describing the behavior of the mixture when base is added is a bit more complicated. Hydroxide ion is a stronger base than acetate so if sodium hydroxide is added to the buffer, H_3O^+ will donate protons to the hydroxide to form water. This loss of H_3O^+ upsets the equilibrium and the acetic acid will donate protons to water in order to replace some of the lost H_3O^+. This nearly restores the original pH as long as there is a sufficient concentration of acetic acid present.

Acid buffer solutions that are "centered" should have pH values that are equal to the pK_a of the weak acid. As an example, using the weak acid "HA" and its conjugate "A^-", the K_a expression for this acid is:

$$K_a = \frac{[H^+][A^-]}{[HA]}$$

which can be rearranged to

$$[H^+] = K_a\frac{[HA]}{[A^-]}$$

Taking the negative log of both sides of the equation on the right provides us with a simple relationship between the pH and the pK_a of an acid-base buffer system. This equation is called the Henderson-Hasselbalch equation:

$$pH = pK_a + \log\frac{[A^-]}{[HA]}$$

Because the acid and its conjugate appear in both the numerator and denominator, if they are of equal size their concentrations will cancel. This tells us that the H_3O^+ concentration in such a mixture is equal to the K_a, or the pH = pK_a. In fact, the pH of a buffer is regulated by some combination of the K value of the weak electrolyte chosen and the ratio of conjugate to acid (or base). This is practical information since it is unlikely that a particular weak acid or base will have exactly the right K value to create a buffer of a desired pH. However, by selecting an acid or base with a pK_a close to the desired pH and altering the ratio of acid (or base) to conjugate, a buffer with any desired pH can be made.

The amount of acid or base a buffer can absorb before a significant pH change is called its capacity. This is governed by the concentrations of the weak acid–or base–and the conjugate. While the pH is supposed to remain reasonably constant, it *does* change, even before one of the concentrations becomes exhausted. The determination of the pH of a buffer solution after the addition of acid or base is a relatively simple process. Using our hypothetical acid, HA, adding an amount of strong acid (x) will shift the equilibrium to the left:

	HA	+	H_2O	⇔	H_3O^+	+	A^-
	+x				x		-x
Eq	HA + x						A^- - x

Note that H_3O^+ readily reacts with A^-, increasing the amount of HA and decreasing the amount of A^-.

It is simpler to use moles in these calculations because the volume of the mixture will typically change since most added acids or bases would be in aqueous form. This volume change would necessitate a recalculation of the molarities of HA and A^-, except that the new volume will cancel in the K_a expression!

$$pH = pK_a + \log\frac{[A^- - x]}{[HA + x]}$$

From the expression above it is clear that the change in the ratio of A^- to HA affects the pH. As long as "x" is small compared to the moles of A^- and HA, the change in pH will be small. The magnitude of x in relation to that amount of acid (or base) and conjugate is what determines the buffer capacity mentioned earlier.

A similar, but opposite, analysis applies to the case of adding strong base (x). HA will react with base in a neutralization reaction. The situation is:

	HA	+	OH^-	⇔	H_2O	+	A^-
	-x		x				+x
Eq	HA - x						A^- + x

To calculate the pH of the solution after base is added the expression would be:

$$pH = pK_a + \log\frac{[A^- + x]}{[HA - x]}$$

Preparing a buffer solution of a particular pH would now seem to be a fairly simple process. After consulting a table of K_a values, an acid (or base) with a pK_a close to the desired pH is selected and then the ratio of conjugate to acid (or base) is adjusted to achieve the desired pH value. When this is done in the lab (as in this experiment), the results are only approximate because the actual solution pH is also affected by both the solution temperature and the pH of the distilled water used to make the buffer. Neither of these factors is accounted for in our original buffer calculation.

The buffer system chosen for this experiment consists solely of ionic species. The equilibrium involved may be written as:

$$\mathbf{H_2PO_4^- + H_2O \Leftrightarrow H_3O^+ + HPO_4^{2-}}$$
$$\mathbf{K_a = 2.3 \times 10^{-7}}$$

Phosphoric acid (H_3PO_4) changes pretty quickly into dihydrogen phosphate, or $H_2PO_4^-$. This dihydrogen phosphate is an excellent buffer, since it can either accept a proton and reform phosphoric acid, or it can give off another proton and become hydrogen phosphate, or HPO_4^{2-}. These ions (which are species derived from phosphoric acid) can be obtained as potassium and sodium salts. The value of K_a (pK_a= 6.64) suggests a reasonable target pH range for this buffer system of about 6.30 to 6.90.

In this experiment you will be assigned a target pH for this system. You can calculate the ratio of buffer component concentrations by using the Henderson-Hasselbalch equation:

$$pH = pK_a + \log\frac{[A^-]}{[HA]}$$

Once the solution is prepared the pH can be measured in the usual fashion. The "capacity" of the buffer can also be checked by calculating the volume of 0.50 M HCl or NaOH needed to change the pH of a 25 mL sample by 0.50 pH units and then observing the result when this is done in the lab. A few additional tests on the buffer solution complete the experiment.

Finally, it should be said that in practice, laboratory chemists do not do these kinds of calculations before preparing buffer solutions. Buffer solutions with integer pH values are readily available commercially. (Someone else has done all of the calculations!) When buffers with non-integer pH values are required, approximate solutions are prepared and additional strong acid or base is carefully added while monitoring the pH of the mixture with a calibrated pH electrode to achieve the desired pH.

Buffer Preparation: An Example

For today's lab both components of the conjugate acid-base pair are to be weighed out separately to obtain the desired ratio and then dissolved in water.

Here's how to do it

Problem: Prepare 1.0 L of a 0.50 M potassium phosphate buffer at pH 6.90, assuming the availability of solid H_3PO_4 (pK_a = 2.00), KH_2PO_4 (pK_a = 6.64), K_2HPO_4 (pK_a = 12.00), and K_3PO_4.

Step 1) Determine the principal components of the buffer system. This is easy for a monoprotic system. For diprotic or polyprotic systems, this can vary depending upon the desired pH. Identify the conjugate acid-base pair and write out the equilibrium.

In this case, the desired pH (6.90) is closest to the pK_a of the second ionization:

$$H_2PO_4^- + H_2O \Leftrightarrow H_3O^+ + HPO_4^{2-} \qquad pK_a = 6.64$$

(You might notice that the pK_a in your text is different from the one given here. This is because the pK_a given here takes into account the activity coefficient of the buffer. Please see A. A. Green, "The Preparation of Acetate and Phosphate Buffer Solutions of Known PH and Ionic Strength", *J. Am. Chem. Soc.* 55, 2331, (1933) for further explanation.)

Step 2) Calculate the desired ratio of the conjugate acid-base pair using the Henderson-Hasselbalch equation:

$$pH = pK_a + \log\frac{[HPO_4^{2-}]}{[H_2PO_4^-]}$$

$$\frac{[HPO_4^{2-}]}{[H_2PO_4^-]} = 10^{pH-pK_a}$$

In this case,

$$\frac{[HPO_4^{2-}]}{[H_2PO_4^-]} = 10^{6.90-6.64} = 10^{0.26} = 1.8$$

We can solve it by using the two statements of what we know about the buffer acid and base concentrations:

$[HPO_4^{2-}] = 1.8$ [H2PO4–]
and
$[HPO_4^{2-}] = 0.50\text{ M} - [H_2PO_4^-]$

If we set the two definitions of $[HPO_4^{2-}]$ equal to each other and solve for $[H_2PO_4^-]$ we get:

$$1.8\,[H2PO4–] = 0.50\text{ M} - [H_2PO_4^-]$$
$$1.8\,[H2PO4–] + [H_2PO_4^-] = 0.50\text{ M}$$
$$2.8[H_2PO_4^-] = 0.50\text{ M}$$
$$[H_2PO_4^-] = 0.50\text{ M}/2.8 = 0.18\text{ M}$$

We can then substitute the $[H_2PO_4^-]$ value back into one of the original two equations to solve for $[HPO_4^{2-}]$:

$$[HPO_4^{2-}] = 0.50\text{ M} - 0.18\text{M} = 0.32\text{ M}$$

Step 3) Determine the most feasible means of obtaining the desired components. In this case, the obvious choice is to weigh out the desired amounts of the potassium salts (KH_2PO_4 and

K_2HPO_4), that will completely ionize upon dissolution, giving both components of the acid-base pair.

Step 4) Calculate the required amount of each material. Multiplying by 1 L, we know that we need 0.18 moles of KH_2PO_4 and 0.32 moles of K_2HPO_4. After calculating the formula weights of these, we can obtain the required masses:

KH_2PO_4: (0.18 mol)(136.1 g/mol) = 25 g KH_2PO_4
K_2HPO_4: (0.32 mol)(174.2 g/mol) = 56 g K_2HPO_4

Step 5) Prepare the buffer. Weigh out 25 g of KH_2PO_4 and 56 g of K_2HPO_4, and dissolve in about 900 mL of distilled water. Check the pH and adjust if necessary. Bring the total volume to 1 L.

Procedure

SAFETY NOTES: Mono- and di-basic phosphate salts are caustic and can cause skin irritation and burns, take all due precaution when handling.

GENERAL INSTRUCTIONS: Each student will prepare and test their own buffer. If interested, you can perform an Acid and Alkaline Self Test by collecting a small piece of litmus paper and applying some saliva to the paper, this will indicate whether your body fluids are too acidic or too alkaline. Litmus paper will change color to indicate pH levels of body fluids. These tests should be performed before eating or one hour after eating.

Part I. Comparing the behavior of the buffer with distilled water

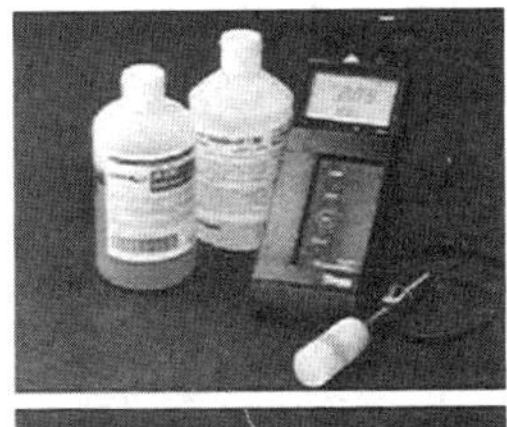

Obtain a pH meter from the stock room. Calibrate the pH meter for data collection.

Place 25 mL of distilled, deionized water in a clean dry 100 mL beaker.

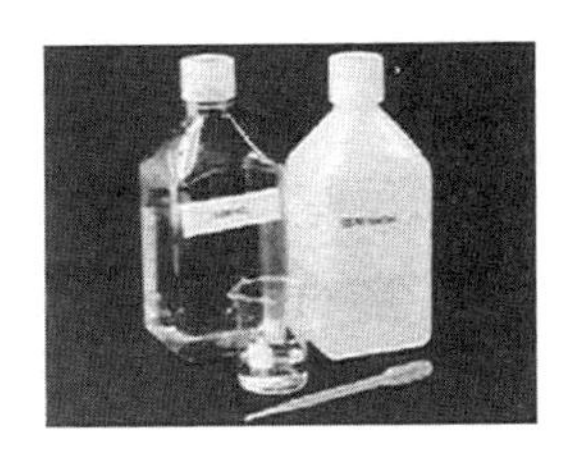

Using a disposable pipet add 1 drop (~0.05 mL) of 0.50 M HCl to the beaker. Record the change in pH.

Discard the solution down the sink with lots of water and place a fresh 25 mL of distilled, deionized water in a clean dry 100 mL beaker.

Using a disposable pipet add 1 drop (~0.05 mL) of 0.50 M NaOH to the beaker. Record the change in pH.

Part II. Preparing and testing of the buffer solution for the assigned pH

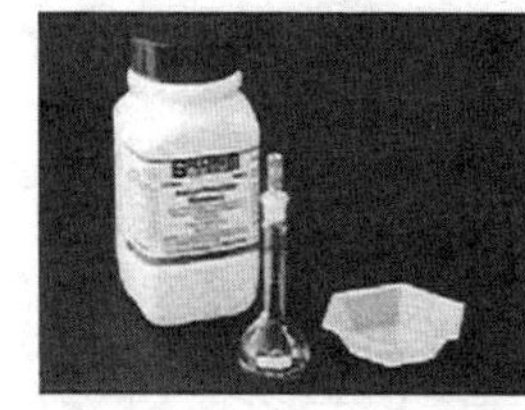

Using the calculations in your pre-lab, prepare 100 mL of your buffer at the assigned pH.

Once complete, measure out a 25 mL aliquot of the prepared buffer and place it in a clean, dry 100 mL beaker. Measure and record the pH.

Part III. Testing the buffer "capacity" with HCl and NaOH

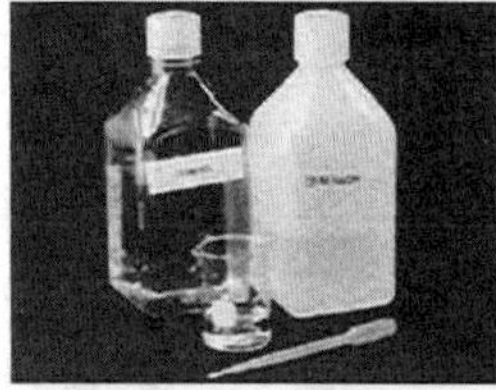

Using the 25 mL of buffer above, add 1 mL of 0.5 M HCl solution to the beaker and mix well. Record the new pH. Repeat the addition of 1 mL of 0.5 M HCl and continue to record the pH until the pH has dropped by 0.5 pH units.

Discard the used buffer solution down the sink with lots of water and place a fresh 25 mL of your prepared buffer in a clean, dry 100 mL beaker.

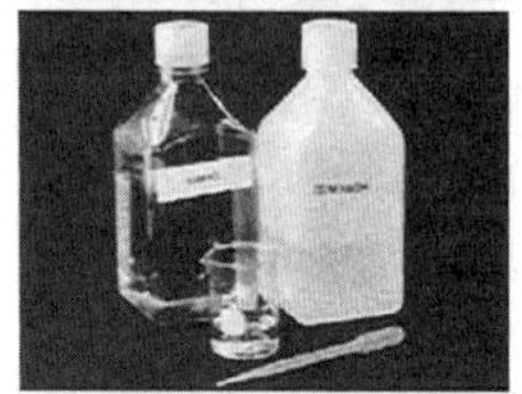

Using the 25 mL of buffer above, add 1 mL of 0.5 M NaOH solution to the beaker and mix well. Record the new pH. Repeat the addition of 1 mL of 0.5 M NaOH and continue to record the pH until the pH has risen by 0.5 pH units.

Part IV. Determining the effect on the pH of diluting the buffer

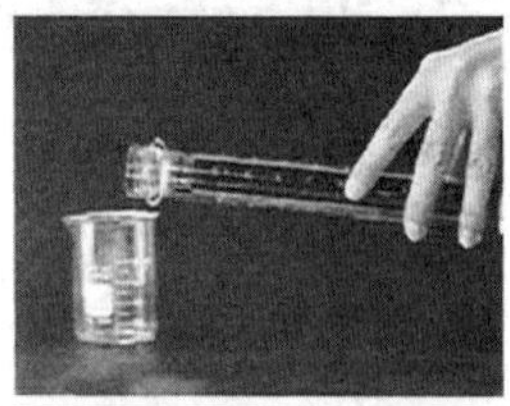

Discard the used buffer solution down the sink with lots of water and place a fresh 12.5 mL of your prepared buffer in a clean, dry 100 mL beaker. Add 12.5 mL of distilled water to the beaker to dilute the buffer.

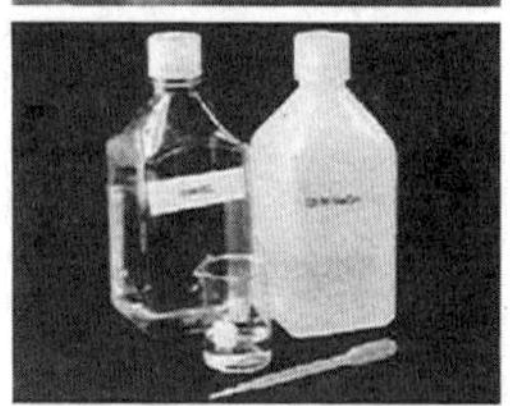

Repeat all of Part 3 to determine the buffer capacity of the diluted buffer.

Experiment 16

Pre-Laboratory Assignment

Name: ______________________ **Date:** ____________

Instructor: ______________________ **Sec. #:** ________

Show all work for full credit.

1. You need to prepare 1.000 L (in a volumetric flask) of 0.50 M phosphate buffer, pH 6.21. Use the Henderson-Hasselbalch equation with a value of 6.64 for pK_2 to calculate the quantities of K_2HPO_4 and KH_2PO_4 you need to add to the flask. Record the steps of these calculations and use them as a guide for the calculation you will have to make in the laboratory. (Calculate the intermediate values to at least one more significant figure than required).

   What is the $[K_2HPO_4]/[KH_2PO_4]$ ratio you will need?

   What concentrations of $[K_2HPO_4]$ and $[KH_2PO_4]$ will you need to make the total concentration 0.50 M? (Units required)

   How many moles of K_2HPO_4 and KH_2PO_4 will you need? (Units required)

   What mass of K_2HPO_4 and KH_2PO_4 will you need? (Units required)
   (FW of K_2HPO_4 = 174.2 g/mol. FW of KH_2PO_4 = 136.1 g/mol.)

2. Buffer capacity refers to the amount of acid or base a buffer can "absorb" without a significant pH change. It is governed by the concentrations of the conjugate acid and base forms of the buffer. A 0.5 M buffer will require five times as much acid or base as a 0.1 M buffer for a given pH change. In this problem you begin with a buffer of known pH and concentration, and calculate the new pH after a particular quantity of acid or base is added. In the laboratory you will carry out some stepwise additions of acid or base and measure the resulting pH values.

   Starting with 60 mL of 0.50 M phosphate buffer, pH = 6.83, you add 1.7 mL of 1.00 M HCl. Using the Henderson-Hasselbalch equation with a pK_2 for phosphate of 6.64, calculate the following values to complete the ICE table.

   What is the composition of the buffer to begin with, both in terms of the concentration and the molar quantity of the two major phosphate species? (Units required)

   What is the molar quantity of H_3O^+ added as HCl, and the final molar quantity of HPO_4^{2-} and $H_2PO_4^-$ at equilibrium?

   What is the new $HPO_4^{2-}/H_2PO_4^-$ ratio, and the new pH of the solution? (Note: You can use the molar ratio rather than the concentration ratio because both species are in the same volume.)

   Now take another 60 mL of the 0.50 M pH 6.83 buffer and add 3.7 mL of 1.00 M NaOH. Using steps similar to those above, calculate the new pH of the solution.

Experiment 16

Laboratory Report

Name: ____________________________ **Date:** _________________

Instructor: _____________________________ **Sec. #:** ___________

PURPOSE: (*The purpose should be several well-constructed sentences describing what your experiment was designed to accomplish and the criteria used to determine success. These sentences should include both concepts and techniques.*)

PROCEDURE: (*The procedure section should reference the lab manual and include any changes made to the procedure during the lab.*)

DATA: (*The data section should include your own personal data and observations.*)

Acid and Base Addition to Distilled Water

Solution	Measured pH	[HA] M	[A$^-$] M	Calculated pH
25mL Distilled Water				
25 mL Distilled Water + 0.05mL HCl				
25 mL Distilled Water + 0.05mL NaOH				

HCl Addition to Buffer

mL of HCl Added	Measured pH	[HA]	[A$^-$]	Calculated pH

NaOH Addition to Buffer

mL of NaOH Added	Measured pH	[HA]	[A⁻]	Calculated pH

HCl Addition to Diluted Buffer

mL of HCl Added	Measured pH	[HA]	[A⁻]	Calculated pH

NaOH Addition to Diluted Buffer

mL of NaOH Added	Measured pH	[HA]	[A⁻]	Calculated pH

CALCULATIONS: (*The calculation section should include sample calculations for each column in the data table. Be sure to include conversions of units.)*

Example calculation of how you prepared the buffer solution at the assigned pH:

Example calculation for determining the acid and base concentration after the addition of HCl or NaOH:

CONCLUSION: (*The conclusion section should include several paragraphs with the following: A summary statement regarding the buffer and pH used, the buffer capacity, and the comparison of the calculated pH values and the experimental pH values recorded in lab. A discussion on the similarities and differences between the buffer and dilute buffer should be included. Any possible errors should be discussed as usual.*)

Experiment 16

Post-Laboratory Assignment

Name: ______________________________ **Date:** __________________

Instructor: ______________________________ **Sec. #:** ____________

Show all work for full credit.

Answer the following question:

1. Human Blood contains a buffering system. What are the key molecules in this buffer and why is it necessary?

2. Using stoichiometry explain why the capacity of a diluted buffer solution is lower than a more concentrated buffer.

Experiment 17

Entropy, Free Energy and Chemical Equilibrium

Introduction

In the first part of general chemistry we learned about solubility as an all or nothing physical property of ionic compounds. Our previous solubility rules simply stated which salts were soluble in water and which were not. As with many concepts introduced in the first semester, that is not the whole story. Now that you have learned more about the concepts of equilibrium and intermolecular forces, we can discuss solubility in more detail.

The solubility of a substance is actually dependent on the forces holding the crystal together (the lattice energy) and on the solvent acting on the crystal. For now, we will only consider water as the solvent. As a solid dissolves, the ions in solution are surrounded by water molecules by a process called hydration. During hydration, energy is released. The extent to which the energy of hydration exceeds the lattice energy determines the solubility. Thus solubility is not all or nothing, but rather is an equilibrium process dependent on the energy availability of the system. Thus almost all substances are soluble to some degree.

Le Châtelier's principle allows us to predict the effects of changes in temperature, pressure, and concentration on a system at equilibrium. It states that if a system at equilibrium experiences a change, its equilibrium will shift to compensate for the change. This is why solubility is affected by changes in temperature. For every reversible reaction, one direction is endothermic and the other is exothermic. A reaction is endothermic if it absorbs heat from its surroundings, and exothermic if it releases heat to the surroundings. If you increase the temperature, then the endothermic reaction will be favored because it will take in some of the excess heat and use it to force the reaction in the forward direction. If you decrease the temperature, the exothermic reaction will be favored because it will replace the heat that was lost.

The purpose of this experiment is to investigate the solubility of $Ca(OH)_2$ at different temperatures. You will also be asked to determine the K_{sp}, ΔG^o, ΔH^o, and ΔS^o for the $Ca(OH)_2$ solution at each temperature. By creating saturated solutions over a range of temperatures and titrating those solutions with a standardized hydrochloric acid solution, the concentration of $Ca(OH)_2$ can be determined. At the conclusion of the experiment, you should be able to discuss the thermodynamics of $Ca(OH)_2$ solubility with respect to Le Chatelier's principle.

Background

The solubility of a compound is defined as the amount of solute that can be dissolved in a particular solvent, normally water. Until now, you have most likely treated solubility as a "yes" or "no" property of a substance. In other words, after memorizing the solubility rules, you know that salts of nitrates, chlorates, and acetates are soluble, while the salts of carbonates, phosphates, and chromates are insoluble. Well, if the truth be told, ALL ionic compounds are soluble in water to some varying degree.

Solubility

If we consider solubility in terms of a reaction where a solid ionic reagent dissolves into its component ions, then the reaction is governed by the same rules of equilibrium that affect all other chemical reactions. And, since dissolving is an equilibrium process, it seems logical that it can be described by an equilibrium constant, referred to as the solubility product constant (K_{sp}). The K_{sp} describes the equilibrium between the soluble and insoluble portion of the solute. The solubility constant expression is derived by following the same rules you would use when writing any other equilibrium expression.

When discussing solubility, it is important to understand the dynamics affecting the process. In particular, the degree of an ionic compound's solubility is dependent upon several factors including: (1) the intermolecular forces between the solute and solvent; (2) the change in entropy accompanying the process; (3) the concentration of the products and the reactants; and the factor which you will be directly observing in this experiment, (4) the temperature.

Calcium Hydroxide

The ionic compound that you will be dealing with in this experiment is unique. For most salts, the solubility of the compound is greater at higher temperatures than lower temperatures. However, the solubility of calcium hydroxide is greatest at lower temperatures, a behavior completely opposite of what is generally expected. In order to observe the effect of temperature upon solubility, and equilibrium, various solutions of calcium hydroxide will be produced at different temperatures. Once equilibrium has been established at each temperature, any excess solid will be removed by filtration; then the filtrate will be titrated with standard HCl to establish the concentration of hydroxide ion.

To further investigate some of the principles just discussed, let's look at some sample data obtained from a similar experiment involving $Ca(OH)_2$. Let's assume that after titrating 35.00 mL of the solution prepared at 25 °C, we found that we had to add 15.10 mL of 0.050 mol/L HCl to reach the endpoint. Recalling that the endpoint occurs when the number of moles of acid equals the number of moles of base, we can easily figure out the concentration of the hydroxide ion in this particular solution by multiplying the volume of acid added by its respective concentration:

$$\text{moles of acid} = 0.01510\text{ L} \times 0.050\ \tfrac{\text{mol}}{\text{L}} = 7.6 \times 10^{-4}\text{ moles}$$

Setting the number of moles of acid equal to the number of moles of base, and then dividing by the volume of $Ca(OH)_2$, we arrive at the concentration of the hydroxide ion:

$$\frac{7.6x10^{-4}\,moles}{0.035L} = 0.022M$$

The equilibrium reaction for dissolving $Ca(OH)_2$ is shown below:

$$Ca(OH)_2(s) \Leftrightarrow Ca^{2+}(aq) + 2OH^{-}(aq)$$
$$K_{sp} = [Ca^{2+}][OH^{-}]^2$$

Since the equilibrium reaction shows a 2:1 molar ratio of OH^{-} to Ca^{2+}, we can easily arrive at the concentration of the calcium ion by dividing the concentration of the hydroxide ion by 2, giving a $[Ca^{2+}]$ of approximately 0.011 mol/L. Taking these values and substituting into the K_{sp} expression, we arrive at a value of about 5.0×10^{-6} for the solubility product constant of $Ca(OH)_2$ at 25 °C. With the same process being carried out at various temperatures, you will be able to see the temperature dependence of solubility.

Thermodynamics of Solubility

Remembering that K_{sp} is a special form of an equilibrium constant, it seems logical that other thermodynamic factors such as enthalpy (ΔH), entropy (ΔS), and Gibbs free energy (ΔG) will also be dependent upon temperature changes. In fact, the equilibrium constant K_{sp} is directly correlated with the Gibbs free energy of an ionic compound's dissociation, as shown in the following equation where R is the ideal gas constant (8.3145 J/mol K) and T is the experimental temperature in Kelvin.

$$\Delta G = -RT\ln(K_{sp})$$

For our example run at 25 °C, 298 K, the value of ΔG comes out to be about 30.1 kJ/mol. After we have calculated ΔG at the various temperatures, we can generate a plot of ΔG versus temperature, in Kelvin, obtain a slope and a y-intercept, and then use the following relationship to calculate ΔH and ΔS for the dissociation process:

$$\Delta G = \Delta H - T\Delta S$$

Note that this equation is in the familiar $y = mx + b$ format with the slope and y-intercept equal to $-\Delta S$ and ΔH, respectively. At this point, it will be your job to figure out the significance of these two parameters!

Procedure

SAFETY NOTES: $Ca(OH)_2$ is caustic and HCl is corrosive. All due precautions should be taken in their handling.

GENERAL INSTRUCTIONS: Students should work in groups and share data for this experiment so that all four titrations can be completed in the time allotted, i.e., each student should complete two of the four different temperature titrations. Be sure to dispose of your calcium hydroxide solutions as directed by your instructor.

Part I: Titration of a room-temperature, saturated $Ca(OH)_2$ solution

Obtain a buret and rinse it well with DI water.

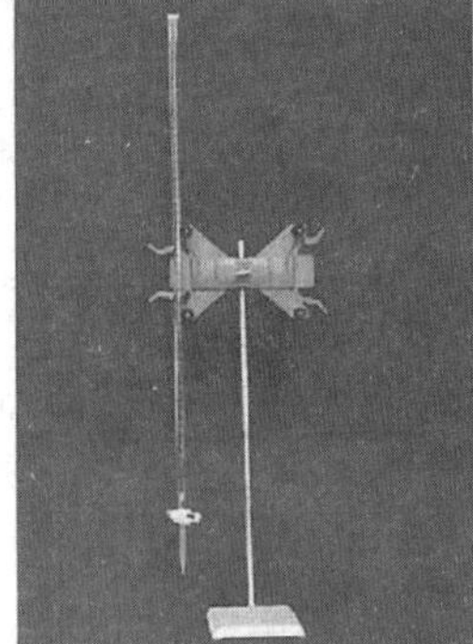

Use a clamp to attach the buret to the ring stand on your bench, and then set a magnetic stirrer (hot plate) underneath it.

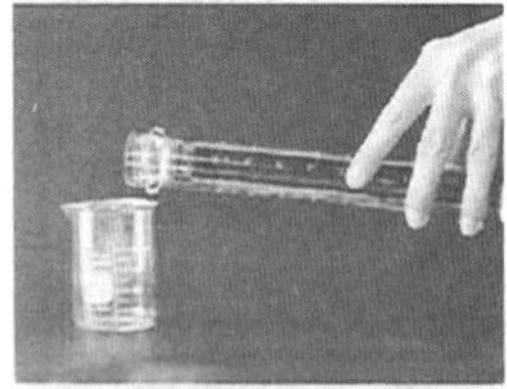

Pour approximately 30 mL of 0.05 M HCl solution into a clean, dry 50 or 100 mL beaker. Record the concentration of the acid.

Rinse the inside of the buret with a few milliliters of the HCl solution, making sure all the walls of the buret are coated. Collect the rinse solution in a beaker labeled "waste."

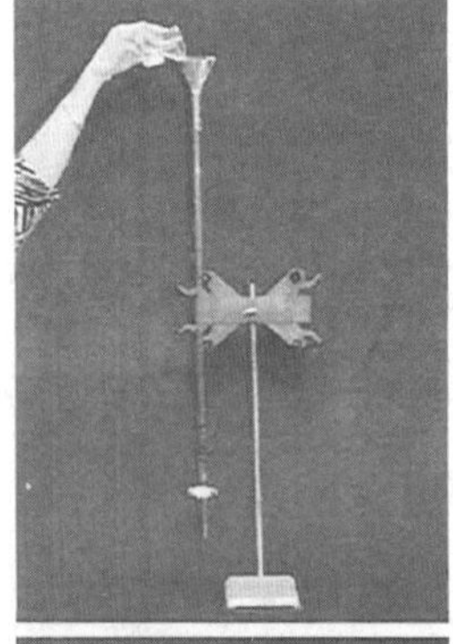

With the stopcock closed, pour the remainder of the HCl solution into the buret. Place a small waste beaker under the buret and slowly open the stopcock until the solution begins to drip out.

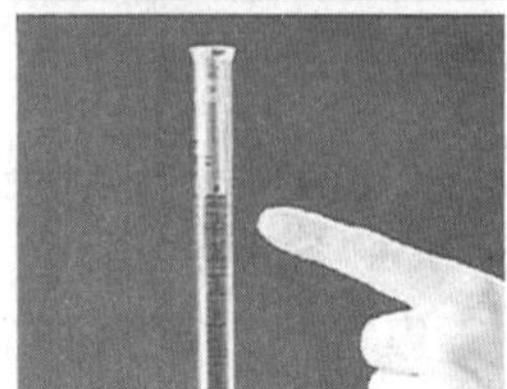

Then close the stopcock and make sure there are no air bubbles in the tip of the buret. Also make sure that the level of liquid in the buret is at or below the 0.00 mL line. Dispose of the wash solution in your waste beaker.

Add about 100 mL of distilled water to a 250 mL Beaker and using a thermometer record its temperature to 0.1 °C. Add 2.0 g of solid $Ca(OH)_2$ with constant stirring until the salt no longer dissolves.

Using a funnel, strain this solution through qualitative filter paper into a clean, dry 125 mL Erlen-meyer filter flask. Note: Only filter small amounts of solution at a time to avoid "swamping" the filter paper. The filtrate solution should be clear when filtration is complete. A "cloudy" filtrate indicates the presence of particulates that will throw off the titration results. Re-filter any "cloudy" filtrates until they are clear.

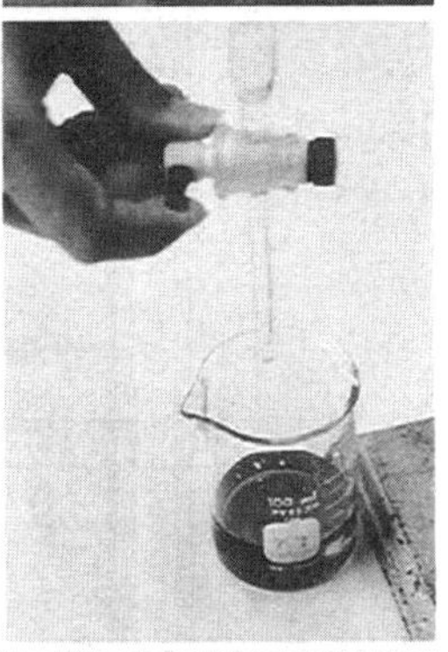

Pipet 10.00 mL of this *clear* solution into another clean, dry 100 mL beaker. Add approximately 25 mL of distilled water and 10–12 drops of the bromothymol blue indicator. Gently slide a magnetic stir bar into the flask.

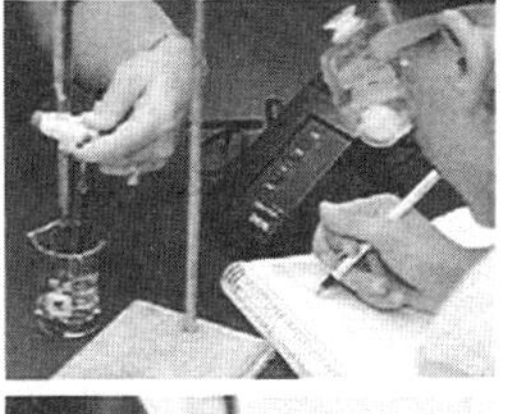

Set the flask on the stirrer and begin stirring at a gentle rate. (Make sure there is no splashing of the solution.) Record to two decimal places the beginning volume level of HCl in the buret.

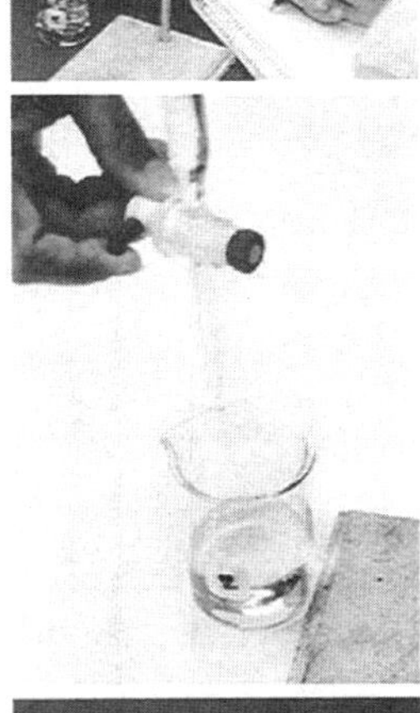

Begin the titration, allowing the titrant to fall from the buret at a rapid drop-by-drop pace. During this addition, the initial blue color will begin to turn green and then yellow. The end point is reached when the entire solution turns yellow. Your goal is to deliver the exact volume needed to reach the endpoint. When you have reached the endpoint, record the final volume in the buret (again to two decimal places) and calculate the volume of titrant delivered.

Pour the titrated solution into the waste container, but don't lose your stir bar.

Part II: Titration of an elevated temperature, saturated $Ca(OH)_2$ solution

Add about 100 mL of distilled water to a 250 mL beaker. Place it on a hot plate and bring the water to between 50 °C and 70 °C. After the water has been at temperature for several minutes, add about 2 g of solid $Ca(OH)_2$ to the water with constant stirring until the salt no longer dissolves. Take a final temperature reading at saturation.

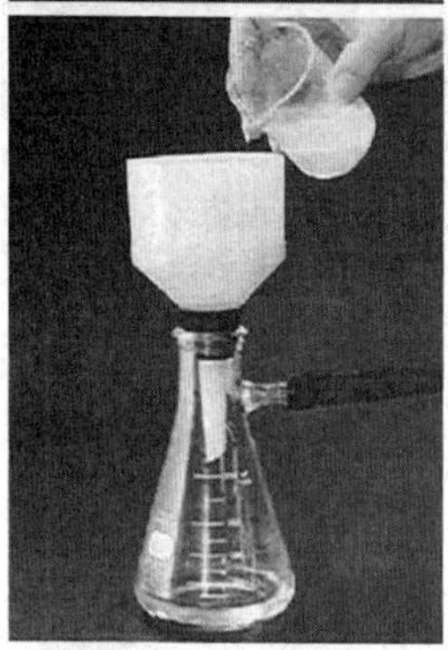

Using a funnel, *quickly* strain this solution through qualitative filter paper into a clean, dry 125 mL Erlen-meyer filter flask. Note: Only filter small amounts of solution at a time to avoid "swamping" the filter paper. The filtrate solution should be clear when filtration is complete. A "cloudy" filtrate indicates the presence of particulates that will throw off the titration results. Re-filter any "cloudy" filtrates until they are clear.

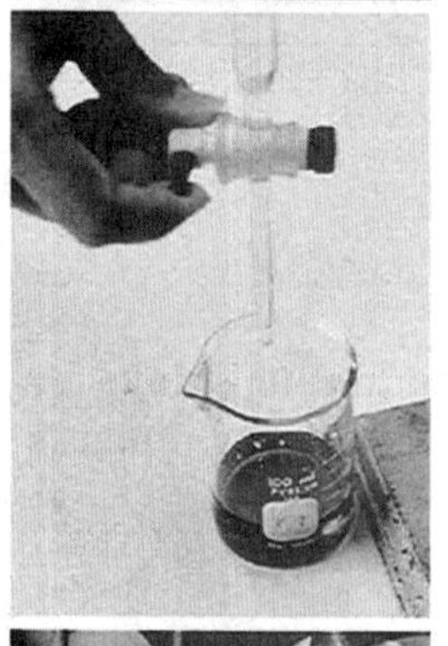

Pipet 10.00 mL of this *clear* solution into another clean, dry 100 mL beaker. Add approximately 25 mL of distilled water and 10–12 drops of the bromothymol blue indicator. Gently slide a magnetic stir bar into the flask.

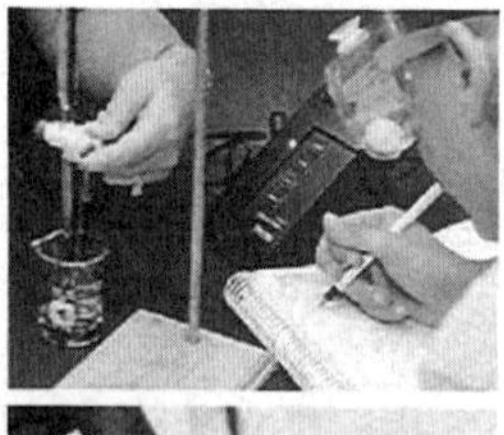

Set the flask on the stirrer and begin stirring at a gentle rate. (Make sure there is no splashing of the solution.) Record to two decimal places the beginning volume level of HCl in the buret.

Begin the titration, allowing the titrant to fall from the buret at a rapid drop-by-drop pace. During this addition, the initial blue color will begin to turn green and then yellow. The end point is reached when the entire solution turns yellow. Your goal is to deliver the exact volume needed to reach the endpoint. When you have reached the endpoint, record the final volume in the buret (again to two decimal places) and calculate the volume of titrant delivered.

Pour the titrated solution into the waste container, but don't lose your stir bar.

Part III: Titration of a low temperature, saturated $Ca(OH)_2$solution

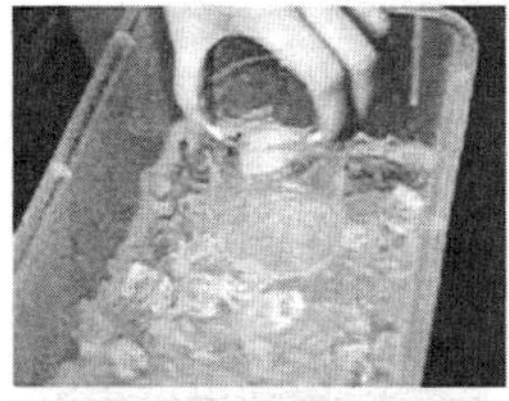

Add about 100 mL of distilled water to a 250 mL beaker. Place it in a salted ice bath for at least 5 minutes. Add solid $Ca(OH)_2$ with constant stirring until the salt no longer dissolves.

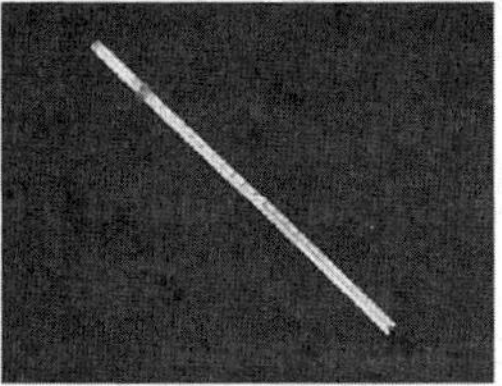

Remove the flask from the ice bath and measure the temperature of the cold solution.

Using a funnel, *quickly* strain this solution through qualitative filter paper into a clean, dry 125 mL Erlen-meyer filter flask. Note: Only filter small amounts of solution at a time to avoid "swamping" the filter paper. The filtrate solution should be clear when filtration is complete. A "cloudy" filtrate indicates the presence of particulates that will throw off the titration results. Re-filter any "cloudy" filtrates until they are clear.

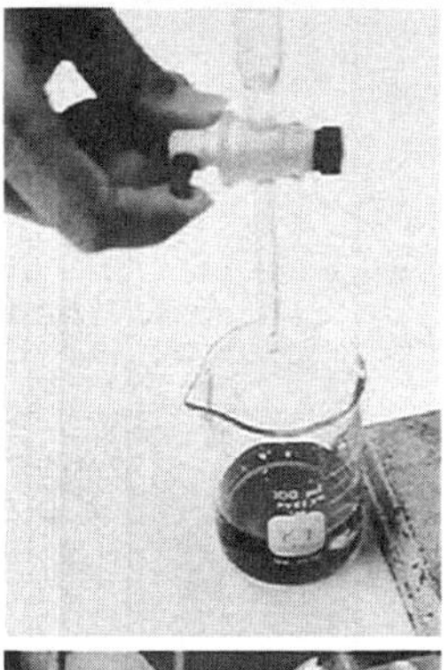

Pipet 10.00 mL of this *clear* solution into another clean, dry 100 mL beaker. Add approximately 25 mL of distilled water and 10–12 drops of the bromothymol blue indicator. Gently slide a magnetic stir bar into the flask.

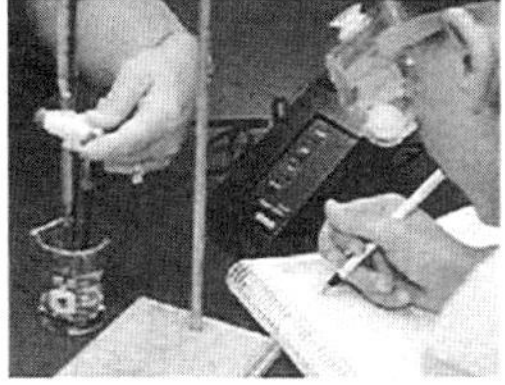

Set the flask on the stirrer and begin stirring at a gentle rate. (Make sure there is no splashing of the solution.) Record to two decimal places the beginning volume level of HCl in the buret.

Begin the titration, allowing the titrant to fall from the buret at a rapid drop-by-drop pace. During this addition, the initial blue color will begin to turn green and then yellow. The end point is reached when the entire solution turns yellow. Your goal is to deliver the exact volume needed to reach the endpoint. When you have reached the endpoint, record the final volume in the buret (again to two decimal places) and calculate the volume of titrant delivered.

Pour the titrated solution into the waste container, but don't lose your stir bar.

Part IV: Titration of a ? temperature, saturated $Ca(OH)_2$ solution

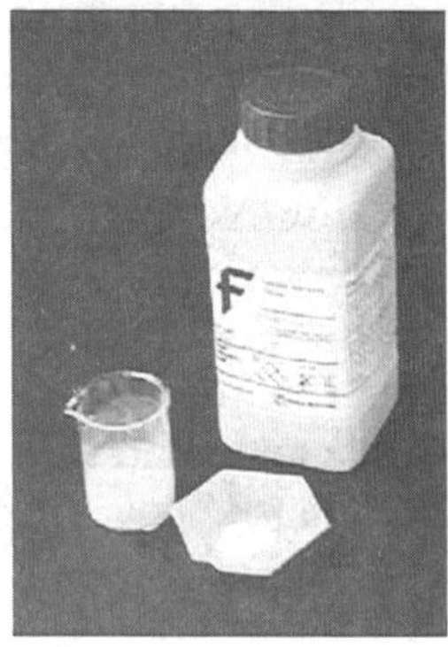

Based on your calculation of the molar solubility of the $Ca(OH)_2$ solution at low, room, and high temperature choose another temperature at which to run the titration. Follow the same procedure as in parts 1–3.

Experiment # 17

Pre-Laboratory Assignment

Name: ______________________ **Date:** __________

Instructor: ______________________ **Sec. #:** __________

Show all work for full credit.

1) Add an excess of $Ca(OH)_2$ to water maintained at a particular temperature, stir until the solution is saturated, filter, then determine the [OH−] in the solution by titration with acid. Titration of 10.0 mL of the calcium hydroxide solution to the endpoint requires 4.86 mL of 0.070 M HCl solution.

What is the molar quantity of OH− in the 10.0 mL of solution?

What are the concentrations of Ca^{2+} and OH^-?

What is the solubility of $Ca(OH)_2$ under these conditions?

What is the Ksp for $Ca(OH)_2$ under these conditions?

2) You measure the solubility of a salt at four different temperatures and calculate the following K_{sp} values:

Temperature (°C):	15	31	51	70
K_{sp}:	2.89×10^{-2}	4.11×10^{-2}	6.08×10^{-2}	8.45×10^{-2}

ΔG^o_{soln} for the solubility process is given by the relationship $\Delta G^o_{soln} = -RT\ln K_{sp}$ where R = 8.314 J/(K*mol) and T is the absolute temperature in degrees K.

Calculate ΔG^o_{soln} (in kJ/mol) at each of these temperatures. As the temperature increases, how does the solubility of the salt change? According to Le Chatelier's principle, dissolving this salt should therefore be which type of process? In other words, what is the sign of ΔH^o_{soln}?

3) It is possible to determine both ΔH^o_{soln} and ΔS^o_{soln} for the solution process from this data. One or both of two graphical methods can be used:

Method 1: From the equation:

$\Delta G^o_{soln} = \Delta H^o_{soln} - T\Delta S^o_{soln}$

a plot of ΔG^o_{soln} versus T will give -ΔS^o_{soln} as the slope,

and ΔH^o_{soln} as the intercept.

Method 2: Alternatively, rearranging the equation relating K_{sp} and ΔG^o_{soln}:

$$\ln Ksp = -\frac{\Delta G^o_{soln}}{RT} = -\frac{\Delta H^o_{soln}}{RT} + \frac{\Delta S^o_{soln}}{R}$$

and plotting $\ln K_{sp}$ versus $\frac{1}{T}$ will give $-\frac{\Delta H^o soln}{R}$ as the slope, and $\frac{\Delta S^o soln}{R}$ as the intercept.

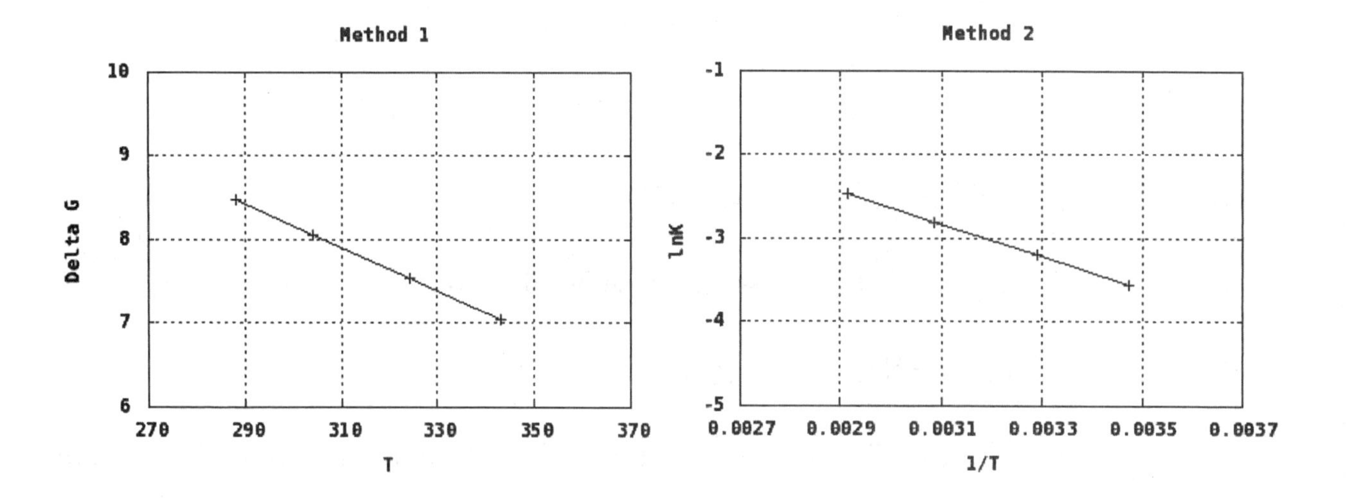

Using one or both of these graphical methods, determine ΔH^o_{soln} (in kJ/mol) and ΔS^o_{soln} (in J/(K*mol)) for this salt. (Be sure to play close attention to your units and signs in both your plots and your calculations).

Experiment 17

Laboratory Report

Name: ______________________ **Date:** ______________

Instructor: ______________________ **Sec. #:** __________

PURPOSE: *(The purpose should be several-well constructed sentences describing what your experiment was designed to accomplish and the criteria used to determine success. These sentences should include both concepts and techniques.)*

____

____

____

____

____

____

____

____

PROCEDURE: *(The procedure section should reference the lab manual and include any changes made to the procedure during the lab. The different temperatures that you ran should be specified. Also include the names of your partners and clearly identify the data contributed by each person.)*

____

____

____

____

____

____

____

____

____

DATA: *(The data section should include your own personal data and observations.)*

Balanced Chemical Equation:

Temperature (°C)	Volume of HCl Used (mL)

Temperature (°C)	$[OH^-]$	Molar Solubility $Ca(OH)_2$ (M)	K_{SP}	ΔG^o (kJ/mol)

Note: the concentration of Ca^{2+} is equal to the molar solubility.

ΔH =

ΔS =

CALCULATIONS: (*The calculation section should include sample calculations for each column in the data table. Be sure to include conversions of units. Example calculations of the following should also be included: [OH^-], molar solubility, K_{sp}, ΔG, ΔH, and ΔS. The values for ΔH and ΔS are determined from the free energy graph where the slope of the graph is $-\Delta S$ and the y-intercept of the graph is ΔH. A graph of ΔG vs. temperature (in Kelvin) AND a graph of molar solubility vs. temperature are required. All graphing guidelines apply. Make sure that the units are correct in all of your calculations and answers.*)

CONCLUSION: (*The conclusion section should be in paragraph form and include [OH^-] for each temperature, and the molar solubility of $Ca(OH)_2$ at each temperature. The trend seen on your molar solubility vs. temperature graph should be explained. Another discussion of the values, trends and discrepancies in the values for K_{sp}, ΔG, ΔH, and ΔS should be included. Finally be sure to discuss any possible experimental errors.*)

____
____
____
____
____
____
____
____
____

Experiment 17

Post-Laboratory Assignment

Name: ______________________________ **Date:** ___________________

Instructor: ________________________________ **Sec. #:** ____________

Show all work for full credit.

1) Why do you have to filter the $Ca(OH)_2$ solution before titrating it?

2) Why don't you have to accurately record the amount of $Ca(OH)_2(s)$ used in the saturated solution?

Experiment 18

Electrochemistry: The Nernst Equation

Introduction

In an earlier experiment we investigated electrochemistry and simple redox reactions by building mock voltaic cells and measuring their potentials. In this lab, we are expanding on that simple experiment by building more sophisticated voltaic cells and measuring their potentials at both room and other temperatures. We have previously discussed the importance of understanding electrochemistry with respect to the development of batteries and other energy storage devices. Another reason we study electrochemistry is more personal and also the reason that we are discussing the concept in relation to non-standard states. Our own bodies run on electrochemical reactions.

Adenosine triphosphate (ATP) is the energy molecule used by the body to fuel cellular reactions. The chemical processes used to produce ATP are oxidation-reduction reactions and are known as the electron transport chain.

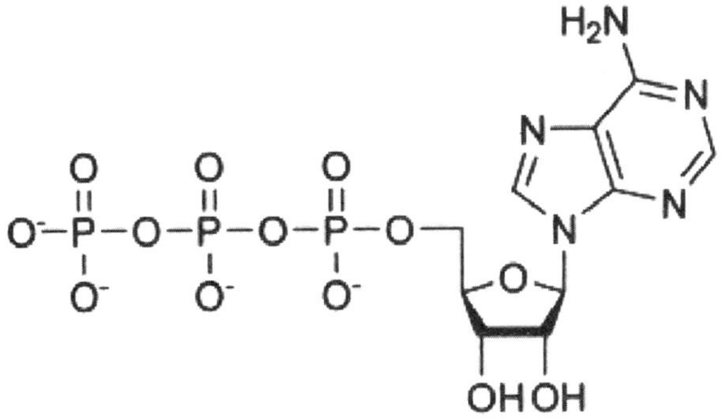

Adenosine Triphosphate (ATP)

Standard state is defined as 1.00 atmosphere of pressure at 25 °C, biological systems are obviously not at standard state. Many of you have plans to be medical doctors or in the medical field and will need to understand this process in great detail since many metabolic diseases are related to the improper function of the enzymes of the electron transport chain. Thus our expansion of this experiment to non-standard temperatures is very important. In this experiment we will determine the potentials of 10 cells at several temperatures and use a version of the Nernst equation to calculate the free energy, enthalpy and entropy of those cells.

Background

In the later years of the 18th century, Luigi Galvani discovered that the nerves of frogs were capable of harboring electrical activity. Shocked by his discovery, Galvani rationalized that such activity could only be found in living tissues. A few years later, a man by the name of Alessandro Volta scientifically refuted Galvani's claim when he observed that electricity could also be produced through inorganic means. By using small metal sheets of copper and zinc separated by pieces of cloth soaked in acidic solution, Volta constructed what is believed to be the first apparatus capable of producing electricity. Through centuries of research and technological advances, electricity is now a crucial part of our everyday lives, an aspect so vital that it has its own chemical course of study—electrochemistry.

Electrochemistry

Electrochemistry is the study of the conversion between electrical and chemical energy. Reactions of this sort are also called redox reactions as they utilize two unique processes, oxidation and reduction, to transfer electrons. Oxidation involves the loss of one or more electrons from a chemical species, while reduction refers to the gain of electrons by a chemical species. These two definitions are easily remembered by the mnemonic "OIL RIG" (Oxidation is Loss, Reduction is Gain).

When an oxidation and a reduction reaction are paired together in a redox reaction, electrons can flow from the oxidized species to the reduced species. For this reason, the oxidized species can also be referred to as the reducing agent or the reductant, while the reduced species is also known as the oxidizing agent or oxidant. The flow of electrons can be observed when an electrochemical cell is constructed. Two types of cells are 1) a galvanic cell, (also known as a voltaic cell), where a spontaneous reaction drives the electron flow, and 2) an electrolytic cell, where an outside source of current is imposed on the cell to drive a chemical reaction.

Electrochemical Cells

When discussing an electrochemical cell, there are several new terms: standard reduction potential (E°_{red}); half-cell; electrode; anode; and cathode. The first and most important term here is the standard reduction potential (E°_{red}), which is a numerical value of a half-reaction for the reduction of a particular metal. The standard reduction potential is a quantitative measure (in volts) of a particular substance's tendency to accept electrons under the standard conditions of 1.0 atm and 1.0 mol/L. A table of standard reduction potentials is in Appendix D in your textbook.

Observing the table, you will notice that the series of half-reactions is listed in descending order from the most positive value to the most negative value. In general, the more positive the standard reduction potential is for a particular substance, the stronger its tendency is to be reduced, or to gain electrons. The standard oxidation potential (E°_{ox}) is equal in value, but opposite in sign, to the standard reduction potential; the value of E°_{ox} measures the tendency of a particular substance to lose electrons, or become oxidized. For example, the dichromate ion ($Cr_2O_7^{2-}$) has a standard reduction potential of +1.33V, so under standard conditions the ion will have a standard oxidation potential of -1.33 V.

The remaining terms deal entirely with electrochemical cells, and are critical components of their construction. The system is composed of two parts commonly referred to as half-cells. A half-cell is the part of the electrochemical cell that houses the electrode. The electrode, as described by Faraday, can be either an anode, the electrode where oxidation occurs, or a cathode, the electrode where reduction takes place. The metal with the greater (more positive) potential will be the anode. For a clearer picture of what we have just described, please carefully note the figure below:

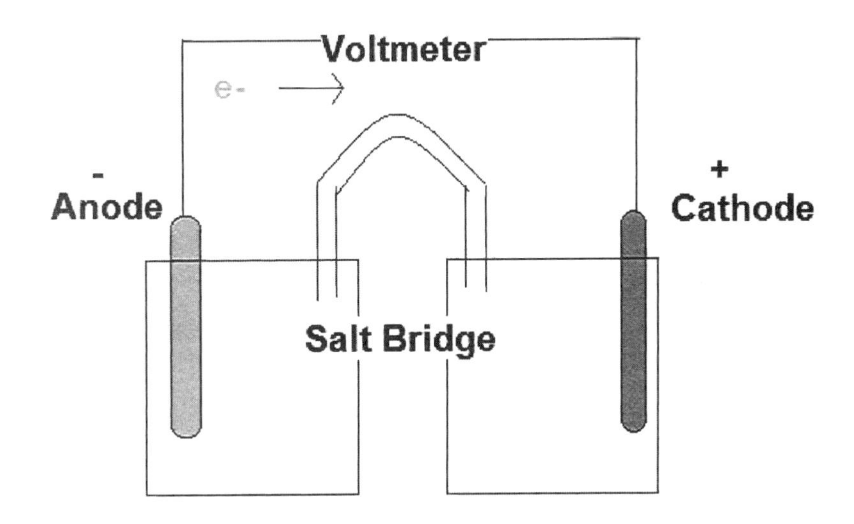

Calculating Cell Potentials

A term not mentioned in the previous section was the overall cell potential (E°_{cell}), which in brief is simply the sum of the standard reduction potential (E°_{red}) and the standard oxidation potential (E°_{ox}). Mathematically, the overall cell potential is expressed as:

$$E^{\circ}_{cell} = E^{\circ}_{red} + E^{\circ}_{ox}$$

Using our diagram above, let's make a reasonable hypothesis as to what we believe the overall potential will be for a cell formed between zinc and copper metals. Use a 1.0 M solution of zinc nitrate and zinc metal as the anode (left,) and copper nitrate and copper metal as the cathode (right). Using the table of the half reaction reduction potentials in Appendix D, we find:

$$Cu^{2+}(aq) + 2e^- \Leftrightarrow Cu^0(s) \qquad E^\circ_{red} = +0.34V$$
$$Zn^0(s) \Leftrightarrow Zn^{2+}(aq) + 2e^- \qquad E^\circ_{ox} = +0.76V$$

Using the equation above for the overall cell potential we can arrive at the estimated value:

$$E^\circ_{cell} = E^\circ_{red} + E^\circ_{ox} = +0.34V + (+0.76V) = +1.10V$$

The actual reading on the voltmeter in the diagram above should confirm the value we have just calculated. However, when you construct this cell in the lab, the voltmeter reading probably will not be exactly +1.10 V. Remember you are not working at standard conditions. However, if the ratio of concentrations of the corresponding salts is 1:1, any differences in the readings can be explained by other sources of error such as resistance in the salt bridge or lack of calibration of the voltmeter.

Free Energy and the Nernst Equation

The value of ΔG for a reaction at any moment in time tells us two things: 1) the sign of ΔG tells us in which direction the reaction must shift to reach equilibrium; and 2) the magnitude of ΔG tells us how far the reaction is from equilibrium at that moment.

The potential of an electrochemical cell is a measure of how far an oxidation-reduction reaction is from equilibrium. The Nernst equation describes the relationship between the cell potential at any moment in time and the standard-state cell potential.

$$E = E^o - \frac{RT}{n\Im} \ln Q$$

Let's rearrange this equation as follows:

$$n\Im E = n\Im E^o - RT \ln Q$$

We can now compare it with the equation used to describe the relationship between the free energy of reaction at any moment in time and the standard-state free energy of reaction.

$$\Delta G = \Delta G^o - RT \ln Q$$

These equations are similar because the Nernst equation is a special case of the more general free energy relationship. We can convert one of these equations to the other by taking advantage of the following relationships between the free energy of a reaction and the cell potential of the reaction when it is run as an electrochemical cell.

$$\Delta G = -n\Im E \text{ and } \Delta G^{\circ} = -n\Im E^{\circ}$$

In the second portion of this experiment you will measure the potential of your galvanic cell at various temperatures, data which will give far more information about a particular reaction that you may be studying. As with any other type of chemical reaction, redox reactions are subject to changes in equilibrium brought about by changes in the overall temperature of the system. However, before we obtain the particular equilibrium constant (K) for our reaction, we need to calculate the change in free energy (ΔG). The free energy (ΔG) of a particular reaction is a measure of the spontaneity of the process, and is defined mathematically by the equation we just derived:

$$\Delta G = -n\Im E$$

In this equation, the variables are as follows: n is the number of moles of electrons transferred; $\Im$ is Faraday's constant; and E is the observed overall cell potential. For our example, the setup would be as follows:

$$\Delta G = -n\Im E = -\left(2\ mol\ e^{-}\right)\left(96.5\frac{kJ}{V \cdot mol}\right)(1.10V) = -212kJ$$

In this particular reaction the ΔG is ~ -212 kJ, meaning that our example reaction is highly spontaneous when both products and reactants are in their standard states.

Now, with an approximate value for ΔG known, we can calculate the reaction's equilibrium constant from the relationship shown below.

$$K = e^{-\frac{\Delta G}{RT}}$$

For this equation, the value of ΔG is calculated as above, R is the ideal gas constant with units of $J\text{-}mol^{-1}\text{-}K^{-1}$, and T is the temperature in Kelvin. Realizing we performed our experiment at room temperature (~ 27 °C) we can quickly calculate the corresponding equilibrium constant.

$$K = e^{-\frac{-212{,}000J}{8.3145\frac{J}{mol \cdot K}(300K)}} = 8.2\ x\ 10^{36}$$

Thus, for our particular example run at room temperature, we get an approximate value of K = 8.2 x 10^{36}! In other words, this reaction is driven completely to product at room temperature.

The second part of this experiment is aimed at furthering your understanding of how temperature can affect the free energy of a reaction as well as its equilibrium. By constructing a graph of ΔG versus temperature, entropy and enthalpy are then easily derived from the slope and intercept of the resulting line, since the equation $\Delta G = \Delta H - T\Delta S$ is of the form y = mx + b.

Procedure

GENERAL INSTRUCTIONS: Because there are 5 metals, there are 10 possible cells you can create. Your Instructor will assign each student one of the possible cells to build and test. Data from each student will then be pooled to allow you to build a potential series of all the metals.

Part I: Building the Cells

The cells will be constructed out of two large test tubes and a salt bridge created by the student.

Each bench can work as a group to make the agar-agar as only a small mL amount is needed for each U-tube.

Place 50 mL of 0.1 M KNO_3 in a 250 mL beaker and using a hot plate bring it to a boil.

Remove the solution from the heat and add 0.50 grams of agar-agar with constant stirring.

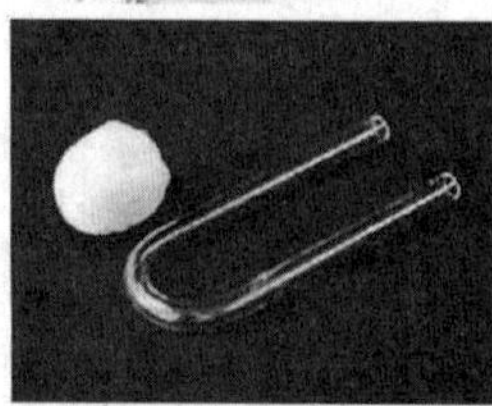

Obtain a U-tube from the front counter.

Use a plastic pipet to fill the agar-KNO_3 gel into the inverted U-tube until the gel fills all but 1/2 inch of the tubes. Allow the gel to cool and solidify.

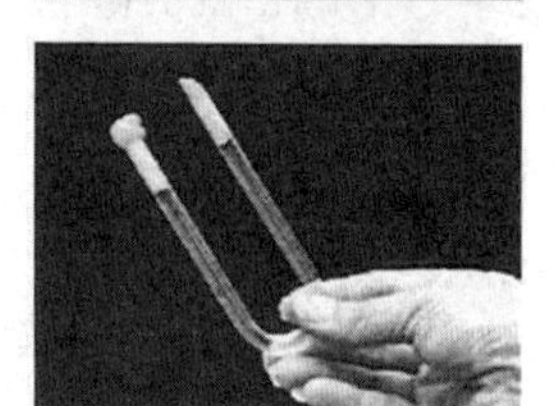

Soak half of some cotton plugs in each of the respective solutions you are using in your cell.

Fill the 1/2 inch spaces in the tubes with cotton plugs, soaked side in. Make sure that the cotton contacts the agar gel and also protrudes from the tubes. Make sure you mark which side of the

cell was soaked in which solvent.

Part II: Testing the Cells at Room Temperature

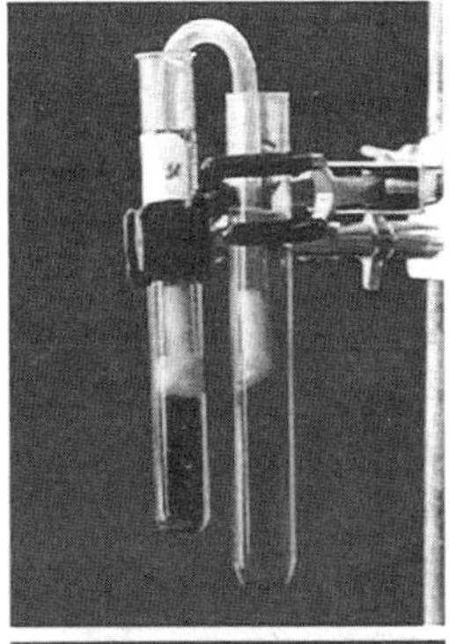

Place 30 mL of each of the salt solutions that correspond to the metals being used into large test tubes. Label the test tubes accordingly. Use a ring stand to hold the two test tubes next to each other.

Insert the appropriate salt bridge into the two tubes. Make sure that you place the correct sides of soaked cotton in the correct solutions. Also make sure that the cotton is fully submerged in the salt solution.

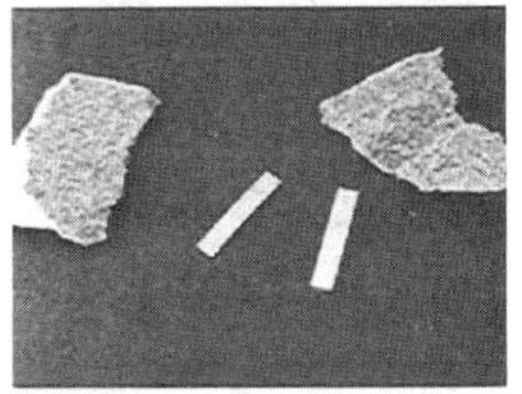

Obtain a strip or piece of the metals being tested and clean the surface with the sand paper provided.

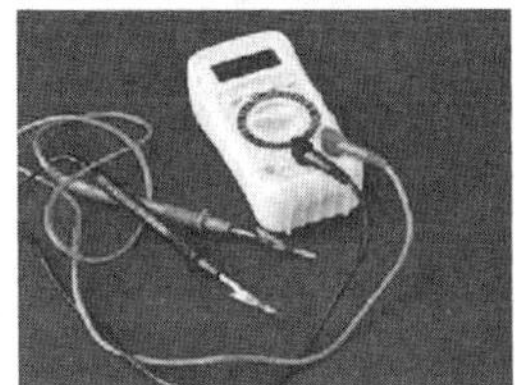

Set the multimeter to read as a voltmeter in direct current volts (V_{DC}).

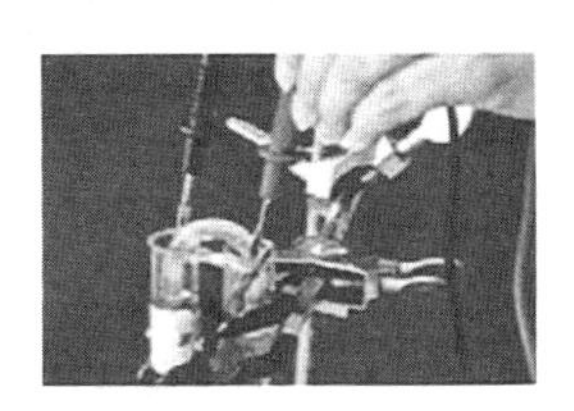

Using the clips on the voltmeter, connect the voltmeter to the metal strips and the sides of the test tubes. Make sure that the metal pieces are in contact with the solution, but do not submerge the clips in the solution.

Measure the potential of the cell. If the potential is negative, reverse the wire connections with the voltmeter. Report your findings to the rest of your class.

Part III: Measuring the potential at different temperatures

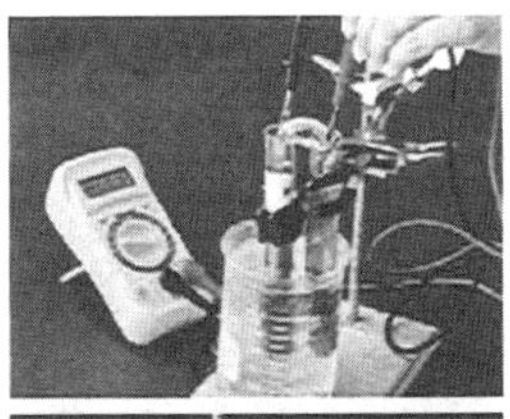

Set up your potential cell as you did in Part 1.

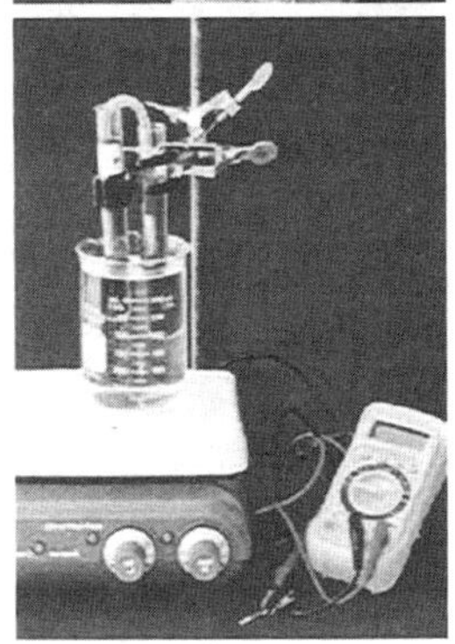

To the set-up add a 400 mL beaker half filled with DI water.

Place the beaker on a hot plate and adjust the cell's position on the ring stand until both test tubes are partially submerged in the water.

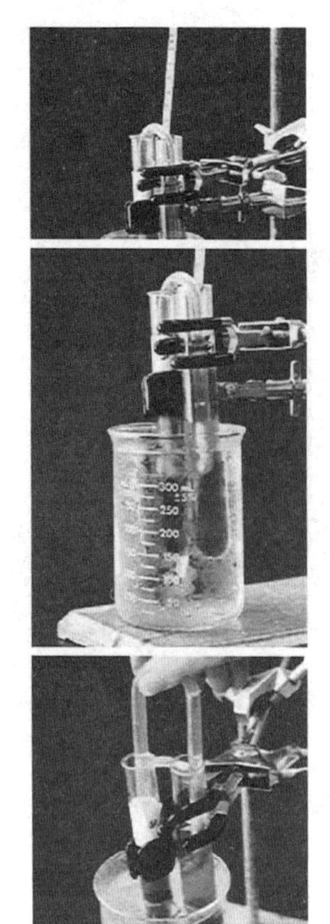

Set the hot plate temperature to 70 °C. Monitor the cell temperature until it reaches and maintains 70 °C. Measure the voltage.

Remove the cell from the hot plate and measure the voltage at 15 °C intervals until the cell cools to room temperature.

Create an ice-water bath in the same beaker as before and submerge the cell as before. Wait 10 minutes and record the temperature and the voltage.

Experiment # 18

Pre-Laboratory Assignment

Name: ______________________________ **Date:** ________________

Instructor: ______________________________ **Sec. #:** __________

Show all work for full credit.

1) Using literature values for the half-cell reactions of the following metals, calculate the total cell potentials for each of the 10 possible cells. Assume complete reduction of the metals. Show all your work please. **Copy the resulting potentials into your lab notebook.**

   Metals: Fe^{3+}, Cu^{+}, Mg^{2+}, Al^{3+}, Pb^{2+}

2) The standard potential of a galvanic cell (**E^{o}_{cell}**) can be determined from thermochemical data such as the free energy of formation (**ΔG^{o}_{f}**) of each of the species in the reaction.

   For example, given the reaction:

   $$Fe^{3+}(aq) + Al(s) \leftrightarrow Al^{3+}(aq) + Fe(s)$$

Species	$Fe^{3+}(aq)$	Fe(s)	$Al^{3+}(aq)$	Al(s)
ΔG^{o}_{f} (kJ/mol)	-10.5	0	-481.2	0

   First use the data to the right to calculate the ΔG^{o}_{rxn}. Then calculate E^{o}_{cell} from ΔG^{o}_{rxn}.

3) Following is the shorthand notation for a Galvanic Cell:

$Al(s)|Al^{3+}(1M)||Cu^{+}(1M)|Cu(s)$

Write the overall balanced equation occurring in this cell. (Indicate the physical states of reactants and products, i.e., Al(s), Cu+(aq), etc.)

Identify the anode and cathode of this cell.

Make the following voltage measurements at four different temperatures.

Temperature (°C):	10	35	55	74
$\mathbf{E^{o}_{cell}}$ **(V):**	2.190	2.176	2.165	2.154

Calculate the ΔG^{o}_{rxn} at each of these temperatures.

Now graph this data using the relationship $\Delta G = \Delta H - T\Delta S$, and determine ΔH^{o}_{rxn} (in kJ/mol) and ΔS^{o}_{rxn} (in J/(K*mol)) for this process. (This assumes that ΔH^{o}_{rxn} and ΔS^{o}_{rxn} do not vary with temperature. Remember to use absolute temperature, and pay attention to units and signs in your plot and your calculations.)

Experiment 18

Laboratory Report

Name: ______________________________ **Date:** __________________

Instructor: ______________________________ **Sec. #:** ____________

PURPOSE: *(The purpose should be several well constructed sentences describing what your experiment was designed to accomplish and the criteria used to determine success. These sentences should include both concepts and techniques.)*

PROCEDURE: *(The procedure section should reference the lab manual and include any changes made to the procedure during the lab. Be sure to note your assigned metals.)*

DATA: *(The data section should include your own personal data and observations.)*

Cell Data from Class Results:

Cell Identity (cathode + anode)	E^{o}_{Cell} Experimental (V)	E^{o}_{Cell} Theoretical (V)	%Error

Balanced Redox Reaction for Your Cell:

Temperature (^{o}C)	E^{o}_{Cell} Experimental (V)	ΔG^{o} (kJ/mol)	K_{eq}

ΔH =

ΔS =

CALCULATIONS: *(The calculation section should include a graph of ΔG vs. Temperature. All graphing guidelines apply. Example calculations of the following should also be included: percent error between the theoretical and experimental potentials for each cell, ΔG, and K. Make sure that the units are correct in all of your calculations and answers.)*

CONCLUSION: *(The conclusion should be several paragraphs with discussions of the values for ΔG, K, ΔH, and ΔS. Also, describe and discuss the potential series you develop from the class data and the discrepancies between the experimental cell potentials and the literature based cell potentials. Finally, discuss the relationship between temperature and cell potential. Remember to use values from your data and calculations to support your conclusions and discussions. As usual, report and explain all errors that may have occurred.)*

Experiment 18

Post-Laboratory Assignment

Name: ______________________________ **Date:** ___________________

Instructor: ______________________________ **Sec. #:** ____________

Show all work for full credit.

1) What is a Watt? How is it related to the potential of a battery?

2) You have probably heard ads about lithium batteries. Why would lithium be a good metal for the construction of a battery?

Experiment 19

Analysis of a Hydrate

Introduction

Most salts have a crystalline structure that appears dry. When heated however these "dry" salts often release a large amount of water. As a result of this dehydration, the salt crystals change form and sometimes even change color as the water molecules are driven off. Such compounds which contain water molecules are known as ***hydrates***. A hydrate that has lost its water is called an ***anhydrous salt or dessicant***. For a hydrate, a simple whole number represents the number of moles of water present per mole of salt. The following are commonly encountered hydrates:

$CuSO_4 \cdot 5H_2O$ $MgSO_4 \cdot 7H_2O$ $BaCl_2 \cdot 2H_2O$

The water in a hydrate is bound loosely, and so is relatively easily removed by heating. Most hydrates lose their water of hydration at temperatures slightly above 100 °C. Sometimes the water is liberated in stages, with one or more lower hydrates being observed during the heating process. Thus, $CuSO_4$ may also be prepared with 3 moles of H_2O or 1 mole of H_2O per mole of ionic solid. If all the hydration is removed, as it will be if the solid is heated sufficiently, the ionic solid is said to be anhydrous (without water).

In this experiment, you will be given a sample of hydrate. You will determine the mass of the water driven off by heating, as well as the amount of anhydrous salt that remains behind. Then, given the mass of one mole of the anhydrous salt, you will determine the empirical formula of the hydrate.

Background

Salt Structure

Salts are comprised of ionic bonds and contain a cation and an anion. Because all metals (M) are cations and all nonmetals (N) are anions, an anhydrous salt is often symbolized ***MN***. Similarly, a hydrate – which consists of an anhydrous salt and water – is often symbolized $\boldsymbol{MN \cdot x\, H_2O}$, where the x indicates the integer number of water molecules for each formula unit of salt. The dot between the MN and the xH_2O means that the water molecules are rather loosely attached to the anhydrous salt rather than attached with an ionic or covalent bond. When referring to an unknown hydrate you should use the notation described above.

Hydrates

Copper(II) sulfate pentahydrate is a good example of a typical hydrate. It is a blue crystal which looks and feels dry, but when heated turns into an ashy white powder as the water is removed. The hydrated salt is bonded to five moles of water. The compound's formula is $\mathbf{CuSO_4 \cdot 5\, H_2O}$. Some anhydrous salts are capable of becoming re-hydrated with exposure to the moisture in their surroundings. These salts are called **hygroscopic** and can be used as chemical drying agents or

desiccants. Some salts are extremely hygroscopic and can absorb so much moisture from their surroundings that they can eventually dissolve themselves. These salts are called **deliquescent.**

The process of removal of water from a hydrate and addition of water to a dessicant can be written in the form of a reversible reaction:

$$CuSO_4 \bullet 5\ H2O \overset{\Delta}{\Leftrightarrow} CuSO_4(s) + 5\ H2O(l)$$

In this example the process is easy to follow experimentally since the copper sulfate pentahydrate has a blue crystalline structure and the anhydrous or desiccant form is a white powder. If water is added to the white anhydrous copper sulfate, the blue color returns indicating that the pentahydrate has been regenerated. True hydrates display this property of reversibility. Other compounds that react when heated to produce water are not hydrates if they are unable to reabsorb the water they have lost.

Calculating Hydration

The molar mass of $\mathbf{CuSO_4 \cdot 5\ H_2O}$ is:

$$63.5\text{ g} + 32.1\text{ g} + 4\,(16.0\text{ g}) + [\,5\,(18.0\text{ g})\,] = 249.6\text{ g}$$

If a 249.6 g sample of $CuSO_4 \cdot 5\ H_2O$ were heated to drive off all the water, the anhydrous salt $CuSO_4$ would weigh 63.5 g + 32.1 g + 4 (16.0 g) = 159.6 g, which is the mass of one mole of $CuSO_4$. The mass of water that has been boiled off into the air is [5 (18.0 g)] = 90.0 g, which is the mass of five moles of water. The formula of the hydrate shows the ratio of the moles of anhydrous salt to the moles of water; in the above case, that ratio is 1:5.

Procedure

SAFETY NOTES: While working with flames, remember to be very cautious of yourself and your surroundings. All due precautions should be taken in handling the apparatus and the experiment.

Part I: Preparation of a dry crucible

Collect a ring stand, ring, clay triangle or wire mesh, crucible and Bunsen burner.

Set up the ring stand. Secure the ring on the stand.

Obtain a clean crucible. Place the crucible on a clay triangle/wire mesh. Heat with a Bunsen burner for 2 minutes, gently at first and then strongly. This will drive off any water that is adsorbed on the walls of the crucible.

Allow the crucible to cool on the triangle/wire mesh for 2 minutes.

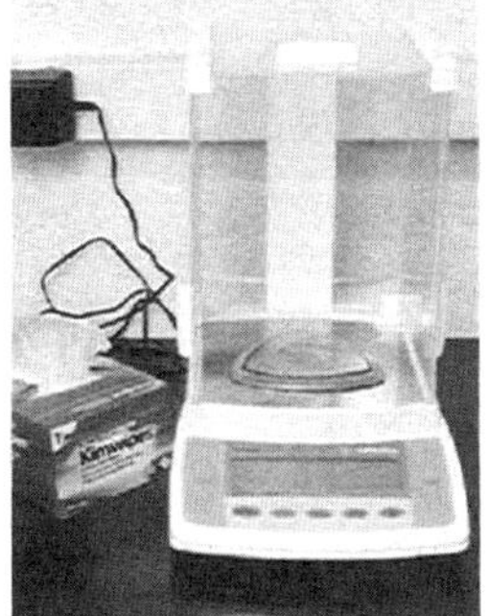

When the crucible is cool, weigh it on a balance. Make this and all successive weights to +/- 0.001 g. Record the weight in your lab notebook.

Part II: Removal of water from a hydrate solution

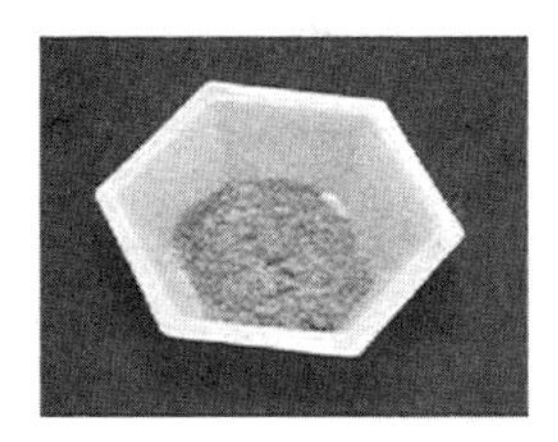

Weigh out ~3.00 g of the unknown hydrate sample in the crucible. Record the mass to the nearest 0.001 g in your lab notebook.

Place the crucible on the clay triangle/wire mesh on the ring stand.

Heat the crucible and its contents, gently at first and then strongly, for approximately 10–15 minutes. The hydrate will probably have a different appearance when the water has been driven off.

Allow the crucible and contents (now an **anhydrous** solid [we hope]) to cool. When they are at approximately room temperature, weigh the crucible and contents and record the mass in your lab notebook.

Repeat the last two steps 2 more times to make sure you have removed all the water from the sample. Record your values for each trial to the nearest 0.001 g in your lab notebook.

Experiment 19

Pre-Laboratory Assignment

Name: ______________________________ **Date:** ___________________

Instructor: ______________________________ **Sec. #:** ____________

Show all work for full credit.

Use the information to answer the questions. Show work, include units, and put your answers in the blanks.

1) Geoffrey weighs an empty beaker and finds it to have a mass of 95.83 g. After putting a spoonful of an unknown hydrate into the beaker, he finds that the mass has increased slightly to 99.87 g. He heats the beaker and its contents twice, and finds that the mass has dropped to 97.22 g. Geoffrey is told by his teacher that the molar mass of the anhydrous salt is 74.10 g.

   a) What mass of hydrate did Geoffrey start with?

   ____

   b) How much water was driven off from the hydrate during the heating process in units of:

      A. **grams**?

      ____

      B. **moles**?

      ____

   c) How much anhydrous salt remained in the beaker in units of:

      A. **grams**?

      ____

      B. **moles**?

      ____

   d) Write down the mole ratio as decimal numbers:

      ______ **moles anhydrous salt :** ______ **moles water**

Write down the mole ratio as whole numbers:

______ moles anhydrous salt : ______ moles water

e) What is the formula of the hydrate? (Use MN to symbolize the anhydrous salt.)

____

f) Based on Geoffrey's data, calculate the percentage of water in the sample of hydrate.

____

g) Why must Geoffrey heat the sample twice instead of just once?

2) In an experiment, a student dehydrated a hydrate unknown and found that the mass of hydrate was initially 4.13 g. After heating, the mass of the desiccant was 3.52 g.
(a) Calculate the mass % of water in the hydrate.

(b) The table below lists the possible hydrates that could be your unknown. Complete the table.

Hydrate	Mass Salt (g)	Mass Water (g)	Mass Percent (%)
$CuSO_4 \bullet 5H_2O$			
$BaCl_2 \bullet 2H_2O$			
$MgSO_4 \bullet 7H_2O$			

(c) Compare the mass % values to determine the identity of your unknown.

Experiment 19

Laboratory Report

Name: ______________________________ **Date:** __________________

Instructor: ______________________________ **Sec. #:** ____________

PURPOSE: *(The purpose should be several well-constructed sentences describing what your experiment was designed to accomplish and the criteria used to determine success. These sentences should include both concepts and techniques.)*

PROCEDURE: *(The procedure section should reference the lab manual and include any changes made to the procedure during the lab. The different temperatures that you ran should be specified. Also include the names of your partners and clearly identify the data contributed by each person.)*

DATA: *(The data section should include your own personal data and observations.)*

Observations:

(Include units, and record all quantities to the nearest 0.01 g.)

Quantity Measured	***Mass***
dry beaker	
beaker and contents before heating	
beaker and contents after first heating	
beaker and contents after second heating	
beaker and contents after third heating	
molar mass of anhydrous salt (obtained from lab)	
percentage of water in hydrate (obtained from lab)	

CALCULATIONS: *(The calculation section should include sample calculations for each column in the data table. Be sure to include conversions of units. Example calculations of the following should also be included: what mass of the hydrate you started with, how much water was driven off from the hydrate during the heating process in moles, how much anhydrous salt remained in the beaker in moles, what the mole ratio of the salt to water is, what the formula of the hydrate is, what percentage of water is in the hydrate and the percent error that occurred in the experiment. Make sure that the units are correct in all of your calculations and answers.)*

CONCLUSION: *(The conclusion section should be in paragraph form and include what the mole ration of the salt to water is, what the formula of the hydrate is, what percentage of water is in the hydrate and the percent error that occurred in the experiment. Finally, be sure to discuss any possible experimental errors.)*

Experiment 19

Post-Laboratory Assignment

Name: ______________________________ **Date:** ___________________

Instructor: ______________________________ **Sec. #:** ____________

Show all work for full credit.

1) Give some reasons why this experiment for finding the formula of hydrates might not work?

2) If the Bunsen burner left some black soot on the bottom of your crucible, how would this change your answer?

3) What are some unavoidable sources of error (like procedural) to explain why you could not achieve a 100% accuracy?

Experiment 20

Qualitative Analysis: Cations, Anions and Complex Ions

Introduction

Qualitative analysis is a process or processes used to determine the presence of a chemical substance based on its reactivity. As the name indicates, qualitative analysis is not concerned with the quantity of substance present, but rather is simply used to confirm its existence. The overall process of a "qualscheme" has become somewhat historical in nature. As modern scientific equipment has improved, qualitative analysis techniques have given way to instruments such as the gas chromatograph and the infrared spectrometer or mass spectrometer for analysis of unknown mixtures. These instruments have the advantage of indicating not only what is present but also the concentration present. That being said, chemical qualitative analysis is still important for two reasons: 1) speed and mobility; 2) every chemist needs a firm understanding of general chemical reactivity.

The first of the two reasons, speed and mobility, refers to the need for fast or "on-site" analysis. The equipment discussed above is laboratory bound and thus inefficient if there is a need to analyze a chemical substance in the field. If you have ever watched CSI (Crime Scene Investigation) television shows, you have probably seen a character spray a solution on a suspect's hands or clothes to confirm gunshot residue. This is qualitative analysis. Another example which you may have experienced is the need to test water samples from lakes or rivers for contaminants.

The second of the two reasons goes to the very heart of what it means to be a chemist. In order to call oneself a chemist you need to be able to predict whether a reaction will take place or not and more importantly the products that will form. A majority of qualitative analysis techniques use the knowledge you have already gained regarding solubility, acid-base, and redox reactions to predict the compounds that will form and under what conditions.

Background

Qualitative analysis refers to the methods used in determining the identity of a chemical compound as opposed to its amount. For example, a majority of this semester has been attributed to determining how *much* of a particular analyte was present through a variety of quantitative techniques. However, in this experiment we switch things up a little. Instead of concentrating upon how *much* of something is present, you will now only be concerned with *what* is present. Overall, the qualitative procedure uses known reactions of a given chemical species and interprets the obtained results using deductive reasoning. In other words, there are a variety of chemical analyses existing for different elements or types of a compound, and combining this with your understanding of acid-base equilibria, redox reactions, and solubility, it becomes fairly elementary to determine *what* particular analyte is found in an unknown mixture.

To increase the accuracy of the analysis process, the procedure for a typical qualitative assay generally breaks down several collections of unknown ions into groups of ions that follow the same chemical pattern. In fact, the **groups** are formed from the basis of the reactivity of the ions, not their groups in the periodic table! Similar ions are then placed in the same group and are then identified by use of **confirmatory** reactions.

Identifying Anions and Cations

Qualitative analysis is used to separate and identify the cations and anions present in an unknown chemical mixture. Ions are grouped according to their reactivity to specific compounds. First, ions are removed in groups from the initial aqueous solution. After each group has been separated, then testing is conducted for the individual ions in each group. What follows here is a common grouping of cations:

Group I cations are those which form insoluble chlorides when reacted with dilute HCl solutions.

Group I cations include Silver (Ag^+), Mercury I (Hg_2^{2+}), and Lead II (Pb^{2+}) and are precipitated out by addition of 1 M HCl to the unknown cation mixture. The **Group I** cations form the precipitates AgCl, Hg_2Cl_2, and $PbCl_2$ that can then be removed from the mixture by centrifugation.

Group II cations, while soluble in dilute HCl, are insoluble when reacted with a dilute H_2S solution at low (acidified) pH <1.

Group II cations include Bismuth III (Bi^{3+}), Cadmium II (Cd^{2+}), Copper II (Cu^{2+}), Antimony III (Sb^{3+}), Antimony V (Sb^{5+}), Tin II (Sn^{2+}) and Tin IV (Sn^{4+}). The **Group II** cations are precipitated out by reaction with 0.1 M H_2S at a low pH, normally around 0.5. The precipitates that are formed are also removed from the remaining mixture by centrifugation.

Group III cations are cations that are soluble in acidic H_2S but are insoluble in basic H_2S.

Group III cations, which include Aluminum (Al^{3+}), Chromium III (Cr^{3+}), Iron II (Fe^{2+}), Iron III (Fe^{3+}), Manganese II (Mn^{2+}), and Zinc II (Zn^{2+}), are precipitated out of the mixture by changing the pH of the previous solution (already saturated with 0.1 M H_2S) to a pH of 9 or greater.

All that is left at this point in the unknown solution should be the **Group IV and V** cations. These are the alkali and alkaline earth metal cations and the ammonium cation. These cations can be separated from each other by their varying reaction to carbonate solution.

Group IV and V cations include Barium (Ba^{2+}), Calcium (Ca^{2+}), and Magnesium (Mg^{2+}) and Potassium (K^+), Sodium (Na^+), and Ammonium (NH^{4+}), respectively. The alkaline earth metal cations Ba^{2+}, Ca^{2+}, and Sr^{2+} are precipitated in 0.2 M $(NH_4)_2CO_3$ solution at pH 10 and then removed from the others by centrifugation. The alkali metal cations K^+ and Na^+ are soluble in most all solutions and thus cannot be separated by precipitation. Rather their presence is confirmed by a colorimetric flame test. The ammonium ion is also soluble and may also have been lost through earlier tests as ammonia. In order to test for the presence of ammonium you should take a small amount of the original mixture and add NaOH until the mixture becomes basic. At that point, the smell of ammonia (NH_3) will confirm the presence of ammonium ion (NH^{4+}).

The portion of qualitative analysis described above simply separates the groups from each other. In order to determine the presence of individual ions in each group a variety of other reactions must be performed. Each reaction is specifically designed to react strictly with only one cation in each group, thereby confirming only its presence. As with the main group separations, the reactions with each cation rely on those chemical properties which are unique to that ion. The discussion of the separation and confirmation of those ions is quite extensive and best broken up into several parts. A more detailed discussion of the confirmation tests for each cation in this experiment is provided here:

Group I Cations

The result of the separation of the Group I cations from the other groups is a precipitate containing Group I cations. In order to analyze the Group I cations further, the precipitate needs to be separated. To do this you must know something about the reactivities of the two Group I cations, silver (Ag^+) and lead(II) (Pb^{2+}).

Lead(II) chloride ($PbCl_2$) is much more soluble than silver chloride (AgCl) and thus can be redissolved by addition of hot distilled water. (Remember, solubility is temperature dependent and most substances are more soluble at higher temperatures.) Lead(II) chloride is almost three times as soluble in hot water as in cold water. Centrifugation of the resulting solution allows you to separate the lead(II) cations from the silver cations. At this point you should have a supernatant containing lead(II) and a precipitate containing silver cations.

For confirmation of the presence of lead(II), we need a reaction that generates a colored precipitate that is characteristic of lead. Addition of several drops of potassium chromate (K_2CrO_4) to the solution containing lead(II) will produce a bright yellow precipitate confirming the presence of the lead. The total ionic reaction is:

$$PbCl_2(aq) + K_2CrO_4(aq) \rightarrow 2KCl(aq) + PbCrO_4(s)\downarrow$$

The net ionic reaction is:

$$Pb^{2+}(aq) + CrO_4^{2-}(aq) \rightarrow PbCrO_4(s)\downarrow$$

The remaining precipitate is then treated with aqueous ammonia. The silver cations in the precipitate react with ammonia to form a complex called the diamminesilver(I) complex:

$$AgCl(s) + 2NH_3(aq) \rightarrow Ag[(NH_3)_2]^+(aq) + Cl^-(aq)$$

This complex is the principal ingredient in Tollen's reagent, the chemical used to make silver mirrors. Addition of strong nitric acid solution (HNO_3) to the resulting solution will break down the diamminesilver(I) complex. Once the complex is broken, the silver can again react with the chloride ions still present in the solution and precipitate as AgCl. The presence of a white precipitate thus confirms the presence of the Ag^+ cation:

$$Ag[(NH_3)_2]^+(aq) + 2H_3O^+(aq) + Cl^-(aq) \rightarrow AgCl(s)\downarrow + 2NH_4^+(aq) + 2H_2O(l)$$

Group II Cations

The supernatant that remained after the removal of the Group I cations contains the Group II through V cations. In order to separate and confirm the Group II cations, copper(II) (Cu^{2+}) and tin(IV) (Sn^{4+}) we only need to know that they, unlike the rest of the remaining ions, are insoluble in acidified sulfide solution.

Note: Hydrogen sulfide (H_2S) is used to precipitate the Group II cations. However, H_2S is a dangerous gas in its pure form. It is much safer to use thioacetamide to form H_2S in solution. When heated, aqueous solutions of thioacetamide hydrolyze to produce H_2S. The formation of H_2S from thioacetamide also provides another level of control in separation of the Group II ions. Below are three reactions based on different solution conditions:

In acid solution:

$$CH_3CSNH_2(aq) + 2H_2O(l) + H^+(aq) \text{ --> } CH_3COOH(aq) + NH_4^+(aq) + H_2S(aq)$$

In basic solution using ammonia:

$$CH_3CSNH_2(aq) + 2H_2O(l) + 2NH_3(aq) \text{ --> } CH_3COO^-(aq) + 3NH_4^+(aq) + S^{2-}(aq)$$

In basic solution using strong base:

$$CH_3CSNH_2(aq) + 2OH^-(aq) \text{ --> } CH_3COO^-(aq) + NH_3(aq) + S^{2-}(aq) + H_2O(l)$$

Each of these reactions produces different concentrations of sulfide ion. An acidic solution produces the least sulfide which will precipitate all of the very insoluble Group II sulfides leaving any slightly insoluble sulfides in solution. A strong base solution produces the most sulfide, capable of precipitating the least insoluble sulfides. Thus, by controlling pH, you can determine which cations precipitate.
In order to precipitate the Group II cations out of the supernatant we must first acidify the solution to a pH of ~0.5. This is done by adding HCl to the supernatant and confirming the pH by use of pH paper. The thioacetamide (CH_3CSNH_2) reacts with the acid and water to form H_2S which in turn precipitates the Group II cations:

$$Cu^{2+}(aq) + S^{2-}(aq) \rightarrow CuS(s)\downarrow$$
$$Sn^{2+}(aq) + S^{2-}(aq) \rightarrow SnS(s)\downarrow$$

The precipitate is removed from solution by centrifugation and the supernatant containing the remainder of the Group III – V ions is saved. The precipitate should be washed with HCl at least once to remove any remaining Group II cations and the supernatant from those washes combined with the first. Once you are satisfied that all of the Group II cations have been removed, you can begin to separate the cations from each other.

Separation and Confirmation of Tin(IV): Unlike copper, tin is soluble in basic solution. Addition of KOH until the solution is basic should dissolve any tin present as the $Sn(OH)_6^{2-}$ complex ion. The solution is heated, centrifuged while still hot, and the supernatant containing the tin complex is removed. Tin(IV) is then confirmed by the addition of HCl until the solution becomes acidic followed by the addition of more thioacetamide. The solution is heated for 5+ minutes and the development of a yellow precipitate (SnS_2) confirms the presence of tin.

$$Sn(OH)_6^{2-}(aq) + 6H^+(aq) + 6Cl^-(aq) \rightarrow SnCl_6^{2-}(aq) + 6H_2O(l)$$
$$Sn^{4+}(aq) + 2S^{2-}(aq) \rightarrow SnS_2(s)$$

The precipitate that was saved from above now contains only the copper(II) (Cu^{2+}) cations.

Confirmation of copper(II): The precipitate is redissolved using nitric acid and heat. If copper is present the solution should be a pale blue color due to the presence of $Cu(H_2O)_4^{2+}$. To confirm the presence of copper(II) add acetic acid and then some potassium ferrocyanide ($K_4[Fe(CN)_6]\cdot 3H_2O$). The formation of a reddish brown precipitate confirms the presence of copper(II).

Group III Cations

Group III consists of manganese(II) (Mn^{2+}) and aluminum (Al^{3+}). Oxidizing the Group III cations with basic H_2O_2 ensures that $Mn(OH)_2$ and $Al(OH)_3$ precipitates are produced. Once the Group III cations are separated by centrifugation, treatment of the precipitate with excess base allows the amphoteric Al^{+3} to form a soluble complex with OH^-. This solubility difference permits the separation of the Mn ions from the Al ions:

$$Mn^{2+}(aq) + 2OH^-(aq) \rightarrow Mn(OH)_2(s)\downarrow \text{ white solid}$$
$$3Al^{3+}(aq) + 4OH^-(aq) \rightarrow Al(OH)_4^-(aq) \text{ clear solution}$$

Analysis of Mn^{2+}: $Mn(OH)_2$ is taken up in HNO_3 and H_2O_2 to dissolve the salt and the resulting aqueous sample tested for the presence of aqueous Mn^{2+} by addition of $NaBiO_3$.

$$Mn^{2+}(aq) + H_2O(l) + BiO_3^-(aq) \rightarrow MnO_4^-(aq) + 2H^+(aq) + Bi(s) \text{ purple solution}$$

The MnO_4^- ion generated in the reaction solution is purple and easily detected. The generation of the color is the confirmation of the presence of the Mn^{2+} ion in the sample.

Analysis of the Al^{3+}: The supernatant separated previously contains $Al(OH)_4^-(aq)$. Acidification with HNO_3 and heating eliminates the basic complexes and generates free Al^{3+} ions, which are amphoteric.

$$Al(OH)_4^-(aq) + 4H^+(aq) \rightarrow Al^{3+}(aq) + 4H_2O(l)$$

Confirmation of Al^{3+}: The aqueous Al^{3+} ion is then precipitated again as the $Al(OH)_3$ -aluminon dye complex.

$$Al(OH)_3(s) + 3CH_3CO_2H(aq) \rightarrow Al^{3+}(aq) + 3H_2O(l) + 3CH_3CO_2^-(aq)$$

$$Al^{3+}(aq) + 3OH^-(aq) + \text{aluminon dye }(aq) \rightarrow Al(OH)_3 - \text{aluminon dye(s) red solution/solid}$$

This step is the confirmation of the presence of Al^{3+}. The dye is added with NH_4Cl solution that is buffered with NH_3. A pH = 5.7 or so ensures that only the correct equivalent amount of OH^- is added. Too much OH^- will cause $Al(OH)_4^{-1}$ (aq) to form and also yield a negative test with the dye. The dye itself is needed to give the $Al(OH)_3$ (s) a color since its presence might be missed otherwise.

Group IV Cations

Once Groups I through III have been removed all that should remain in an unknown sample are the Group IV and Group V cations. The Group IV cations are the alkaline earth metals Ba^{2+} and Sr^{2+}. In order to separate the Group IV cations from those of Group V, we note that all of the alkaline earth metals form insoluble carbonates when treated with ammonium carbonate ($(NH_4)_2CO_3$). The precipitate that forms is removed by centrifugation and the supernatant saved as it contains the Group V cations.

$$Ba^{2+}(aq) + CO_3^{2-}(aq) \rightarrow BaCO_3(s) \downarrow$$

$$Sr^{2+}(aq) + CO_3^{2-}(aq) \rightarrow SrCO_3(s) \downarrow$$

The precipitate containing the Group IV cations is redissolved by acidification and heating. Addition of acetic acid and heat will dissolve the precipitates of both ions.

Test for Barium: Addition of potassium chromate (K_2CrO_4) will precipitate any barium present as a bright yellow solid. Centrifugation removes the precipitate and the supernatant is saved to test for strontium.

$$Ba^{2+}(aq) + CrO_4^{2-}(aq) \rightarrow BaCrO_4(s) \downarrow \text{ bright yellow precipitate}$$

Test for Strontium: The supernatant is used to perform a flame test for strontium. A crimson-red flame confirms the presence of strontium.

Group V Cations:

The supernatant removed from the separation of the Group IV cations contains the Group V cations (Na^+, K^+ and NH_4^+). These cations are soluble in most common reagents and thus cannot be confirmed by precipitation. Flame tests are therefore used to confirm both sodium and potassium. The sodium flame is a strong, yellow colored flame and the potassium emits a weaker violet color. The potassium flame is observed through a piece of cobalt glass to remove the stronger sodium wavelengths which might otherwise mask its presence.

A sample of the original unknown mixture, not the supernatant above, is used to test for ammonium ion because several of the separation techniques used previously may have removed the ammonium as NH_3. A small amount of the original sample is made basic by use of 1M NaOH solution and heated. The odor of ammonia indicates the presence of the ammonium ion.

$$NH_4^+(aq) + OH^-(aq) \xrightarrow{\Delta} NH_3(g) + H_2O(l)$$

A piece of moistened red litmus paper held over the mouth of the test tube confirms the presence of the base.

Anions

As with cations, the anions in a mixture are similarly determined using confirmation reactions based on the unique chemical reactivity of the anion. The most common anions found are: Carbonate (CO_3^{2-}), Sulfate and Sufite (SO_4^{2-}, SO_3^{2-}), Phosphate (PO_4^{3-}), the Halogens (Cl^-, Br^- and I^-), Nitrate and Nitrite (NO_3^-, NO_2^-), Acetate (CH_3COO^-), and Chromate and Dichromate (CrO_4^{2-}, $Cr_2O_7^{2-}$). A list of the reactions used to confirm the presence of each of these anions is given below:

A) CO_3^{2-}

Only the alkali metal and ammonium carbonates are water soluble. Heating (at Bunsen burner temperatures) decomposes all but the alkali and alkaline earth metal carbonates, giving the oxide and carbon dioxide:

$$CuCO_3(s) \rightarrow CuO(s) + CO_2(g)$$

Dilute hydrochloric acid gives vigorous effervescence with carbonates, evolving carbon dioxide:

$$CO_3^{2-}(aq\ or\ s) + 2H^+(aq) \rightarrow H_2O(l) + CO_2(g)$$

B) SO_4^{2-}

$BaSO_4$, $SrSO_4$ and $PbSO_4$ are insoluble. $CaSO_4$ is only slightly soluble. Barium chloride solution added to sulfate solution acidified with dilute hydrochloric acid produces a white precipitate of barium sulfate:

$$Ba^{2+}(aq) + SO_4^{2-}(aq) \rightarrow BaSO_4(s)$$

Adding lead(II) acetate solution gives a precipitate of white lead sulfate:

$$Pb^{2+}(aq) + SO_4^{2-}(aq) \rightarrow PbSO_4(s)$$

C) PO_4^{3-}

A test for phosphate ion in solution is performed by adding nitric acid followed by a solution of ammonium molybdate, $(NH_4)_6MoO_{24}$. **In acidic solution, the ammonium molybdate forms molybdic acid, which reacts with the phosphate ion.** If phosphate ion is present, a canary-yellow precipitate of triammonium dodecamolybdophosphate will form when heated:

$$Ca_3(PO_4)_2(s) + 6\ H^+(aq) \rightarrow 3\ Ca^{2+}(aq) + 2H_3PO_4(aq)$$

$$PO_4^{3-}(aq) + 12\ H_2MoO_4(aq) + 3\ NH_4^+(aq) \rightarrow (NH_4)_3[PO_4 \cdot 12\ MoO_3] \cdot 12\ H_2O(s)$$

D) I^-

AgI, PbI_2, Hg_2I_2 and CuI are all insoluble in water. When added to solid iodides, concentrated sulfuric acid gives a mixture of hydrogen iodide, iodine, hydrogen sulfide, sulfur and sulfur dioxide; the HI produced is oxidized by sulfuric acid. The mixture evolves purple acidic fumes, turns into a brown slurry, and is a mess, but does confirm the presence of iodide:

$$NaI(s) + H_2SO_4(l) \rightarrow NaHSO_4(s) + HI(g)$$

$$2\ HI + H_2SO_4 \rightarrow I_2 + SO_2 + 2\ H_2O$$

$$6\ HI + H_2SO_4 \rightarrow 3\ I_2 + S + 4\ H_2O$$

$$8\ HI + H_2SO_4 \rightarrow 4\ I_2 + H_2S + 4\ H_2O$$

NOTE: There are no state symbols in these equations because the mixture is such a mess, and is more sulfuric acid than water.

Addition of silver nitrate solution to a solution of an iodide that has been acidified (test with blue litmus paper) with dilute nitric acid gives a yellow precipitate of silver iodide. The precipitate is insoluble even in concentrated ammonia:

$$Ag^+(aq) + I^-(aq) \rightarrow AgI(s)$$

$$AgI(s) + NH_3(aq) \rightarrow \text{No reaction}$$

Acidification with nitric acid is necessary to eliminate carbonate or sulfite, both of which interfere with the test by giving spurious precipitates. (An alternative test for iodine, introduced in the halogens experiment, is to use an oxidizing agent to oxidize iodide to iodine, which is brown in aqueous solution. A suitable oxidizing agent is sodium hypochlorite; this is added to the test solution, followed by a little dilute hydrochloric acid and a few mL of hexane:

$$OCl^-(aq) + 2\ H^+(aq) + 2\ I^-(aq) \rightarrow I_2(aq) + Cl^-(aq) + H_2O(l).$$

Iodine can be extracted from the solution by shaking with an immiscible organic solvent, such as hexane, and noting the purple color of the organic layer.

E) Cl^-

The chloride salts AgCl, $PbCl_2$, Hg_2Cl_2 and CuCl are all insoluble in water. Concentrated H_2SO_4 produces steamy acidic fumes of HCl from solid chlorides:

$$NaCl(s) + H_2SO_4(l) \rightarrow NaHSO_4(s) + HCl(g)$$

Addition of silver nitrate solution to a solution of a chloride that has been acidified (test with pH paper) with dilute nitric acid gives a white precipitate of silver chloride. The precipitate is readily soluble in dilute ammonia or in sodium thiosulfate solution:

$$Ag+(aq) + Cl–(aq) \rightarrow AgCl(s)$$

$$AgCl(s) + 2NH3(aq) \rightarrow [Ag(NH3)2]+(aq) + Cl–(aq)$$

$$AgCl(s) + 2\ S2O32–(aq) \rightarrow [Ag(S2O3)2]3–(aq) + Cl–(aq)$$

Acidification with nitric acid is necessary to eliminate carbonate or sulfite, both of which interfere with the test by giving spurious precipitates.

Concentrated solutions of sulfates can give a precipitate of silver sulfate in this test; its appearance is wholly different from AgCl. The latter is truly white; the sulfate is a pearly white, rather like pearlescent nail varnish.

F) CH_3COO^-

Acetates on heating with dilute hydrochloric acid give acetic acid, recognizable by its vinegary smell. Neutral iron(III) chloride solution added to neutral solutions of acetate ion gives a deep red color owing to formation of iron(III) acetate.

$$NaCH_3COO(aq) + HCl \rightarrow CH_3COOH(aq) + NaCl(aq)$$

$$NaCH_3COO(aq) + FeCl_3(aq) \rightarrow Fe(CH_3COO)_3(aq) + NaCl(aq)$$

G) Br^-

AgBr, $PbBr_2$, Hg_2Br_2 and CuBr are all insoluble in water. Concentrated sulfuric acid gives a mixture of hydrogen bromide, bromine and sulfur dioxide with solid bromides; the HBr produced is oxidized by sulfuric acid. The mixture evolves steamy brownish acidic fumes:

$$NaBr(s) + H_2SO_4(l) \rightarrow NaHSO_4(s) + HBr(g)$$

$$2\ HBr(aq) + H_2SO_4(aq) \rightarrow Br_2(aq) + SO_2(aq) + 2H_2O(l)$$

Addition of silver nitrate to a solution of a bromide that has been acidified (test with litmus paper) with dilute nitric acid gives a cream precipitate of silver bromide. The precipitate is readily soluble in concentrated ammonia:

$$Ag^+(aq) + Br^-(aq) \rightarrow AgBr(s)$$

$$AgBr(s)+2\ NH_3(aq) \rightarrow [Ag(NH_3)_2]^+(aq)+Br^-(aq)$$

Acidification with nitric acid is necessary to eliminate carbonate or sulfite, both of which interfere with the test by giving spurious precipitates. Oxidizing agents oxidize bromide to bromine, which is yellow or orange in aqueous solution. Bromine can be extracted from the solution by shaking with an immiscible organic solvent, such as hexane, and noting the orange color of the organic layer.

A suitable oxidizing agent is sodium hypochlorite; this is added to the test solution, followed by a little dilute hydrochloric acid and a few mL of hexane:

$$OCl^-(aq) + 2\,H^+(aq) + 2\,Br^-(aq) \rightarrow Br_2(aq) + Cl^-(aq) + H_2O(l)$$

Bromine can be extracted from the solution by shaking with an immiscible organic solvent, such as hexane, and noting the orange color of the organic layer.

H) NO_3^-

Since all nitrates are water soluble, there is no precipitation reaction for this ion. In chemical analysis, a test for nitrates involves the addition of a solution of ferrous sulfate to the substance to be tested, followed by the addition (without mixing) of a few drops of concentrated sulfuric acid; the presence of a nitrate is indicated by the formation of a brown ring where the sulfuric acid contacts the test mixture. The brown ring is a complex containing the $Fe(NO)^{+2}$ ion.

I) SO_3^{2-}

A sulfite solution acidified with dilute HNO_3 produces sulfur dioxide gas on warming; bubbling of SO_2 gas through potassium dichromate(VI) solution produces a green colored solution.

$$SO_3^{2-}(aq) + 2\,H^+(aq) \rightarrow H_2O(l) + SO_2(g)$$

J & K) CrO_4^{2-} and $Cr_2O_7^{2-}$

These ions are related through the equilibrium

$$Cr_2O_7^{2-}(aq) + 2\,OH^-(aq) \Leftrightarrow 2CrO_4^{2-}(aq) + H_2O(l)$$

In alkaline solution the yellow chromate dominates and in acidic solution orange dichromate dominates. All dichromate salts are soluble; addition of dichromate ions to solutions of ions of metals that have insoluble chromate salts leads to the precipitation of chromates.

Addition of barium chloride solution to a chromate or dichromate solution precipitates bright yellow barium chromate:

$$Ba^{2+}(aq) + CrO_4^{2-}(aq) \rightarrow BaCrO_4(s)$$

Acidified potassium dichromate solutions oxidize primary alcohols to aldehydes and then acids, and secondary alcohols to ketones. Ethanol can be used to test for dichromate, turning the solution green and the apple smell of ethanal being evident.

$$Cr_2O_7^{2-} + 8\,H^+ + 5\,H_2O + CH_3CH_2OH \rightarrow CH_3CH{=}O + 2[Cr(H_2O)_6]^{3+}$$

The orange dichromate(VI) ion is reduced to hexaquachromium(III) in this reaction.

The Experiment

Qualitative analysis today is predominantly performed by machines and is thus somewhat historical in nature. As such, we have decided not to perform experiments running for weeks in order to cover all of the separations described above. Rather, the experiment you will be performing will only take a single lab period but should give you experience in the types of reactions and techniques you would typically perform during a qualitative analysis scheme.

In your experiment you will be investigating seven cations (Na^+, Mg^{2+}, Ni^{2+}, Cr^{3+}, Zn^{2+}, Ag^+, and Pb^{2+}) as well as four anions (NO_3^-, Cl^-, I^- and SO_4^{2-}). You will use the reactivities of these ions to both separate and then identify them. The first part of this experiment asks you to investigate the cations and anions with the goal of gathering enough information about their reactivity and physical properties to use that data as a reference guide later in the experiment when given an unknown mixture of some of the same ions. Thus it is extremely important that you take meticulous notes with respect to every aspect of a solution or precipitate's appearance. For example, don't just write down "a white precipitate", comment on the structure of the precipitate, e.g., is it flaky or crystaline or powdery, etc., as there may be several white precipitates in the experiment before you are through.

Part I: Reactions of Cations with NaOH and NH_3:

Most hydroxide salts are insoluble and thus precipitate out of solution when formed by reaction. The precipitates that form from these reactions often have distinctive colors or textures lending to their identification. Some hydroxide salts undergo further reaction to form oxides. For example:

$$2Ag^+(aq) + 2OH^-(aq) \leftrightarrows \{2AgOH(s)\} \leftrightarrows Ag_2O(s) + H_2O(l)$$

If an excess of hydroxide is added, however, there are some hydroxide salts that will exhibit amphoteric character and re-dissolve, forming complex ions.

For example, aluminum in dilute hydroxide solution forms a white gelatinous powder, $Al(OH)_3$. But if more hydroxide is added the aluminum hydroxide dissolves and the soluble complex ion $Al(OH)_4^-$ forms instead.

The simplest way to determine the structure of a complex ion is to look it up in a reference book or your text. In this experiment, all of the complex ions containing OH^- will contain four OH^- ions. This generalization should be used strictly for this experiment. The overall charge on the complex ion will therefore be the charge on the metal cation minus the number of OH^- ions.

In addition to aluminum, you will also be looking at the reaction of hydroxide with aqueous ammonia, NH_3.

NH_3 solutions are basic enough to create insoluble hydroxides, but do not supply an excess of OH- ions in the solution. If you add a small amount of ammonia, insoluble hydroxides will precipitate.

Ammonia can also form complex ions with some metals. These complex ions are soluble and will generally have four NH_3 ligands bound to the metal ex. ($Zn(NH_3)_4^{2+}$). Silver is the only exception to the 4 ligand rule for this experiment, it has only 2 ammonia groups in its complex,

$Ag(NH_3)_2^+$.

A Stability Sequence

A stability sequence is a sequence of salts and ions for a given cation ranked on their stability from more stable to less stable. In the following series the most stable salts are the ones farthest to the right. While you can perform reactions to make more stable salts (left to right), you cannot go in the opposite direction.

For Ag^+: $Ag_2O(s) < AgCl(s), Ag(NH_3)_2^+(aq) < AgI(s)$

For Pb^{2+}: $PbCl_2(s) < PbSO_4(s) < PbI_2(s) < Pb(OH)_2(s) < Pb(OH)_4^{2-}(aq)$ in xs OH^-

For Zn^{2+}: $Zn(OH)_{2(s)} < Zn(NH_3)_4^{2+}(aq) < Zn(OH)_4^{2-}(aq)$ in xs OH

For Ni^{2+}: $Ni(OH)_2(s) < Ni(NH_3)_4^{2+}(aq) < Ni(OH)_2(s)$ in xs OH-

What these sequences indicate is that if, for example, you were to add aqueous sulfate ions to solid lead chloride, $PbSO_4$ would form. You could not, however, go in the other direction. But you could add iodide to lead sulfate or lead chloride and get lead iodide to form. Any species above shown as an ion is soluble in water. The neutral compounds are all solids and will precipitate.

Anion Precipitation Reactions

In this experiment only two cations will form precipitates when combined with the anions we are using. Silver forms two precipitates and lead forms 3 precipitates. You will need to test these anions (NO_3^{1-}, Cl^{1-}, I^{1-}, and SO_4^{2-}) with each of the two reactant cations and take careful notes to allow you to determine the identity of the anions in your unknown.

Procedure

SAFETY NOTES: There are several strong acids and bases used in this lab. Use great care when handling them and wash hands immediately if you are splashed.

GENERAL INSTRUCTIONS: Each student should work independently for this lab.

Part 1: Reactions of Cations with NaOH and NH_3

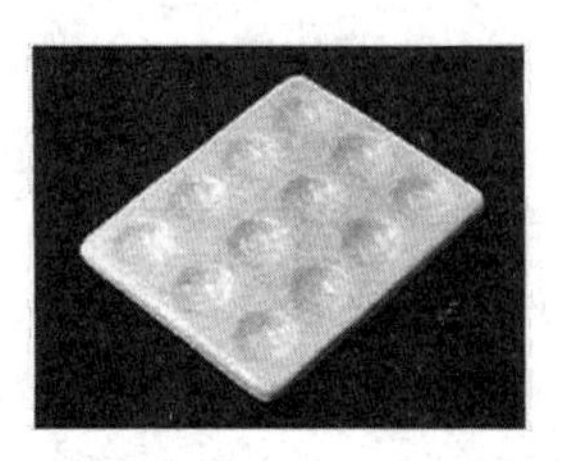

Obtain a well plate from the front counter. Make sure it is clean and dry.

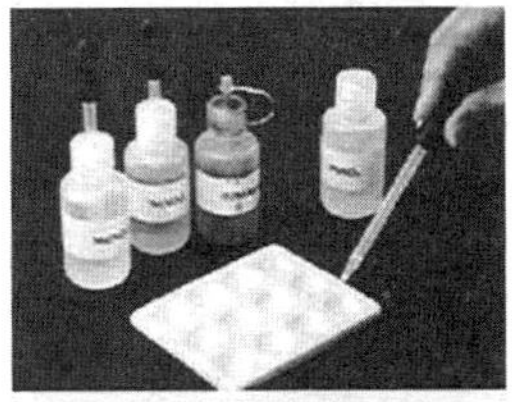

In separate wells, place 10 drops ~ 1/2 mL of each of the cations except for Na^+. (All Na^+ salts are soluble so we know that they will not react.)

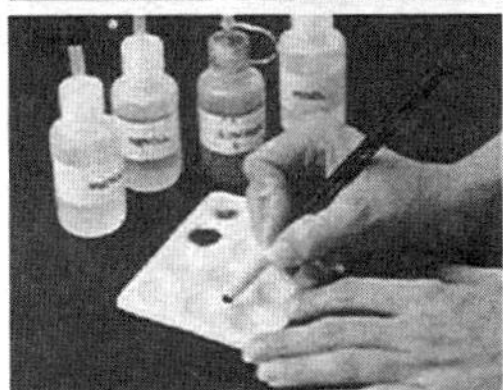

Use a wax pencil to label the identity of each cation in the well to avoid confusion.

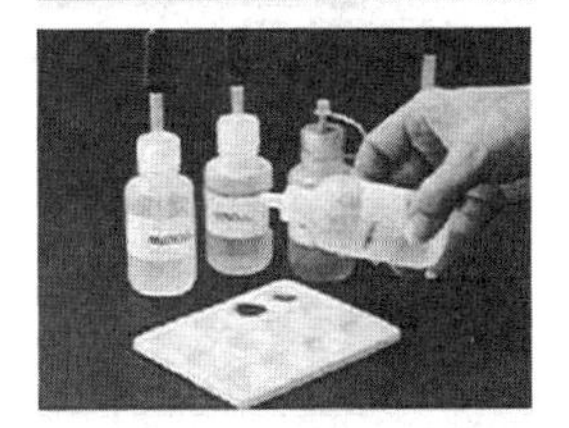

To each of the cation wells, add 1 drop of 6M NaOH. Tap the plate gently to mix and carefully record observations of each cation.

Excess OH Addition: Now add 10 more drops of 6M NaOH to the same wells. Tap the plate gently to mix and carefully record observations of each cation.

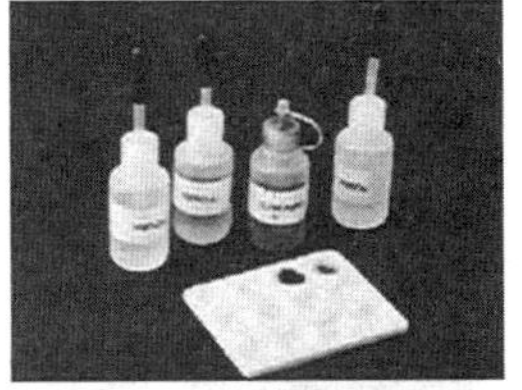

Discard the used solutions in the well plate in the designated waste container and rinse well. Remove any excess water by shaking or Kimwipes.

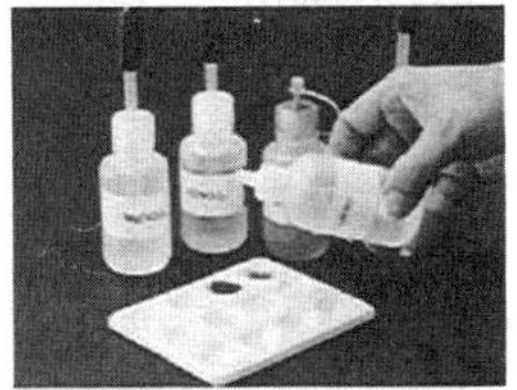

Repeat the process above adding NH_3 instead of OH to each of the cations. Tap the plate gently to mix and carefully record observations of each cation.

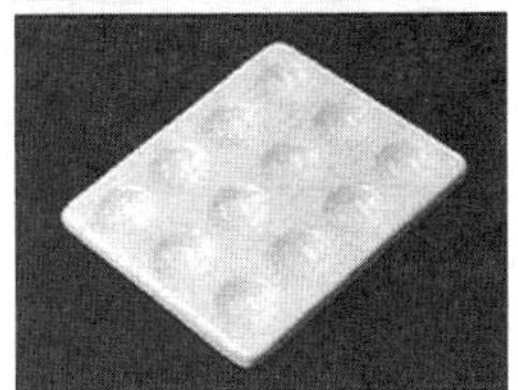

Discard the used solutions in the well plate in the designated waste container and rinse well. Remove any excess water by shaking or Kimwipes.

Part 2: A Stability Sequence

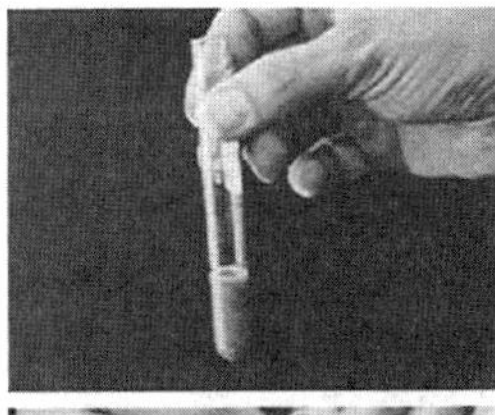

Combine 1 mL of AgNO3 and 1 mL of NaOH in a small test tube. Make and record your observations.

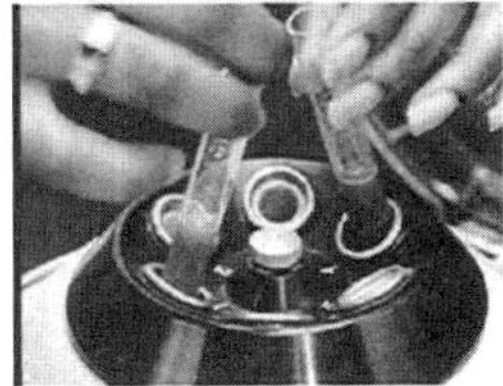

Centrifuge and discard the supernatent. Record all of your observations.

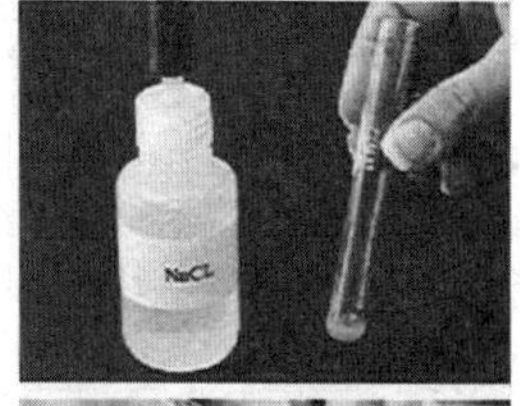

Add several eyedroppers full of aqueous NaCl solution to the test tube and mix well. Record all of your observations.

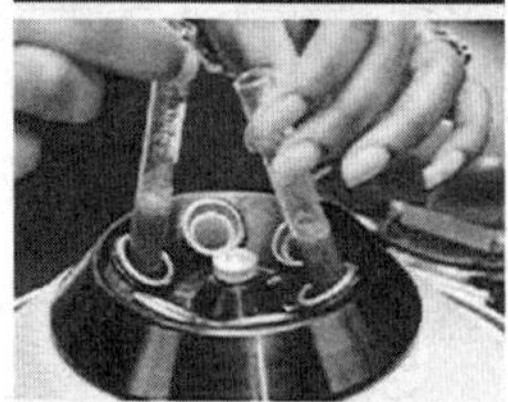

Centrifuge and discard the supernatent. Record all of your observations.

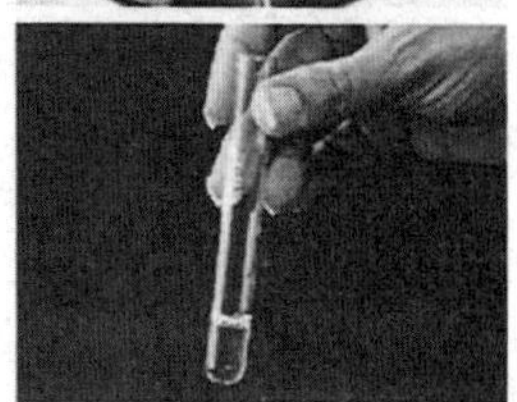

Add NH_3(aq) until the solid dissolves. Record all of your observations.

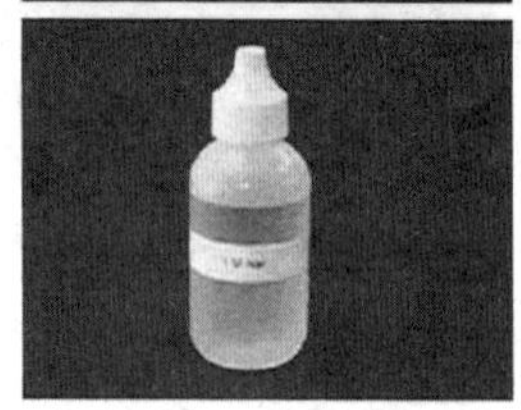

Add NaI until a reaction is seen. Record all of your observations.

Part 3: Anion Precipitation Reactions

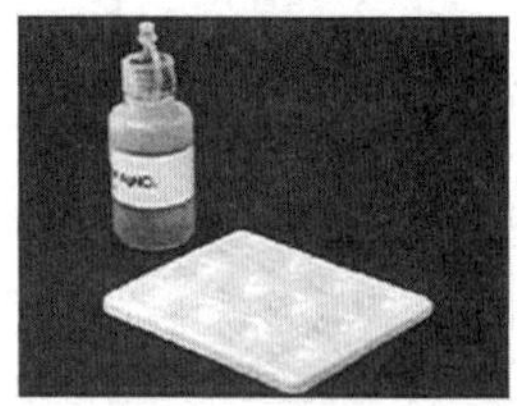

Place 10 drops of Ag^+ into four wells on your well plate.

Place 10 drops of Pb^2+ into another four wells on your well plate.

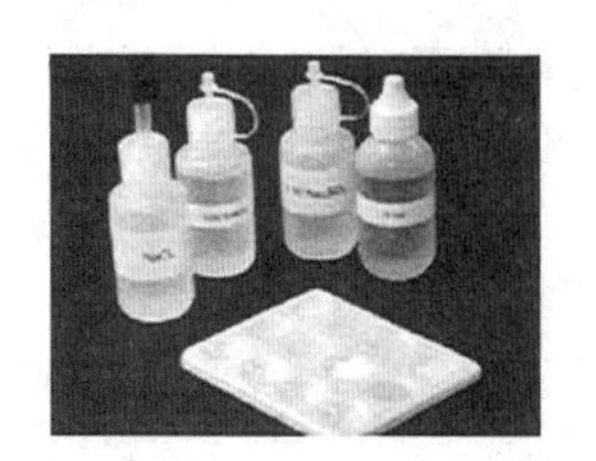

Add 1 drop of Cl^- to one Ag^+ well and one Pb^{2+} well. Record your observations.

Add an additional 10 drops of Cl^- to the same Ag^+ and Pb^{2+} wells. Record any changes.

Repeat the above process for the remaining 3 anions (I^-, SO_4^{2-}, and NO_3^-).

Part 4: Sodium Ions

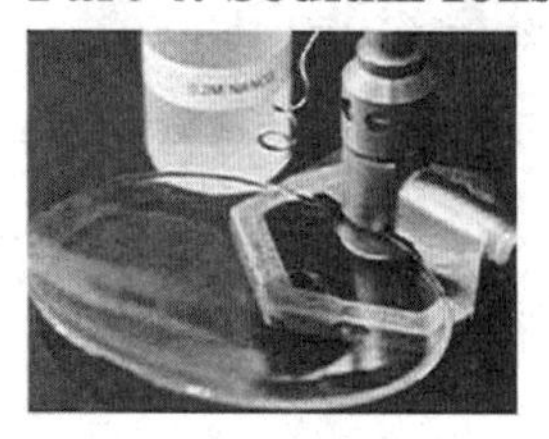

Place a small amount of $NaNO_3$ solution on a watch glass.

Vaporize the $NaNO_3$ under a Bunsen burner in the hood. An intense yellow flame confirms the presence of sodium. See video example.

Part 5: Nitrate Ions

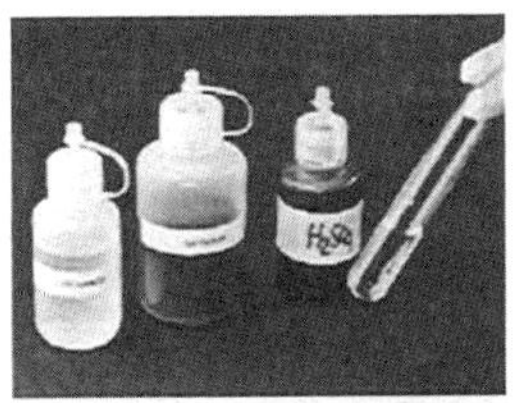

Place 20 drops of aqueous $NaNO_3$ in a test tube.

Note: If this was an unknown you would need to remove any iodide from the sample first by precipitating it out using several drops of saturated Ag_2SO_4. Otherwise you would get a false positive.

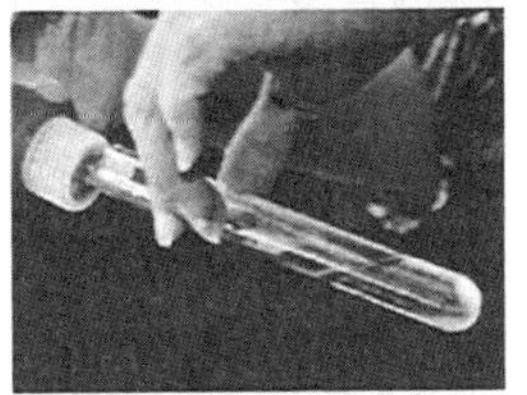

Slowly and very carefully add 20 drops of concentrated sulfuric acid.

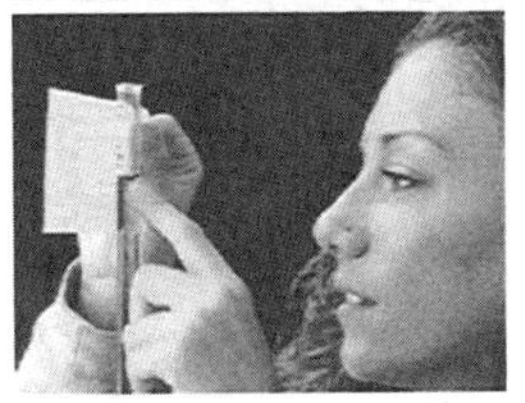

Now add ~ 5 drops of iron II sulfate heptahydrate by allowing the drops to roll gently down the inside of the test tube. DO NOT MIX!

The formation of a smokey brown ring at the solution interface confirms the presence of nitrate ion.

Part 6: The Unknown

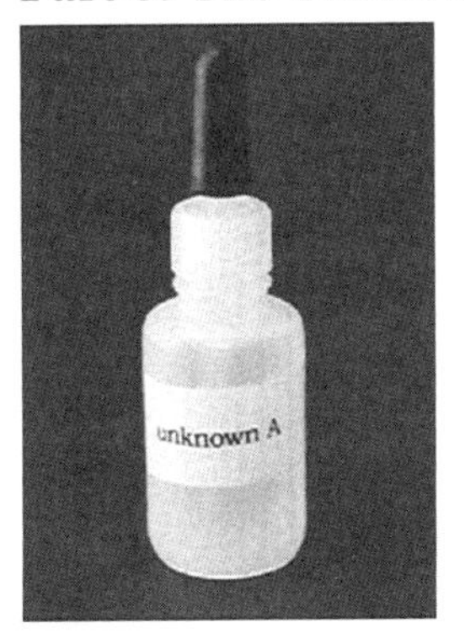

Collect ~ 10 mL of one of the unknowns provided from the front counter. Record the code of the unknown in your notebook.

Using the same tests and your recorded observations from above, determine the identity of the cation(s) and anion(s) present in your unknown.

Experiment # 20

Pre-Laboratory Assignment

Name: ______________________ **Date:** ____________

Instructor: ______________________ **Sec. #:** __________

Show all work for full credit.

1) The qualitative analysis scheme depends on the relative solubilities of the salts of the cations with various anions. The aqueous solubility of a salt can be estimated from K_{sp}, the solubility product constant, describing the equilibrium reaction between the solid salt and the ions in solution:

$$\mathbf{M_xA_y(s) \leftrightarrows xM^{y+}(aq) + yA^{x-}(aq), \quad K_{sp} = [M^{y+}]^x[A^{x-}]^y}$$

(M_xA_y does not appear in the expression because the pure solid is considered to be in its standard state with an activity of 1.)

The actual solubility may differ slightly from the calculated value, though, if the undissociated salt itself is slightly soluble, and if there are intermediate ion-pair structures formed in solution. Nevertheless, the calculated value is still very useful in comparing solubility relationships of various ions and predicting whether or not a precipitate will form when two ionic solutions are mixed.

For the salt BaSO4, K_{sp} = 1.1×10-10.
What is $[Ba^{2+}]$ for a saturated solution of $BaSO_4$?

What is the calculated solubility for $BaSO_4$?

Addition of a soluble sulfate, such as Na_2SO_4, will suppress the solubility of the Ba^{2+} ion because of the common ion effect.
What will $[Ba^{2+}]$ be in the presence of 0.15 M Na_2SO_4?

2) For the salt SrF_2, $K_{sp} = 4.3\times10^{-9}$.
What is $[Sr^{2+}]$ for a saturated solution of SrF_2?

What is the calculated solubility for SrF_2?

What will $[Sr^{2+}]$ be in the presence of 0.10 M NaF?

3) To estimate whether a salt will precipitate under a particular set of conditions, a **reaction quotient, Q_{sp}**, can be calculated from given concentrations of the cation and anion from the equation:

$$Q_{sp} = [M^{y+}]^x[A^{x-}]^y$$

If $\mathbf{Q_{sp}}$ is less than or equal to $\mathbf{K_{sp}}$, no precipitate will form, but if $\mathbf{Q_{sp}}$ is greater than $\mathbf{K_{sp}}$ a precipitate will form and the ion concentrations will decrease until they reach values that satisfy the $\mathbf{K_{sp}}$ expression.

In the case of metal hydroxides, the anion is the hydroxide ion, and its concentration depends on the pH of the solution. Whether a precipitate occurs will therefore depend both on the K_{sp} of the hydroxide and on the pH.

For the reaction: $Mg(OH)_2 \Leftrightarrow Mg^{2+} + 2OH^-$, $K_{sp} = 5.6\times10^{-12}$.
You prepare a solution of 0.20 M Mg^{2+} and adjust the pH to 6.0.
What is the value of the reaction quotient, Q_{sp}, under these conditions? Will a precipitate form?
What will be the equilibrium concentration of the Mg2+ ion?

4) Insoluble salts can sometimes be brought into solution by the formation of a soluble complex ion. For example AgCl(s) can be dissolved in ammonia solutions by the formation of the soluble $Ag(NH_3)_2^+$ complex ion. The ammonia competes with the Cl^- ion for reaction with Ag^+ as follows:

$$AgCl(s) \leftrightarrow Ag^+(aq) + Cl^-(aq) \quad K_{sp} = [Ag^+][Cl^-] = 1.8\text{x}10^{-10}$$

$$Ag^+(aq) + 2NH_3(aq) \leftrightarrow Ag(NH_3)_2^+(aq) \quad K_{formation} = \frac{[Ag(NH_2)_2^+]}{[Ag^+][NH_2]^2} = 1.6\times10^{\partial}$$

Sum: $AgCl(s) + 2NH_3(aq) \leftrightarrow Ag(NH_3)_2^+(aq) + Cl^-(aq)$

$$K_{net} = K_{formation} \times K_{sp} = \frac{\left[Ag(NH_{\partial})_2^+\right]\left[Cl^-\right]}{\left[NH_{\partial}\right]^2} = 2.9 \times 10^{-\partial}$$

Assuming all the Cl^- comes from the dissociation of AgCl, then $[Ag(NH_3)_2^+] = [Cl^-]$, and

$$\left[Ag(NH_{\partial})_2^+\right]^2 = \left[NH_{\partial}\right]^2 \times 2.9 \times 10^{-\partial} \text{ or } \left[Ag(NH_{\partial})_2^+\right] = \left[NH_{\partial}\right] \times \sqrt{2.9 \times 10^{-\partial}}$$

What is the maximum $[Ag(NH_3)_2^+]$ that can be obtained by dissolving AgCl in 1.5M ammonia?

5) For another silver salt, $AgBrO_3$, $K_{sp} = 5.4 \times 10^{-5}$.

What is K_{net} for the reaction $AgBrO_3 + 2NH_3 \leftrightarrow Ag(NH_3)_2^+ + BrO_3^-$?

What is the maximum $[Ag(NH_3)_2^+]$ that can be obtained by dissolving $AgBrO_3$ in 4.1 M ammonia?

Experiment 20

Laboratory Report

Name: ______________________________ **Date:** ____________________

Instructor: ______________________________ **Sec. #:** _____________

PURPOSE: *(The purpose should be several well constructed-sentences describing what your experiment was designed to accomplish and the criteria used to determine success. These sentences should include both concepts and techniques.)*

PROCEDURE: *(The procedure section should cite the lab manual and include any changes made to the procedure during lab work. The procedure section should also include a step-by-step procedure on how you determined the identity of the unknown.)*

Attach Step by Step Procedure.

DATA: *(The data section should include your own personal data and observations.)*

Part 1: Observations

Cation	Initial Observation	OH^- Addition	XS OH^- Addition	NH_3 Addition	Anion PPT

Part 2: Observations

Initial:

After Centrifugation:

After NaCl Addition:

After Centrifugation:

After NH_3 Addition:

After NaI Addition:

Part 3: Observations

Anion	Ag^+	Pb^{2+}
Cl^-		
XS Cl^-		
I^-		
XS I^-		
SO_4^{2-}		
XS SO_4^{2-}		
NO_3^-		
XS NO_3^-		

Part 4: Observations

Part 5: Observations

Part 6: *(Attach a sheet containing all of your observations for your unknown.)*

Unknown #: ______________________________

CALCULATIONS: *(The calculation section for this experiment will include balanced net ionic equations. Write the balanced net ionic equation for each reaction from Part 1. If you observed a precipitate re-dissolve in excess OH^-, write two separate balanced equations: The first showing the formation of the precipitate from the metal ions and the hydroxide ion; and the second with the precipitates and additional hydroxide ions as the reactants. NH_3 was also reacted with all the cations except Na^+. If a precipitate formed and remained, a neutral hydroxide salt formed when the cation reaction with NH_3 and H_2O. If a precipitate was not observed or immediately re-dissolved, the cation reacted with NH_3 to form a complex ion. Formulas of any complex ions can be determined from the information provided in the Background section to this experiment. Also write the balanced net ionic equations for the anion precipitation reactions you observed in Part 3.)*

CONCLUSION: *(The conclusion section should include several detailed paragraphs which include the following information: A paragraph explaining how a stability series works; include how the observations you made specifically illustrate the stability series for silver ions (this is from part II). The identity of the ions in your unknown and how you reached these conclusions; this should include a discussion of the known reactions you did in parts I–V and how they compare to the unknown reactions. Be sure to include a description of both supporting and eliminating experiments that you conducted. Finally include all possible errors found in this experiment.)*

Experiment 20

Post-Laboratory Assignment

Name: ______________________ **Date:** ______________

Instructor: ______________________ **Sec. #:** __________

Show all work for full credit.

1) Which cation forms a 2-ligand complex with NH_3?

2) Which anion does not form a precipitate with lead?

3) Write the balanced equation for the dissociation of the compound in question 2 in water.

Experiment 21

Isomerism in Coordination Chemistry

Preparation of a Coordination Compound: Tetraamminecopper(II) Sulfate Monohydrate

Introduction

Coordinate covalent bonds are chemical bonds in which the two electrons are donated by one of the participating atoms. This type of interaction is a common mode of bonding in transition metal compounds. These same bonds can be considered as Lewis acid-base interactions; a situation where there is an electron acceptor (acid) and an electron donor (base) that form a Lewis acid/base adduct.

This lab uses the tendency for metal ions to accept electrons from multiple Lewis bases to form a new compound. The change in physical properties of the material is often easy to see. The compound made in this experiment is a blue crystalline product that is easily isolated by filtration.

Background

By accepting electron density from multiple Lewis bases, transition metal ions form *complex ions*. These electron donors are called *ligands*. The number of ligands attached to the metal ions is called the *coordination number*. The most common coordination numbers are 4 and 6, but it can range from 1 to 12 depending on the metal ion.

Forming coordination complexes is easy. Indeed, the simple act of dissolving metal salts in water forms one type of complex ion. An example of this is the dissolution of anhydrous nickel chloride in water shown in equation 1. The drawing in Figure 1 shows the structure of the $[Ni(H_2O)_6]^{2+}$ ion.

$$\underset{\text{(yellow)}}{NiCl_2} \xrightarrow{H_2O} \underset{\text{(pale green)}}{[Ni(H_2O)_6]^{+2}\,(aq)} + 2Cl^-\,(aq) \qquad \text{Eq.1}$$

$$\left[Ni^{+2}(OH_2)_6 \right]^{+2}$$

Figure 1: The hexaaquanickel(II) complex ion.

Equation 2 below shows how the addition of a new ligand can replace the water and form a new complex ion.

$$[Ni(H_2O)_6]^{2+}\ (aq) + 6\ NH_3\ (aq) \rightarrow [Ni(NH_3)_6]^{2+}\ (aq) + 6\ H_2O\ (l) \qquad \text{Eq. 2}$$

NH3
H3N NH3
Ni+2
H3N NH3
NH3
+2

In this experiment, students will have the opportunity to make a coordination compound by reacting copper ions with a common ligand, ammonia (NH_3). The reaction is very fast, and is complete shortly after mixing the two reactants.

As with many chemical syntheses, the method used to isolate the product is critical to the success of the experiment. In this case, the isolation takes advantage of the ionic properties of the product. It is very soluble in water, and only slightly soluble in non-polar solvents. After the reaction is complete, ethanol is added to the mixture and the new coordination complex precipitates as blue crystals. These crystals are isolated by suction filtration. Once dried, the material can be massed and the percent yield for the reaction determined.

Procedure

SAFETY NOTES: While working with these samples, remember to be very cautious of yourself and your surroundings. All due precautions should be taken in handling the apparatus and the experiment. Dispose of crystals, filtrates and leftover reagents in the waste container provided. Wash all equipment and return it to its original location.

Part I: The reaction

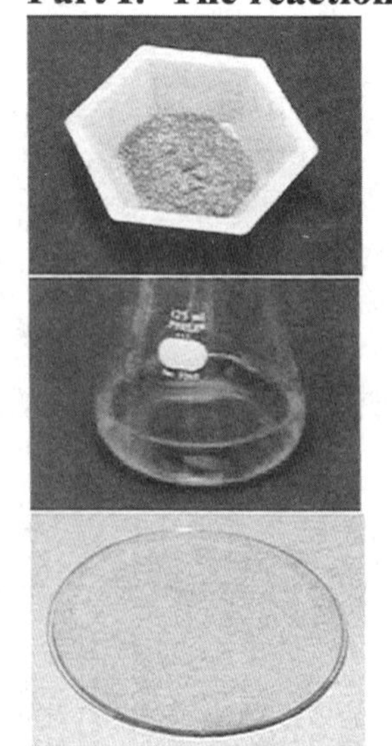

Mass 1.5 g of copper sulfate pentahydrate to the nearest milligram in a 50 mL beaker. Record the mass.

Add 5 mL of water and swirl gently to dissolve the solid. In the hood, add 5 mL of 15 M ammonia and stir well with a glass rod.

Cover the beaker with the watch glass and return to the work station.

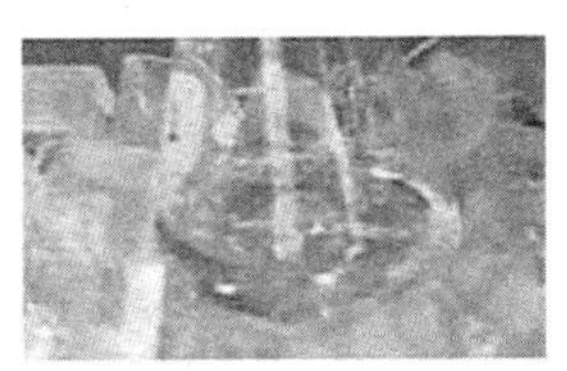

At the bench, add 8 mL of 95% ethanol. Stir well with a glass rod. Cool the reaction beaker in an ice bath for 10 minutes.

Part II: Isolation of product

Using vacuum filtration, collect the solid product on the funnel.
Once the solid looks dry, stop filtration and pour the filtrate into a waste beaker.
Record your observations. What is the color of the solid? What is the color of the filtrate?

Part III: Wash the filtered product

Obtain 10 mL of 1:1 ethanol/ammonia solution from the hood.

Use the Pasteur pipet to add small portions of the wash solution to the solid product.

Wait for the wash to go all the way through the filter before adding another portion.

Dipose of the washing in the waste jar provided in the hood.

Next, wash the solid with 3–6 mL of 95% ethanol added in small portions (one pipet-full at a time).

To completely dry the solid, continue suction for another 5 minutes.

Part IV: Yield and Percent Yield

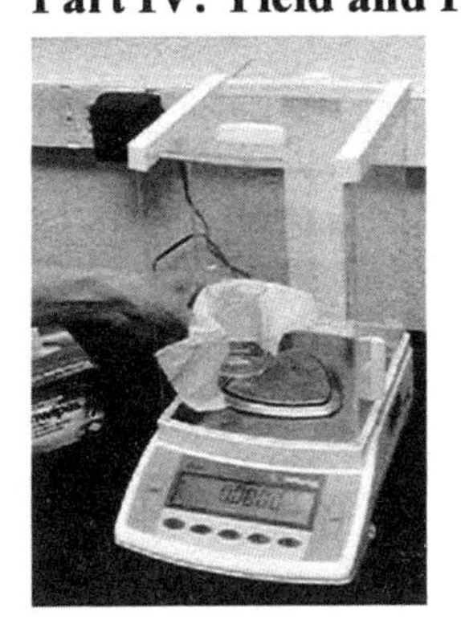

Once the product is dried, determine the mass of the product to the nearest milligram (0.001 g).

Mass a clean and dry 50mL beaker. Remove the dry solid from the funnel by carefully loosening the powder with a spatula. Transfer the solid to the beaker to determine the mass of product.

Make sure to record to all your values in your lab manual.

Experiment 21

Pre-Laboratory Assignment

Name: ______________________ **Date:** ______________

Instructor: ______________________ **Sec. #:** __________

Show all work for full credit.

Use the information to answer the questions. Show work, include units, and put your answers in the blanks.

1. Write a balanced chemical equation for the reaction of $CuSO_4 \cdot 5H_2O$ with ammonia to form $[Cu(NH_3)_4]\ SO_4 \cdot H_2O$.

2. How many moles of ammonia are present in 5 mL of a 15 M NH_3 solution?

3. How many moles of Cu^{2+} are present in 1.5 grams of $CuSO_4 \cdot 5H_2O$?

4. Calculate the molar mass of $[Cu(NH_3)_4]\ SO_4 \cdot H_2O$.

Experiment 21

Laboratory Report

Name: ______________________________ **Date:** ___________________

Instructor: ______________________________ **Sec. #:** ____________

PURPOSE: *(The purpose should be several well-constructed sentences describing what your experiment was designed to accomplish and the criteria used to determine success. These sentences should include both concepts and techniques. Write a paragraph consisting of several sentences that describe the nature and goals of the experiment. This section should concisely answer the following questions: What (and how) data are being collected during the lab? How are the data used to reach a result? (Or what is the relationship between the data and the calculated value of interest?... a brief description of the calculations.) What concepts and techniques are illustrated in the experiment? Spelling and grammar should be correct.)*

PROCEDURE: *(The procedure section should reference the lab manual and include any changes made to the procedure during the lab. The different temperatures that you ran should be specified. Also include the names of your partners and clearly identify the data contributed by each person.)*

DATA: *(The data section should include your own personal data and observations)*

(Include units, and record all quantities to the nearest 0.01 g.)

Mass of $CuSO_4 \cdot 5H_2O$ (g)	
Moles $CuSO_4 \cdot 5H_2O$ (g)	
Mass of beaker and product (g)	
Mass of product beaker (g)	
Mass of product (g)	
Molar mass $[Cu(NH_3)_4]\ SO_4 \cdot H_2O$ (g/mol)	
Moles of $[Cu(NH_3)_4]\ SO_4 \cdot H_2O$	
Limiting reagent	
% yield	

CALCULATIONS: *(The calculation section should include sample calculations for each column in the data table. Be sure to include conversions of units. Example calculations of the following should also be included: the theoretical yield calculating the number of grams of $[Cu(NH_3)_4]\ SO_4{\cdot}H_2O$ that could be made from the starting amount of $CuSO_4{\cdot}5H_2O$, the actual yield by using the mass-by-difference to determine the grams of product isolated from the synthesis, and the percent yield by determining the ratio of actual yield to theoretical yield, multiplied by 100 gives the percent yield of the reaction. Make sure that the units are correct in all of your calculations and answers.)*

CONCLUSION: *(The conclusion section should be in paragraph form and include what the percent yield of the reaction is. What does the yield imply with respect to the actual and theoretical values? What caused the yield to be so off from being 100%? Finally be sure to discuss any possible experimental errors.)*

Experiment 21

Post-Laboratory Assignment

Name: ______________________________ **Date:** ____________________

Instructor: ______________________________ **Sec. #:** ____________

Show all work for full credit.

1) Based on the color of the filtrate, is it possible to obtain a 100% yield on this reaction? Explain why/why not.

Experiment 22

Light Emitting Diode

Introduction

When a complex ion is sandwiched between two metal electrodes a circuit can be created. If a current is passed through this chemical circuit an organic light emitting diode may be formed.

In this experiment the coordination compound tris(2,2bipyridine) ruthenium(II) tetrafluoroborate is used as an intermediate in the formation of an organic diode. To synthesize the complex ion, Ru^{3+} found in the form of $RuCl_3$ is first reduced to Ru^{2+} using sodium hypophosphite (NaH_2PO_2) as a reducing agent. The Ru^{2+} is then systematically introduced to 2,2'dipyridyl and sodium tetrafluoroborate to form $[Ru(bpy)_3](BF_4)_2$. The 2,2'dipyridyl is a bidentate ligand (aka chelating agent) in that one molecule has two pairs of electrons it can donate to the central atom, therefore forming two coordinate covalent bonds.

$$2H_2O + NaH_2PO_2 + 12\ bpy + 4RuCl_3 \rightarrow 4[Ru(bpy)_3]Cl_2 + 3HCl + H_3PO_4 + NaCl$$

$$[Ru(bpy)_3]Cl_2 + 2NaBF_4 \rightarrow [Ru(bpy)_3](BF_4)_2 + 2NaCl$$

Background

Photoluminescence

When a current is passed through a semiconductor, electrons from the valence band can be promoted across the energy gap into the conduction band, which allows the semiconductor to conduct electricity. When an electron is promoted into the conduction band, it is said to leave behind a hole in the valence band. As with electrons, holes can move in the substance, and they act as a positively charged particle. Holes are known as p-carriers for carrying a positive charge, while electrons are known as n-carriers for carrying a negative charge. When an electron in the conduction band looses energy and falls back into a hole it is said to recombine. This electron-hole recombination can produce photons and give off energy in the form of light through a process called photoluminescence.

Diode Formation

Semiconductors can be doped to increase the number of p-carriers or n-carriers. A p-type semiconductor is one that has had additional p-carriers added, while an n-type semiconductor has had additional n-carriers added. When an n-type semiconductor is brought into contact with a p-type semiconductor of the same material, a p-n junction is formed. The holes on the p-side of the junction move toward the n-side, while the electrons on the n-side move toward the p-side due to attraction between the positive and negative charges.

An external voltage can be applied across a p-n junction, causing it to be biased. If the magnitude of the potential difference between the n-side and p-side is reduced, it is called forward bias, and if it is increased it is called reverse biased. Applying a voltage across the p-n junction allows it to behave as a diode. A diode is a device that allows current to only travel in one direction. A light-emitting diode (LED) is created by inducing forward bias, and when the electrons and holes recombined on both sides of the p-n junction light is given off.

Procedure

SAFETY NOTES: While working with these samples, remember to be very cautious of yourself and your surroundings. All due precautions should be taken in handling the apparatus and the experiment. Dispose of crystals, filtrates and left over reagents in the waste container provided. Wash all equipment and return it to its original location.

Part I: Preparing "$RuCl_3$" from commercial $RuCl_3 \cdot 3H_2O$

NOTE: *Parts I through III may be completed by lab personnel in advance to shorten the time required.*

Place the commercial $RuCl_3 \cdot 3H_2O$ (~3 g) in a mortar and pestle.

Place this in an oven (120 to 130 °C) for 3 hours, allow it to cool, and grind it. Place the ground material back in the oven overnight prior to the laboratory experiment. There is most likely some water remaining, as anhydrous $RuCl_3$ is insoluble.

Part II: NaH_2PO_2 preparation from H_3PO_2

NOTE: *Hypophosphorous acid is monoprotic (two of the hydrogens are attached directly to phosphorus). It does not fume, but with a pKa of 1.1, it is stronger than phosphoric acid. Use caution when handling, and keep it away from oxidizers (do not store it in a cabinet where the usual mineral acids are kept!)*

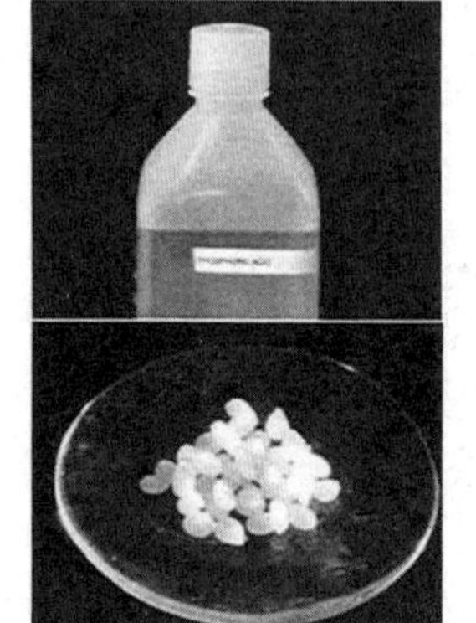

Measure 10 mL of 50% H_3PO_2 into a beaker with a stirbar, add 6 mL of water, and begin stirring.

Slowly add pellets of NaOH until the pH is about 6-8 (wait until the previous pellet dissolves before adding the next). This will require about 3.9 g of NaOH. The final solution is about 6 M in NaH_2PO_2.

Part III: Synthesis

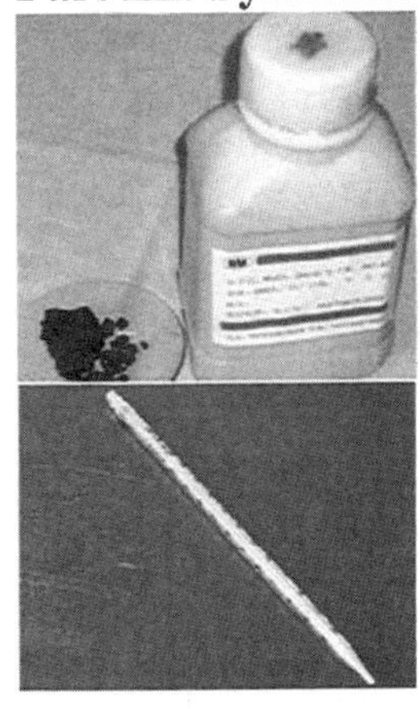

Dissolve 0.083 g (0.40 mmol) "$RuCl_3$" in 8 mL of water in a 30-mL beaker equipped with a Teflon-coated stirbar.

Add 0.188 g (1.204 mmol) 2,2'-dipyridyl and, using a syringe or Mohr pipet, 0.44 mL of NaH_2PO_2. The liquid will be deep greyish-green.

Mark the beaker to indicate the level of liquid, place a watch glass on top of the beaker, and heat so it boils for 30 min. Add water as necessary to bring the level to the mark. The liquid will now be deep orange. Allow the solution to cool to room temperature.

Dissolve 0.333 g of $NaBF_4$ in 1.5 mL of water and add the resulting solution to the orange solution. Solid should form. Heat the mixture until the solid dissolves, adding water as needed.

Cool in ice; crystals should form. Collect these by filtration and wash with ethanol. Draw air through the orange solid to dry.

Part IV: OLED fabrication

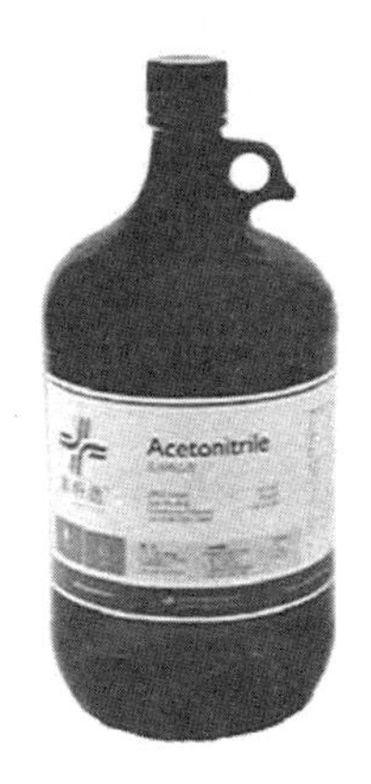

Dissolve about 0.01 g of the $[Ru(bpy)_3](BF_4)_2$ in 1 mL of acetonitrile.

Determine which side of a piece of ITO coated glass is conductive and make sure it is face up. With a pipet, place about 3 drops of the orange solution on the ITO coated glass and spread it with the pipet tip so that it covers about half of the area.

The solution will dry within a minute or so. Place it in the oven (120 °C) for about 10 min. Have the gallium/indium eutectic ready, take the coated glass out of the oven, and spread a few mm2 of the eutectic onto the dried $[Ru(bpy)_3](BF_4)_2$ film with a cotton swab; this is most easily done while the coated glass is still warm.

In a fairly dark spot, place the assembly on a white sheet of paper, coated side up. Touch the positive lead of a 6-volt power supply directly to the ITO coating (not on the $[Ru(bpy)_3](BF_4)_2$ film). Gently touch the negative lead to the gallium/indium eutectic. The light emission occurs downward from the eutectic, the reflection of which can be observed through the paper. Alternatively, hold the assembly up, back side facing you, and have a partner touch the leads. Observe what part glows.

Make sure to record to all your findings in your lab notebook.

NOTE: *An alternative to spreading the eutectic onto the* $[Ru(bpy)_3](BF_4)_2$ *film is to coat the negative lead with the Ga/In eutectic and gently touch it onto the film. Move it over the surface as needed to maximize brightness, but be sure not to let it touch the tin oxide coated glass itself.*

Ref.:

LED fabrication procedure modified from Frank G. Gao and Allen J. Bard, "Solid-State Organic Light-Emitting Diodes Based on Tris(2,2'-bipyridine)ruthenium(II) Complexes", *Journal of the American Chemical Society*, **122**(30), 7426-7427 (2000)

$[Ru(bpy)_3](BF_4)_2$ synthesis modified from John A. Broomhead and Charles G. Young, *Inorg. Syn.* **28**, 338-340 (1990).

Sevian, H., Müller, S., Rudmann, H., & Rubner, M.F. (2004). Using organic light-emitting electrochemical thinfilm devices to teach materials science. *Journal of chemical education, 81*.

Experiment 22

Pre-Laboratory Assignment

Name: ______________________________ **Date:** __________________

Instructor: ______________________________ **Sec. #:** ____________

Show all work for full credit.

Use the information to answer the questions. Show work, include units, and put your answers in the blanks.

1. Calculate the theoretical yield of $[Ru(bpy)_3](BF_4)_2$ (MM = 743.29 g) recovered. Use the $RuCl_3$ as the limiting reactant.

2. Identify the central atom, ligand(s), and coordination number in the coordination compound $[Ru(bpy)_3](BF_4)_2$. Draw a sketch of the complex ion.

3. Would the photoluminescence in this reaction be classified as fluorescence or phosphorescence? Why?

4. When the negative and positive leads are switched no light is emitted. Why is this and what does this prove?

Experiment 22

Laboratory Report

Name: ____________________________ **Date:** _________________

Instructor: _____________________________ **Sec. #:** ___________

PURPOSE: *(The purpose should be several well-constructed sentences describing what your experiment was designed to accomplish and the criteria used to determine success. These sentences should include both concepts and techniques. Write a paragraph consisting of several sentences that describe the nature and goals of the experiment. This section should concisely answer the following questions: What (and how) data are being collected during the lab? How are the data used to reach a result? (Or what is the relationship between the data and the calculated value of interest?... a brief description of the calculations.) What concepts and techniques are illustrated in the experiment? Spelling and grammar should be correct.)*

PROCEDURE: *(The procedure section should reference the lab manual and include any changes made to the procedure during the lab. The different temperatures that you ran should be specified. Also include the names of your partners and clearly identify the data contributed by each person.)*

DATA: *(The data section should include your own personal data and observations.)*

Observations made:	

CALCULATIONS: *(The calculation section should include sample calculations for each column in the data table. Be sure to include conversions of units. Example calculations of the following should also be included: the theoretical yield calculating the number of grams of* $[Ru(bpy)_3](BF_4)_2$ *that could be made from the starting material, the actual yield by using the mass-by-difference to determine the grams of product isolated from the synthesis, and the percent yield by determining the ratio of actual yield to theoretical yield, multiplied by 100 gives the percent yield of the reaction. Make sure that the units are correct in all of your calculations and answers.)*

CONCLUSION: *(The conclusion section should be in paragraph form and include what the percent yield of the reaction is. What does the yield imply with respect to the actual and theoretical values? What caused the yield to be so off from being 100%? Finally be sure to discuss any possible experimental errors.)*

Experiment 22

Post-Laboratory Assignment

Name: ____________________ **Date:** ______________

Instructor: ____________________ **Sec. #:** __________

Show all work for full credit.

1. Calculate the percent yield of $[Ru(bpy)_3](BF_4)_2$ (MW = 743.29 g) recovered. Use the $RuCl_3$ as the limiting reactant.

2. Does the $[Ru(bpy)_3]^{2+}$ complex exhibit a small or large crystal-field splitting energy? Explain. (Hint: think about the color of the pure compound.)

Experiment 23
Amino Acid Analysis by Ninhydrin

Introduction

As we all know, amino acids are the building blocks of proteins, and as such, serve as one of the more important biochemical molecules you will work with in this lab. The identification and characterization of these molecules can be done numerous ways depending upon what you are trying to prove. One such method to determine the presence of amino acids within a sample is accomplished with ninhydrin.

Ninhydrin, which is originally yellow, reacts with amino acids and turns deep purple, and it is this purple color that confirms the presence of amino acids in unknown biochemical samples.

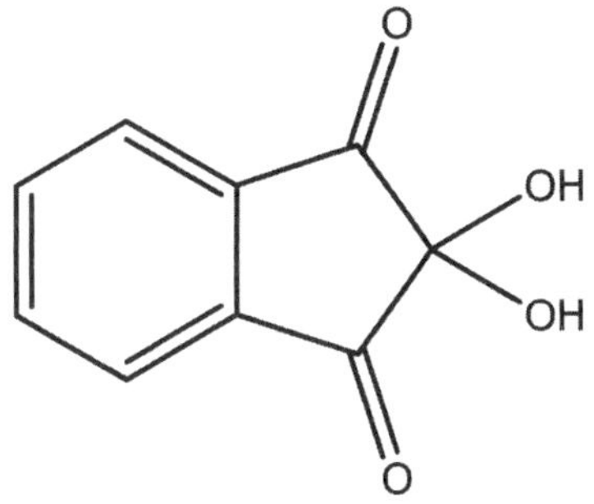

O
OH
OH
O

Ninhydrin

Despite being successful for indicating the presence of amino acids, the ninhydrin technique cannot uniquely identify individual amino acids. A second method is required for this purpose. Thin-layer chromatography (TLC), a method commonly used for amino acid identification will be coupled with the Ninhydrin technique to both confirm and identify the amino acids in an unknown mixture.

Background

Ninhydrin

The reaction between an α-amino acid and ninhydrin can be summarized by the following steps:

Step 1:

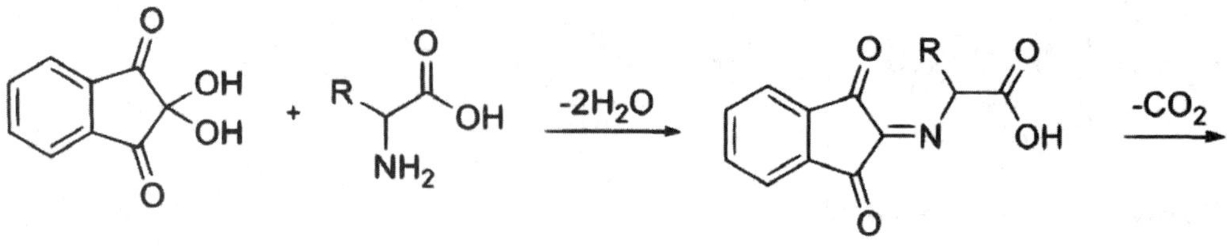

In this initial step of the reaction, ninhydrin removes two hydrogens from the α-amino acid to yield α-imino acid. The ninhydrin molecule is reduced during the process and loses an oxygen atom. As a result a molecule of water is formed from the two hydrogen and oxygen atoms.

Under heated conditions, the α-keto acid is decarboxylated to form an aldehyde, with one less carbon than the original amino acid, in addition to a molecule of carbon dioxide.

Step 2:

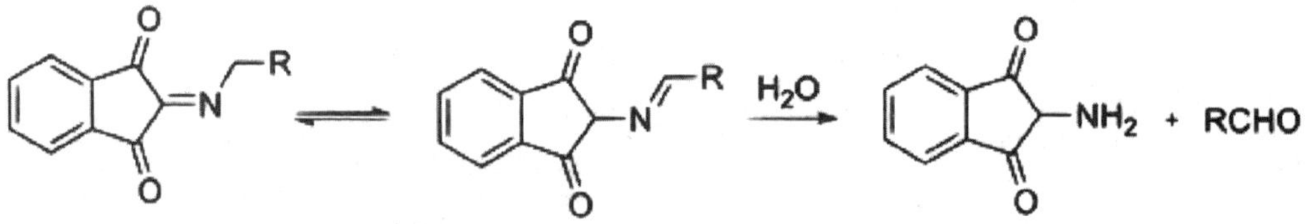

In this step, the –NH group of the α-imino acid is rapidly hydrolyzed forming an α-keto acid concurrent with the production of a carbonyl molecule.

Step 3:

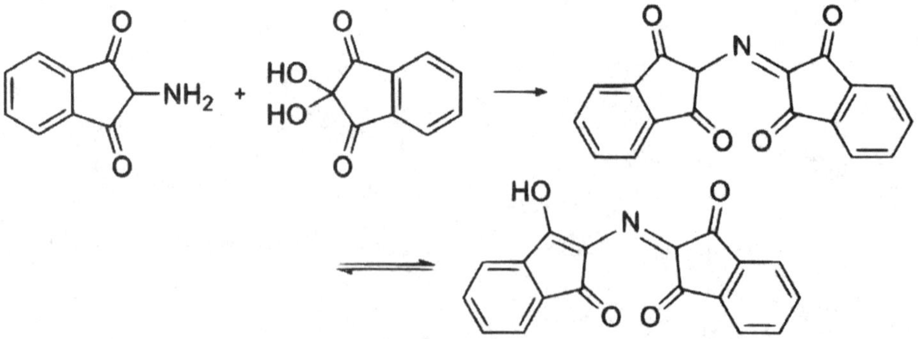

In the final reaction steps the reduced ninhydrin and ammonia react with another nonhydrin compound to create the purple chromophore.

Thin Layer Chromatography

A relatively simple and inexpensive technique, thin-layer chromatography (TLC) is often used to judge the purity of a synthesized compound or to indicate the progress of a chemical reaction. In this technique, a small quantity of a solution of the mixture to be analyzed is deposited as a small spot on a TLC plate that, acting as the stationary phase, consists of a thin layer of silica gel (SiO_2) or alumina (Al_2O_3) coated upon a glass or plastic sheet. When the plate is placed in a chamber containing a small amount of solvent, generally referred to as the mobile phase, the solvent gradually moves up the plate via capillary action. As the mobile phase moves up the plate, it carries along the components within the mixture to be separated at varying rates. The desired result is that each component of the deposited mixture is moved a different distance up the plate by the solvent, with the components appearing as a series of spots at different locations along the plate. The differences in migration allow for the calculation of a substance's Rf value, a unique ratio of the distance that the component of a mixture travels to the distance that the solvent front travels up the plate. For example, an Rf value of 0.5 indicates that the spot corresponding to this particular component travels exactly half as far as the solvent travels along the plate. This value will aid in your task of identifying the amino acids contained in an unknown sample.

Performing a TLC analysis consists of a number of steps, each of which will be briefly introduced here.

1) **Preparing the TLC Plate** - After obtaining a TLC plate of the correct size for the analysis, the plate is generally marked with a pencil (pencil must be used rather than pen because inks are moved by many developing solvents) by drawing a light straight line parallel to and about 1-cm from the end of the plate. Be very careful not to press too hard since that would create a groove in the plate that would not allow the mobile phase to move through it. The second step is to make light pencil marks in the line you just drew to set the spacing for the spots of unknown. These marks will subdivide the plate, serve as guides for placing the substance spots, and most importantly act as a point from which to measure Rf values. The last step is to draw another light line, parallel to the first, on the opposite end of the plate so that it is easy to discern where the solvent front should rise to during developing.

2) **Spotting the TLC Plate -** a capillary is used to spot the plate. The capillary is dipped into a container of the solution mixture and capillary action will cause a small amount of sample to be drawn into the tube. To release the solution and spot the plate, hold the capillary over one of the lines made when preparing the plate and touch it gently to the plate. Make sure that the spot remains small <1mm. Allow the spot to dry completely. You can also repeat the spotting if you need to increase the concentration of the unknown.

3) **Running the TLC Plate** – Construct a TLC chamber out of a tall beaker and either saran wrap or paraffin. The volume of mobile phase should be such that it will not rise above the 1cm line drawn on the TLC plate. Once the mobile phase is added to the beaker, cover with saran wrap or paraffin to allow the atmosphere in the chamber to saturate. Quickly but carefully add the TLC plate to the chamber, being sure to have the plate as vertical as possible and handling it strictly with gloves or tweezers. The plate is then allowed to run for either a specified time or to a specific distance on the plate. When the time or distance has elapsed, remove the TLC plate from the chamber and immediately mark the solvent line with pencil.

4) **Drying the TLC Plate** - The drying step of this type of analysis is accomplished by placing the TLC plate on a clean dry surface so that the solvent may completely evaporate. In some cases where the solvent is not highly volatile, this step can be facilitated by placing the plate on a flat surface in an oven set at a temperature of 50 °C to 60 °C (NOTE: When working with a plastic substrate material, higher temperatures may melt the plate.) Once the plate is completely dry, it is ready for the visualization step.

5) **Visualizing the TLC Plate** - The ninhydrin technique will be used in this experiment to visualize the amino acid spots. In this method, a solution containing 0.2% ninhydrin in ethanol is sprayed on the dry plate. Once in contact with an amino acid, ninhydrin displays a purple coloration that is easily visible once the plate has been sprayed and allowed to sit for several minutes. Once the spots have developed they should be outlined in pencil.

6) **Analyzing the TLC Plate** - To calculate each component's unique Rf value, the first measurement that must be obtained is the distance the solvent advanced, and is simply recorded as measuring the distance between the two pencil lines. Next, the distance the components traveled is required and is ascertained by measuring from the bottom pencil line to the center of the spot corresponding to a particular component. With both values obtained, the ratio of the distance traveled by a component to the distance traveled by the solvent is then resolved. For example, the figure below shows a developed plate for a mixture consisting of 4 components with Rf values of about 0.05, 0.20, 0.50, and 0.90.

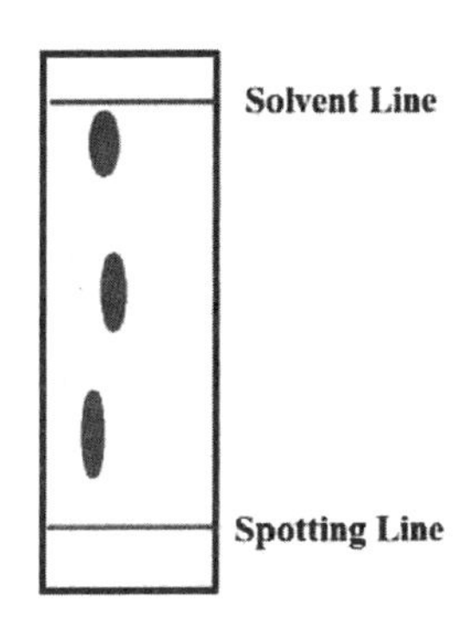

In this experiment you will prepare and develop TLC plates spotted with known amino acids and a sample composed of unknown amino acids. After successfully developing your plate, visualizing it with ninhydrin, and determining the Rf values of each spot, you should be able to identify which amino acids were in the unknown sample.

PROCEDURE

Safety Notes: Always wear gloves when working with the organic solvents and ninhydrin. The mobile phase may cause irritation to those with respiratory issues. Notify your instructor if you suffer from asthma or similar so that you can work at the hood.

Part I: Preparing the TLC Plate

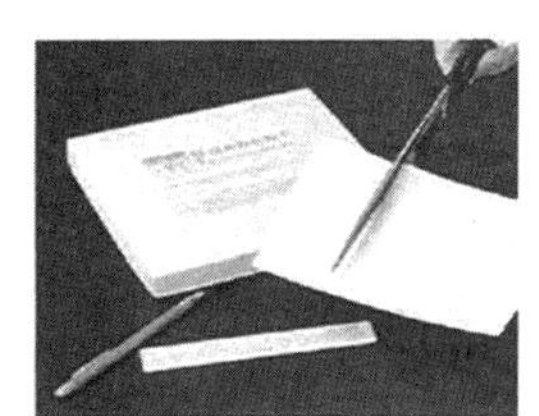

Collect 2 samples of the unknown amino acid mixture, samples of amino acid standard solutions (small volumes of aqueous solutions are provided), a thin-layer chromatography plate, latex gloves, a pencil, and micropipettes.

NOTE: *Always handle your TLC plate with gloved hands, because the staining technique we will use is so sensitive that fingerprints will show!*

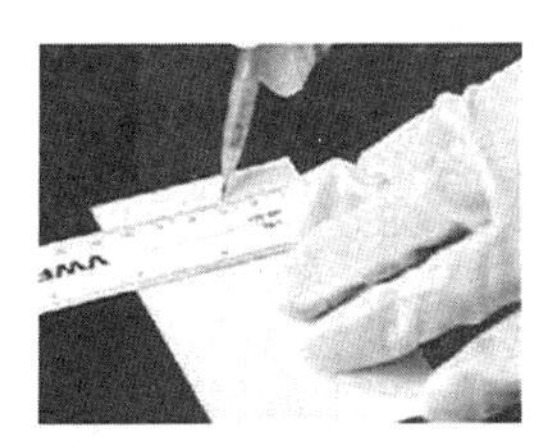

Mark your TLC plate with your pencil, being careful to make a light line and not to remove any of the silica coating that makes the thin layer. Mark the positions of the spots of the standard and the unknowns that you will make with your pencil, too.

Carefully apply samples of the standard amino acid solutions and your unknown mixture to the origin of your TLC plate. Be sure to make note of what material has been applied to which place on the plate. Also be sure to record the volumes of the solutions that you apply to the plate, because when small amounts of amino acids are present, the staining intensity of a spot is roughly proportional to the amount of amino acid that is present. This will allow you to make a semi-quantitative estimate of the amounts of amino acids present in your unknown mixture.

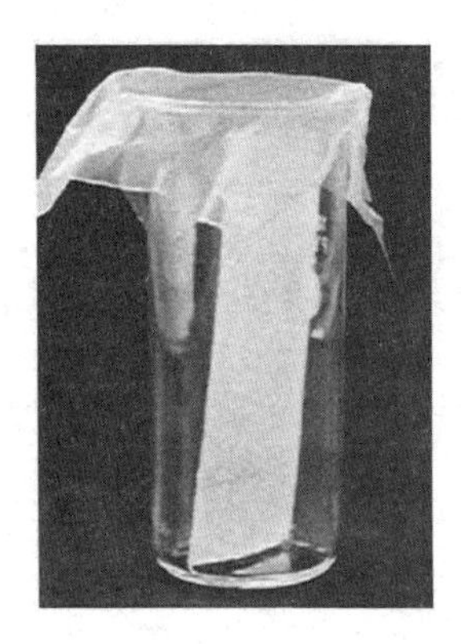

Prepare the chromatography chamber in the fume hood by pouring ~200ml of the mobile phase into it. After you have applied your samples and the spots are completely dry, place your plate into the chromatography chamber. With the TLC plate leaning against a side of the chamber (2 plates may be put in one chamber) cover the chamber with parafilm so that evaporation is minimal. The mobile phase we will use is a mixture of 1-propanol:H2O(1:1 v/v/).

Allow the solvent to move up the plate about 10-cm before removing it from the chamber. Previous experiments have shown that this will take between 1 to 1.5 hours to complete.

Part II: Analyzing the TLC Plate

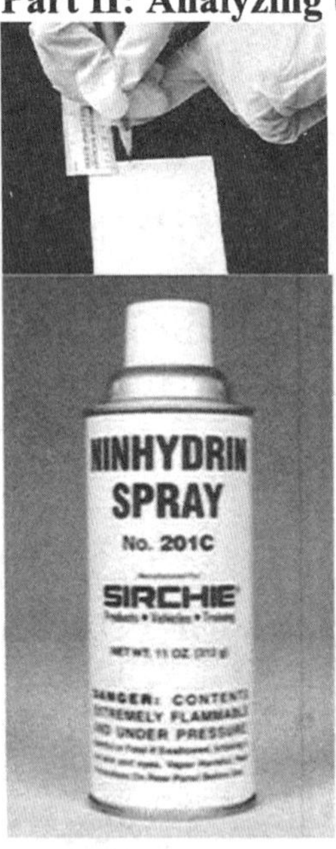

After removing the plate, indicate with pencil marks the position of the solvent front.

Place your plate in the fume hood to dry.

When the solvent has evaporated, spray the plate with an even, light mist of the ninhydrin reagent.

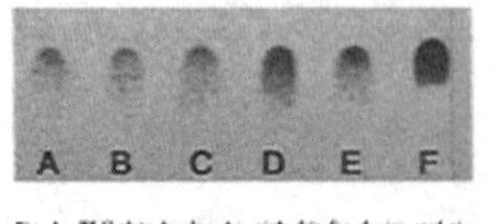

Fig. 1 - TLC plate developed in ninhydrin for glycine analysis. A, B, and C are glycine samples synthesized from 50, 100, and 150 ml ammonia, respectively. Sample D refers to ^{15}N-enriched glycine. E and F correspond to the glycine p.a. standard.

Heat the plate for about 10 minutes at ~110 °C (or with a heat gun). The amino acids will be visualized as bluish or purple spots (for primary amino acids) or yellow spots (for secondary or imino acids like proline). These spots may fade with time and exposure to light, so you may want to make a tracing of the plate for your records.

You should also note carefully any slight differences in color of the spots and their relative staining intensity.

Experiment 23

Pre-Laboratory Assignment

Name: ____________________________ **Date:** __________________

Instructor: _____________________________ **Sec. #:** ____________

Show all work for full credit.

1) How would you expect the amino acids glycine, alanine, and phenylalanine to travel relative to each other on the silica plates using a mixture of chloroform/methanol/ammonia? Explain your answer.

2) You spot an unknown sample on a TLC plate and elute it with chloroform. After visualizing the plate, you detect only one spot with an Rf of 0.97. Does this indicate that the unknown material is a pure compound? What could be done to verify the purity?

3) You spot a sample of an unknown pure compound on a TLC plate and elute it with hexane. Its Rf is 0.25. Would you expect its Rf to increase, decrease, or remain the same if you elute with acetone? Why?

Experiment 23

Laboratory Report

Name: ______________________________ **Date:** ___________________

Instructor: _______________________________ **Sec. #:** ____________

PURPOSE: *(The purpose should be several well-constructed sentences describing what your experiment was designed to accomplish and the criteria used to determine success. These sentences should include both concepts and techniques.)*

PROCEDURE: *(The procedure section should reference the lab manual and include any changes made to the procedure during the lab.)*

DATA: *(The data section should include your own personal data and observations.)*

Observations:

Analysis of TLC Plate

Lane	Sample Identity	Rf 1	Rf 2	Rf 3	Rf 4
1					
2					
3					
4					
5					
6					
7					
8					
9					
10					

Input a picture of your TLC plate here.

CALCULATIONS: *(The calculation section should include sample calculations for how an Rf value was determined. Be sure to include conversions of units.)*

CONCLUSION: *(The conclusion section should include several paragraphs with the following: A summary statement regarding the identity of the amino acids in your unknowns. A discussion of how your TLC analysis supports your conclusions. Any possible errors should be discussed as usual.)*

Experiment 23

Post-Laboratory Assignment

Name: ______________________________ **Date:** ________________

Instructor: ______________________________ **Sec. #:** ___________

Show all work for full credit.

Answer the following questions:

1) Why is it important that the spots you made on the TLC plate not be more than 1mm in diameter? What might have happened if the spots had been larger?

2) We saw in the mechanism for the ninhydrin reaction that the molecule reacts with amino acids via their amine group. What other molecules might give a "false positive" reaction with ninhydrin?

Appendix A
Atomic and Molecular Structure Review

In the atom's first philosophy both the structure of the atom and the molecules built from those atoms is introduced at the beginning of the term. A laboratory to compliment this approach assumes a level of knowledge of structure that students may or may not have achieved. This appendix summarizes the main points of both atomic and molecular structure to act as a reference for those that have covered the material and as a brief introduction for those that have not.

A1 The Atom

Basic Structure = Nucleus (Protons + Neutrons) surrounded by Electrons

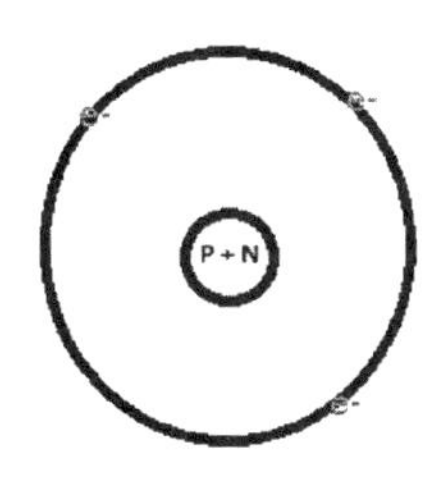

Component of the Atom	Mass	Charge
Proton	$1.6726216 \times 10^{-27}$ kilograms	Positive
Neutron	1.6749286×10^{-27} kilograms	Neutral
Electron	9.1093897×10^{-31} kilograms	Negative

The **Atomic Number** for an element is equal to the number of protons in its nucleus and therefore the element is defined by this number. If you change the number of protons you change the identity of the element.

The **Atomic Mass Number** of an element is given by the total of its protons and its neutrons. The number of neutrons in an element can change and therefore the Atomic Mass Number for an element can change. Each of these different **Atomic Mass Numbers** for an element are called **Isotopes**.

The **Charge** on an element is calculated as the total of its protons and its electrons. A charged element is called an **Ion**. **Ions** can be positive (more protons than electrons) or negative (more electrons than protons). A positive ion is called a **Cation**. A negative ion is called an **Anion.**

A2 Electronic Structure of the Atom

The electronic structure of an atom is described by **Quantum Theory** which maps the location of the electrons around the nucleus.

Quantum Theory uses four quantum numbers, n, ℓ, m_ℓ, and m_s to set coordinates for each electron found in an atom. No two electrons in the same atom can have the same set of four quantum numbers. This is the **Pauli Exclusion Principle.**

The four quantum numbers designed to describe the location of an electron in an atom are:

n = The Principle Quantum Number

- can be any whole number from 1–∞
- measures the average distance of the e^- from the nucleus
- different values of n mean different energy levels
- different values of n mean relatively large differences in the energies of the e^-s
- the smallest average distance and the lowest energy occurs when n = 1; each increase in n increases those quantities.
- the number of e^- possible in a level is $2n^2$

ℓ = Azimuthal Quantum Number

- ℓ goes from 0 to n-1 within an energy level
- ℓ values = 0 (for s), 1 (for p), 2 (for d), 3 (for f) sublevels
- indicates the shape of the orbital (set of probable locations of the e^-)
- different values of ℓ mean different sublevels. In a sublevel all the e^- have nearly the same energy.
- different sublevels within the same level may have moderately large differences in energy
- within any level, the lowest energy sublevel is s, then p, then d, then f
- the number of possible values of ℓ for a level is equal to the value of n

m_ℓ = Magnetic Quantum Number

- values of m_l go from + ℓ to - ℓ, which gives 2 ℓ + 1 number of values
- identifies the direction the e^- orbital has around the nucleus
- specifies the e^- orbital in which the e^- is located within a sublevel
- different values of m_ℓ mean little difference in energies of the e^-
- the number of possible values of m_ℓ within a sublevel identifies how many e^- pairs that the sublevel can hold

m_s = Spin Quantum Number

- has 2 values: +1/2 (spin up) and -1/2(spin down)
- identifies the "spin" or rotation of the e^- about its own axis
- shows that each orbital can contain only 2 e^-
- the direction of spin is either in one direction or the other
- when 2 e^- (in an atom) have the same set of quantum numbers except for m_s, then these e^- are called an e^- pair
- these e^- within an e^- pair have essentially the same energy

Orbitals

There are four orbital types that are used in general chemistry to describe the electronic structure of the atom. These are s, p, d and f orbitals. Each has a distinct shape and can hold 2 electrons. The table below summarizes the number of these orbitals for each atomic shell (n value):

Orbitals and Electron Capacity of the First Four Principle Energy Levels				
Principle energy level (n)	**Type of sublevel**	**Number of orbitals per type**	**Number of orbitals per level (n^2)**	**Maximum number of electrons ($2n^2$)**
1	**s**	**1**	**1**	**2**
2	**s**	**1**	**4**	**8**
	p	**3**		
3	**s**	**1**	**9**	**18**
	p	**3**		
	d	**5**		
4	**s**	**1**	**16**	**32**
	p	**3**		
	d	**5**		
	f	**7**		

Electron Configurations

Electrons are added to the shells and subshells surrounding the nucleus by increasing energy. The lowest energy levels fill first. The order of fill is therefore not always numerically in shell order since as the atoms get larger the subshells start to overlap in energies.

The order in which electrons should be added to an atom is given by the handy tool below:

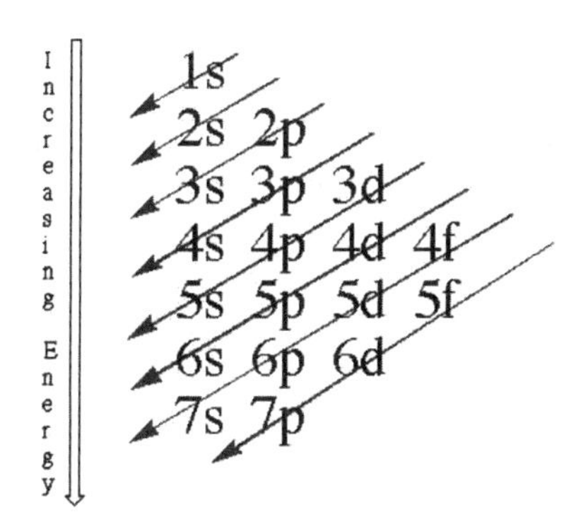

The order of fill follows down the arrows.

$1s^2 2s^2 2p^6 3s^2 3p^6 4s^2 3d^{10} 4p^6 5s^2 4d^{10} 5p^6 6s^2 4f^{14} 5d^{10} 6p^6 7s^2 5f^{14} 6d^{10} 7p^6$

s orbitals contain a maximum of 2 electrons
p orbitals contain a maximum of 6 electrons
d orbitals contain a maximum of 10 electrons
f orbitals contain a maximum of 14 electrons

A3 Bonding and Lewis Structures

There are two main types of bonds—**Ionic** and **Covalent**. **Ionic Bonds** form between two ions of opposite charge. The ratio of positive to negative ions is determined by the number of each required to form a neutral molecule. Covalent bonds form between uncharged elements. The bonds form based on the sharing of electrons that provides an **Octet** of electrons for each.

The Octet Rule

Bonding and formation of ions is promoted by each atom's desire to reach an energetically stable form. The more energetically stable an atom, the less reactive it is. **Noble gases** (Group 8A), also referred to as *inert gases,* are the most energetically stable, least-reactive atoms in the periodic table. The electron configurations of the noble gases are marked by the completion of an outer shell.

Helium	He	$1s^2$	Full 1s orbital
Neon	Ne	$[He]\mathbf{2s^2 2p^6}$	Full 2p orbitals
Argon	Ar	$[Ne]\mathbf{3s^2 3p^6}$	Full 3p orbitals
Krypton	Kr	$[Ar]3d^{10}\mathbf{4s^2 4p^6}$	Full 4s & p orbitals
Xenon	Xe	$[Kr]4d^{10}\mathbf{5s^2 5p^6}$	Full 5s & p orbitals
Radon	Rn	$[Xe]4f^{14}5d^{10}\mathbf{6s^2 6p^6}$	Full 6s & p orbitals

It should also be noticed that the completion of the outer shell has a total of 8 electrons (bold). This is the origin of the **Octet Rule.** If all atoms wish to be energetically stable and the most stable atoms have 8 electrons in their outermost shell, then all atoms should strive to have 8 electrons in their outermost shell as well.
Elements other than the Noble gases can accomplish the Octet Rule by either forming ions or by sharing electrons with other atoms.

Valence Electrons

Valence electrons are the electrons found in the highest energy shell of an atom, i.e., highest principal quantum number, n, value. **Core electrons** are those corresponding to the closest previous Noble gas configuration.

The number of valence electrons can be determined from the group number of the element for the main group elements.

Group 1A = 1 valence electron	Form +1 ions
Group 2A = 2 valence electrons	Form +2 ions
Group 3A = 3 valence electrons	Form +3 ions
Group 4A = 4 valence electrons	Do not generally form ions
Group 5A = 5 valence electrons	Form -3 ions
Group 6A = 6 valence electrons	Form -2 ions
Group 7A = 7 valence electrons	Form -1 ions
Group 8A = 8 valence electrons	Do not generally form ions (already have an octet)

Gain or loss of valence electrons creates ions. Ions form in order to satisfy the Octet Rule by gaining or losing electrons in the easiest manner possible to create an electron configuration similar to that of the nearest noble gas.

Group 4A elements do not generally form ions because the energy required to add 4 or lose 4 electrons is too great.

Lewis Structures

For those elements such as the ones in Group 4A, the only method by which they can satify the Octet Rule is to form covalent bonds and share the electrons they need. Lewis structures are constructed in order to satisfy the octet rule for each of the atoms in a molecule.

Bonds are represented by "–" and lone pairs of electrons are represented by ":".

There are simple steps to creating a correct Lewis structure:

1. Calculate the total number of valence electrons available.
2. Determine which atom will be central in the molecule.
3. Arrange atoms symmetrically around the central atom.
4. Place bonds/electrons around the atoms until the octet rule is satisfied for each atom. Use double or triple bonds if necessary.
5. Show any charges on the molecule using brackets [ ] and place the charge in the upper right hand corner just outside the brackets

Step1: Calculate the total number of valence electrons available.

Let's use CH_4 as our example.

We need to know how many electrons are available to make the bonds for methane.

Carbon is in group IVA so it has 4 valence electrons and hydrogen is in group IA so each hydrogen has 1 valence electron. Total valence electrons = 4 + 4(1) = 8.

This means we have 8 electrons to distribute into bonds to create methane.

Step 2: Determine which atom will be central in the molecule.

- The central atom in a molecule is usually the least electronegative atom.
- This is often the atom which will allow you to create the most symmetrical molecule.

For methane, CH_4, we know that hydrogen can only make 1 bond so none of the hydrogen atoms can be the central atom. By deduction we know that carbon is the central atom.

Step 3: Arrange atoms symmetrically around the central atom.

Most of nature's creations are symmetrical and the same holds true for most chemical compounds.

When writing Lewis structures, the most symmetrical arrangement of atoms around the central atom is best.

For methane:

```
  H
H C H
  H
```

The four hydrogen atoms are placed symmetrically around the carbon.

Step 4: Place bonds/electrons around the atoms until the octet rule is satisfied for each atom. Use double or triple bonds if necessary.

```
   H
   |
H—C—H
   |
   H
```

For methane:

Step 5: Show any charges on the molecule using brackets [ ] and place the charge in the upper right hand corner just outside the brackets.

This is inapplicable here as methane is uncharged.

A4 Hybridization and VSEPR

Hybrid Orbitals:

What is a hybrid? Well, when you combine two things into one that is a hybrid. Scientists hybridize plants all the time to give them better taste, more resilience to disease, etc. When we talk about hybrid orbitals we are visualizing what we believe must occur within a molecule's bonding structure to result in the molecular structures we can see.

For example: Carbon has an electron configuration of $1s^22s^22p^2$. There are four valence electrons in carbon's outermost shell that can bond: 2 s orbital electrons and 2 p orbital electrons. Now, remembering back to the atomic theory, we know that s orbitals are of lower energy than p orbitals and thus that should mean that when they bond to other atoms, the p orbital electrons would form stronger (higher energy) bonds than the s orbital electrons. So in a molecule of CH_4 you should see two long bonds between the s-s orbital overlaps, and two shorter bonds between the p-s orbital overlaps. So the structure would look like this:

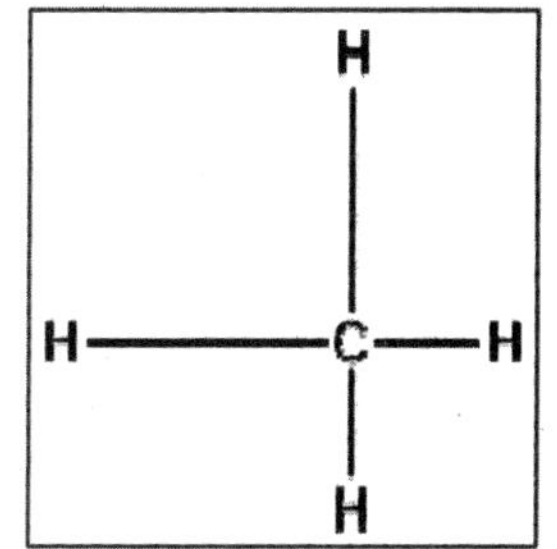

But we know this is not what methane (CH_4) actually looks like. All the bond lengths and strengths in methane are roughly the same. So even though the bonds are made up of different energy orbitals they make all the same type of bonds, how can this be? Well, the way we explain it is hybridization.

We take the two higher energy p orbital electrons and the two lower energy s orbital electrons and meld them into four equal energy sp3 (1s + 3 p orbitals = sp3) hybrid orbitals. When these sp3 hybrid orbitals overlap with the s orbitals of the hydrogens in methane, you get four identical bonds, which is what we see in nature.

There are two ways to form sp^2 hybrid orbitals that result in two types of bonding.

Hybridization of an element with three valence electrons in its outer shell, like boron will yield three full sp^2 hybrid orbitals and no left over electrons.

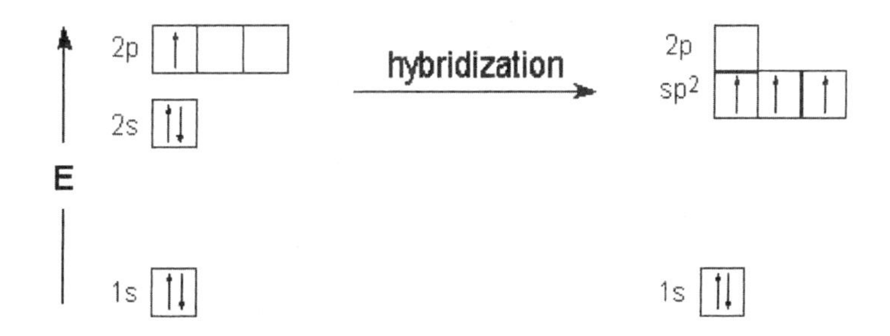

If the atom has more than three valence electrons in its outer shell three of the electron orbitals hybridize and one of the p orbitals remains unhybridized:

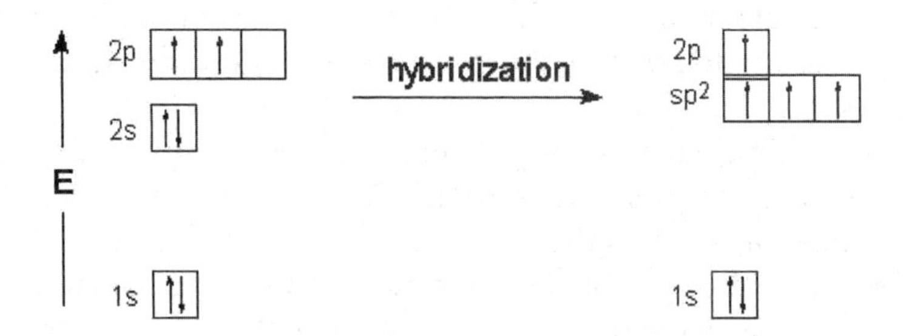

It is the unhybridized p orbitals that then form pi bonds for double bonding:

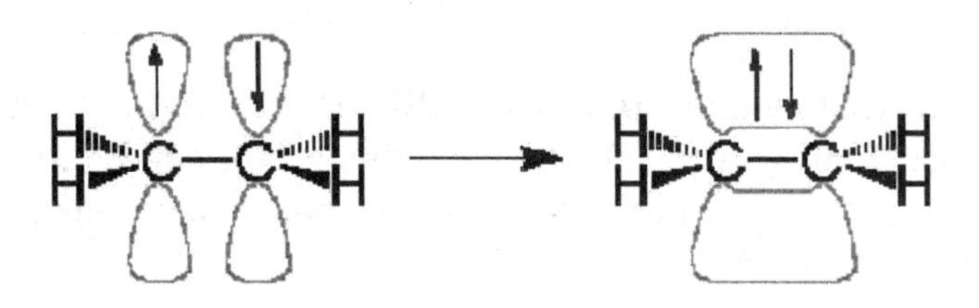

The table below summarizes the relationship between valence bond theory (hybridization) and electron pair geometry. Both of these designations can be assigned simply by counting the number of groups (bonds or lone pairs) attached to a central atom.

Number of Groups Attached to a Central Atom	Description and 3-Dimensional Shape
Two Groups...sp	2 groups = sp hybridization 180 degree bond angle linear electron-pair geometry
Three Groups...sp^2	3 groups = sp^2 hybridization 120 degree bond angles trigonal planar electron-pair geometry
Four Groups...sp^3	4 groups = sp^3 hybridization 109.5 degree bond angles tetrahedral electron-pair geometry
Five Groups...sp^3d	5 groups = sp^3d hybridization 120 and 90 degree bond angles trigonal bipyramidal electron-pair geometry
Six Groups...sp^3d^2	6 groups = sp^3d hybridization 90 degree bond angles octahedral electron-pair geometry

VSEPR

Valence Shell Electron Pair Repulsion is the theory of molecular structure that assumes the atoms of a molecule will spread out around a central atom and conform to a structure that minimizes the repulsion between their electrons.

In molecular structures we see this best when there are lone pair electrons on the molecule. The electron pair geometry which indicates the location of electrons does not change as bonds are removed; only the appearance of the molecular geometry changes. The table on the next page summarizes the changes as bonds are removed for each form of central atom.

Hybridization	Electron Pair Geometry	Molecular Geometry	Bonds to Central Atom (Lone Pairs)
s	Linear	Linear	1
sp	Linear	Linear	2
sp	Linear	Linear	1 (1)
sp^2	Trigonal Planar	Trigonal Planar	3
sp^2	Trigonal Planar	Bent	2 (1)
sp^2	Trigonal Planar	Linear	1 (2)
sp^3	Tetrahedral	Tetrahedral	4
sp^3	Tetrahedral	Trigonal Pyramidal	3 (1)
sp^3	Tetrahedral	Bent	2 (2)
sp^3	Tetrahedral	Linear	1 (3)
sp^3d	Trigonal Bipyramidal	Trigonal Bipyramidal	5
sp^3d	Trigonal Bipyramidal	See-Saw	4 (1)
sp^3d	Trigonal Bipyramidal	T-Shaped	3 (2)
sp^3d	Trigonal Bipyramidal	Linear	2 (3)
sp^3d	Trigonal Bipyramidal	Linear	1 (4)
sp^3d	Octahedral	Octahedral	6
sp^3d	Octahedral	Square Pyramidal	5 (1)
sp^3d	Octahedral	Square Planar	4 (2)
sp^3d	Octahedral	T-Shaped	3 (3)
sp^3d	Octahedral	Linear	2 (4)
sp^3d	Octahedral	Linear	1 (5)

Appendix B
Excel Tutorial

The competent use of a spreadsheet program such as Excel is an important skill for a scientist to master. In addition to simplifying repetitive computations, it allows the user to produce publication quality graphics. In science more so than any other area, a picture can indeed be worth a thousand words, but only if it is accurately created. What follows here is a very brief summary of how to use Excel to calculate some simple algebraic formulas and common statistical values, and to create graphs from typical experimental data.

Getting Started

When you open Excel you will see a spreadsheet with alphabetically labeled columns and numerically labeled rows. This means that each cell in the spreadsheet has a unique alphanumeric label. It is these labels that can be input into formulas:

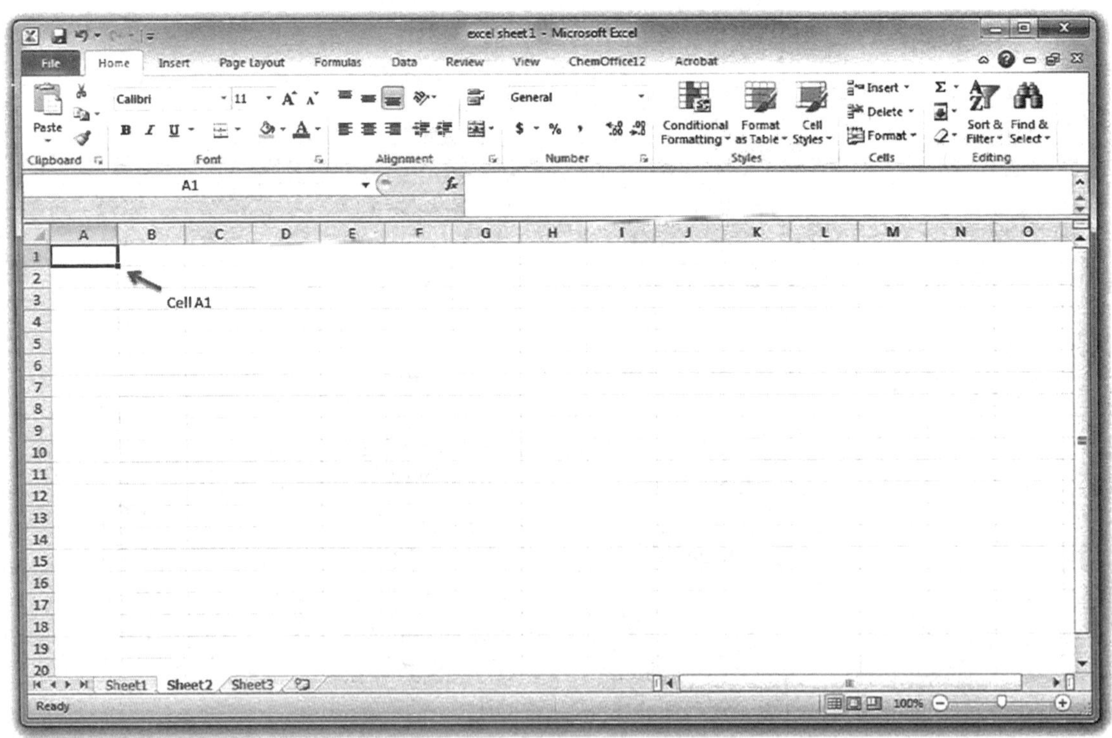

Note: Your screen may vary a bit from the one shown here. Commands not shown on your toolbar can usually be accessed from one of the menus. If you will be using that command frequently, the button can be added to your toolbar.

You can enter data into each cell, in either alphabetic or numeric form, by simply typing whatever you need into each cell. You can move from cell to cell using the Enter (down one cell) or Tab (across one cell) key after you type in your entry. The number of decimal places displayed in the data can be changed by highlighting the data, selecting "Cells" under the format menu, and choosing the "Number" tab. From the new menu shown, select number and you will see a display that allows you to select the number of decimal places shown. This is especially useful for displaying calculation results with the correct number of significant figures.

Basic Math

Once you have your data in the columns as you want it, you need to be able to use simple expressions to manipulate the data. A mathematical equation in Excel is always preceded by an equal sign, "=".

For basic addition, subtraction, multiplication and division of cells the formulae are:

Operation	Example Input
Addition	=A1+B2
Subtraction	=A1-B2
Multiplication	=A1*B2
Division	=A1/B2

Note: Combinations of these operations and more complex expressions can be entered using parentheses between operations.

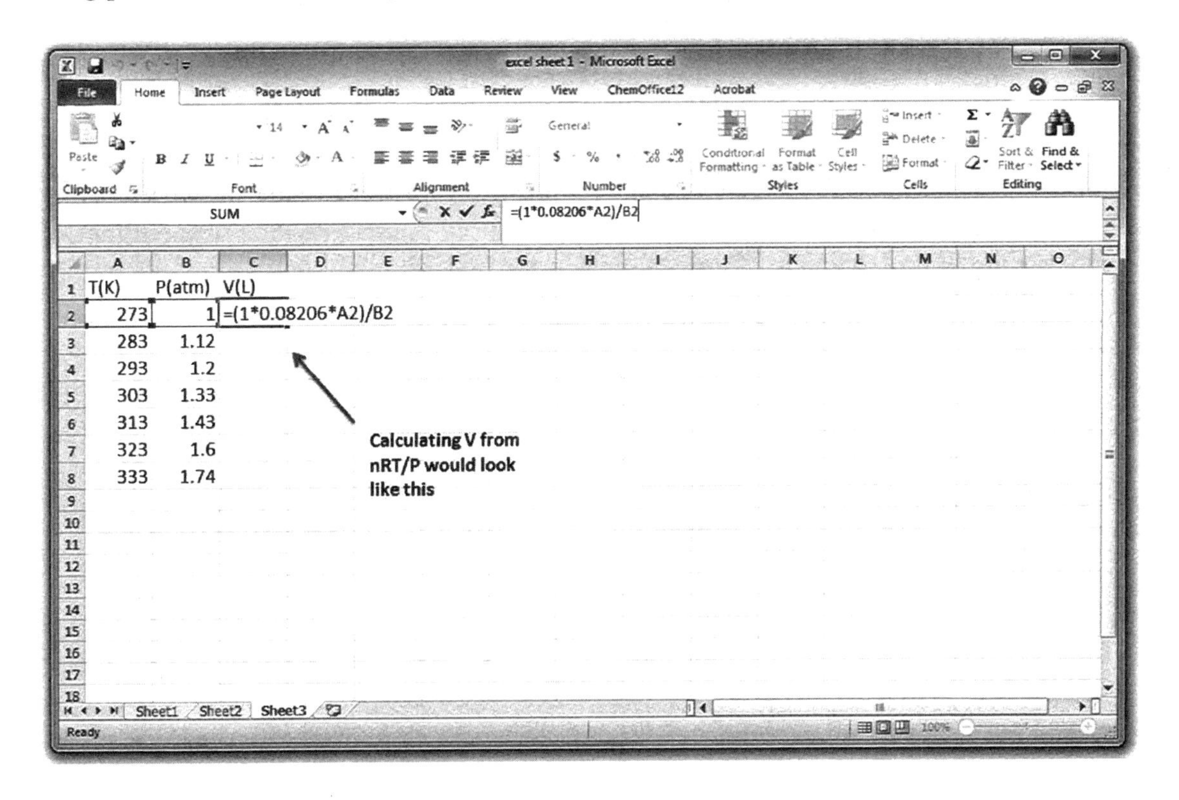

Now if you want to continue with the calculation but don't want to type the equation in several times, you can copy the equation simply by double-clicking your mouse on the small box at the bottom right of the cell you want to copy. You should notice that the expressions remain the same but the cell values change running down the column.

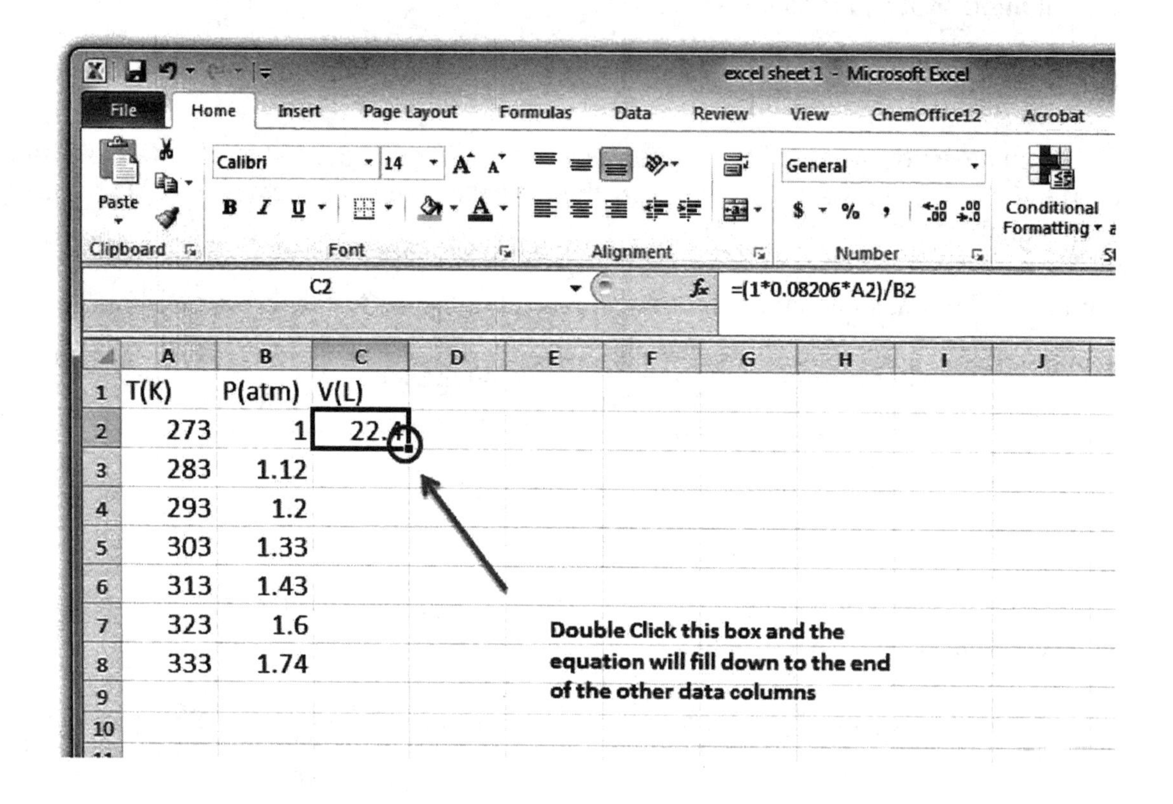

If you want a value to remain constant and not change as you copy values down a column, this can be done by simply placing a $ in front of the number and letter of the cell containing the value you do not want to change. For example, if cell E2 contains a constant in your formula, enter it as E2. Notice that the cell only contains the actual number. The identifying words are in other cells.

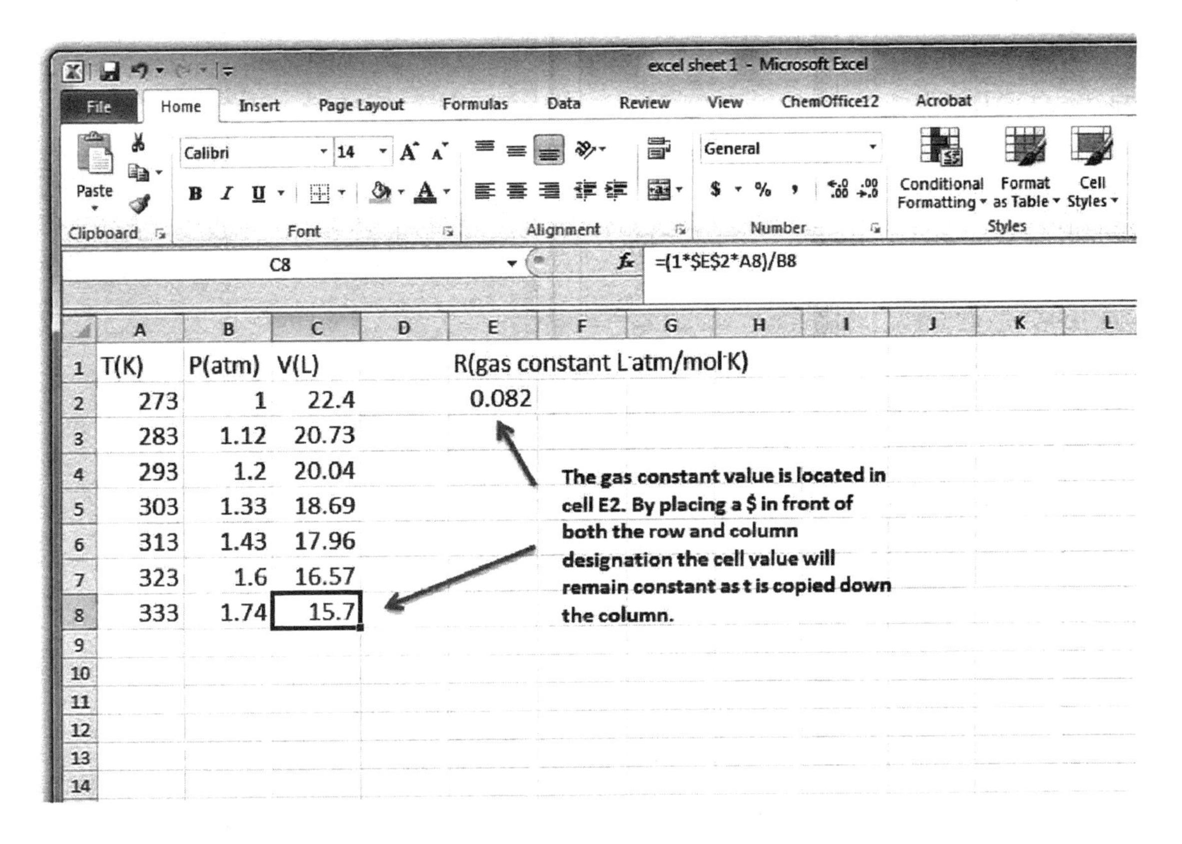

Graphing

In order to create a graph in Excel, first input the data you want to graph with the x values in the column on the left and the y values in the column to its right.

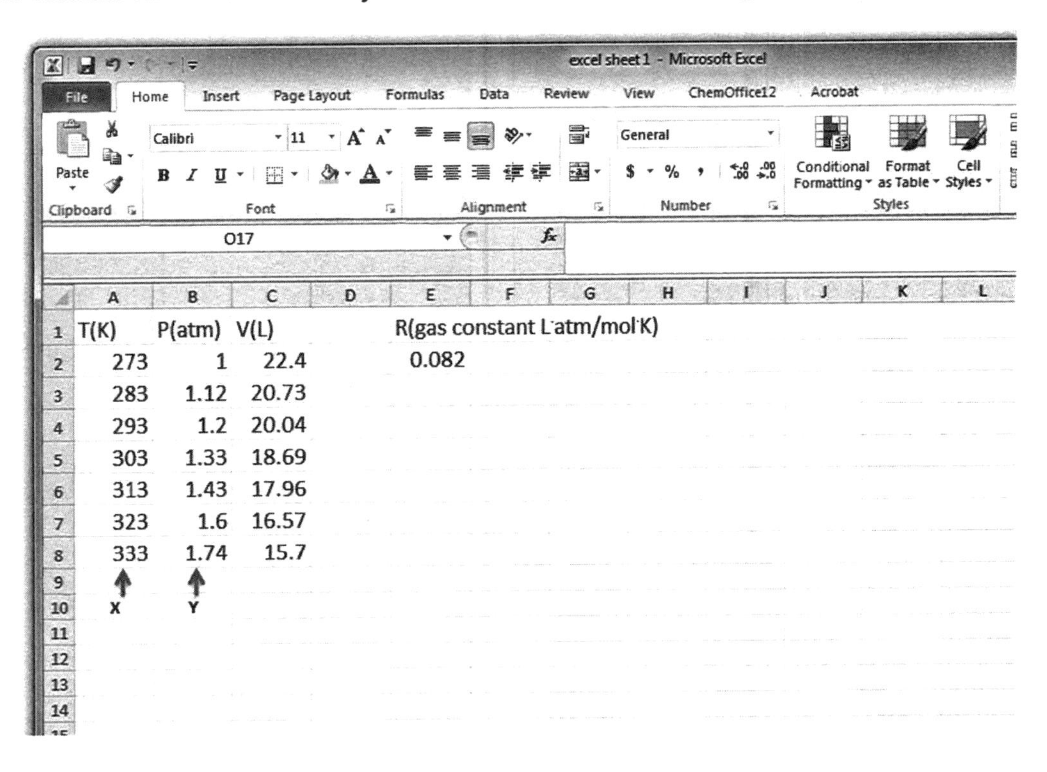

Highlight the columns you wish to graph, with headings, and press the graph button on the toolbar. If the graph button is not on your toolbar, use the Insert menu, select Chart, and continue with the instructions below.

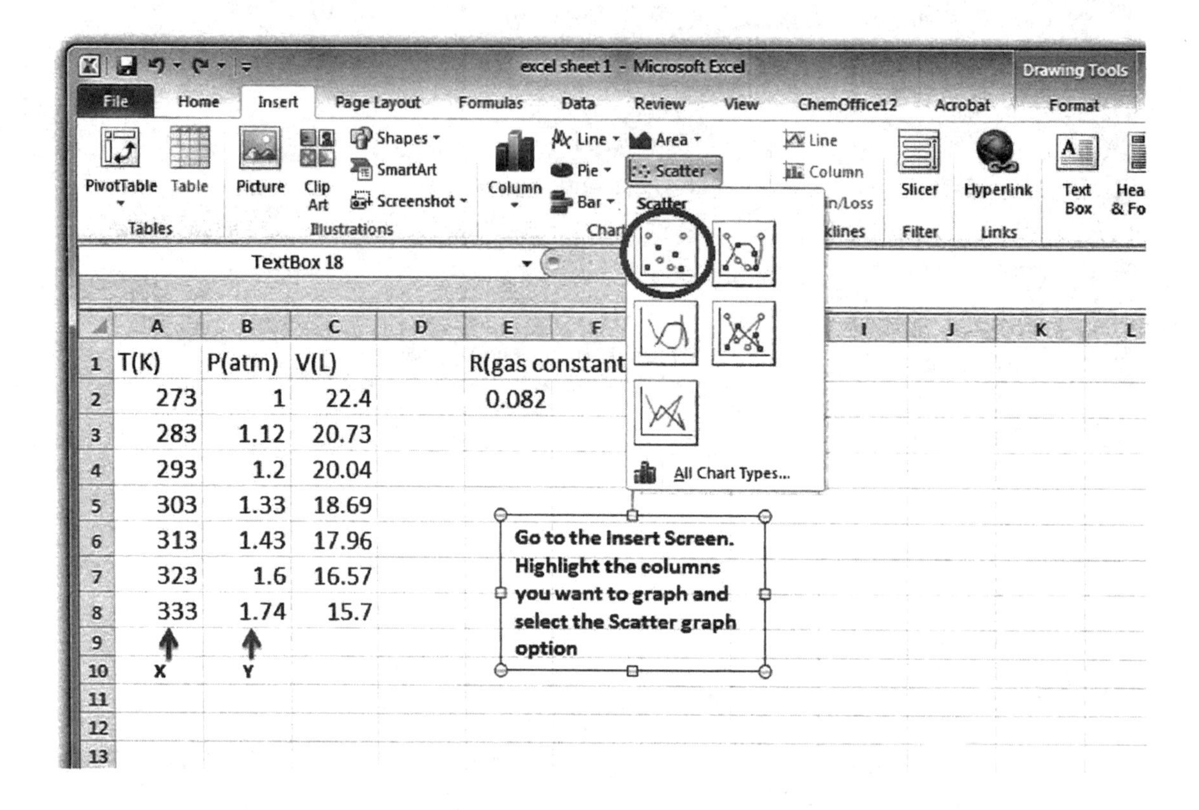

Select the type of graph you want to make. Most graphs for the laboratories in this manual will be XY scatter plots. Do not select the "dot-to-dot" option. If you need to add a line to the graph it should be a "best-fit" trendline.

Once you select the type of graph, the graph will automatically appear in your spreadsheet. It will not, however, have titles or necessarily be a "good" graph as you can see from the picture below. You will need to modify the X and Y axis ranges and Titles, as well as provide a title for the entire graph. To add titles, click on the graph and then select the Layout section of the Chart Tools. You can then input the appropriate titles for each axis (don't forget the units) and the overall chart title.

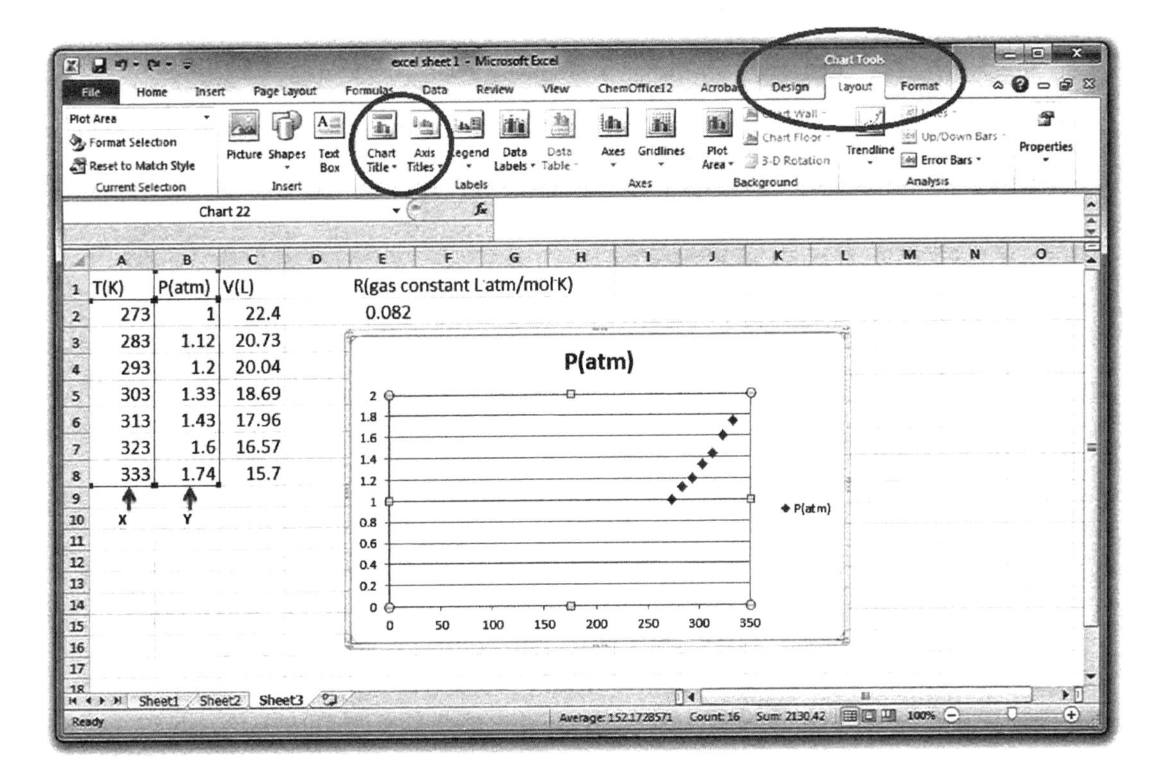

In order to reset the ranges on the axis to more fully center the data on the graph double click the numbers on the axis you wish to adjust. A box will appear that will allow you to alter the minimum and maximum range for the axis. If you plan to use your graph to predict a value of x or y, be sure your scale goes a bit beyond that value.

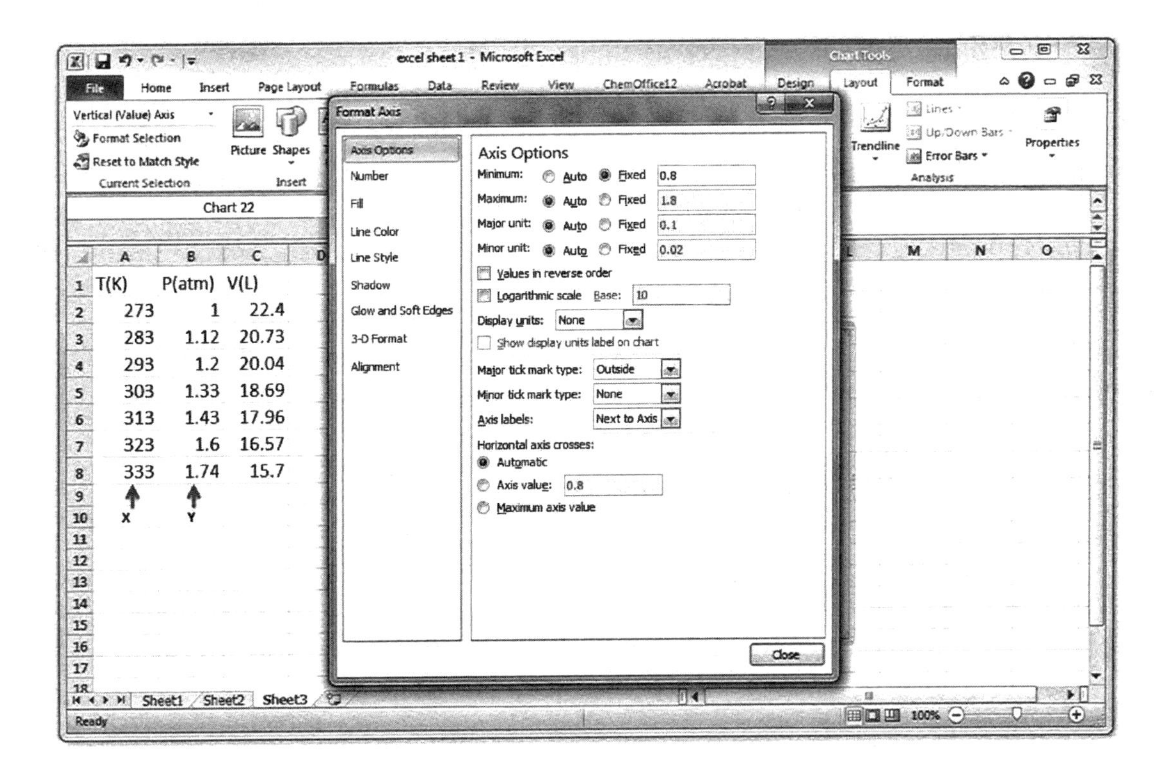

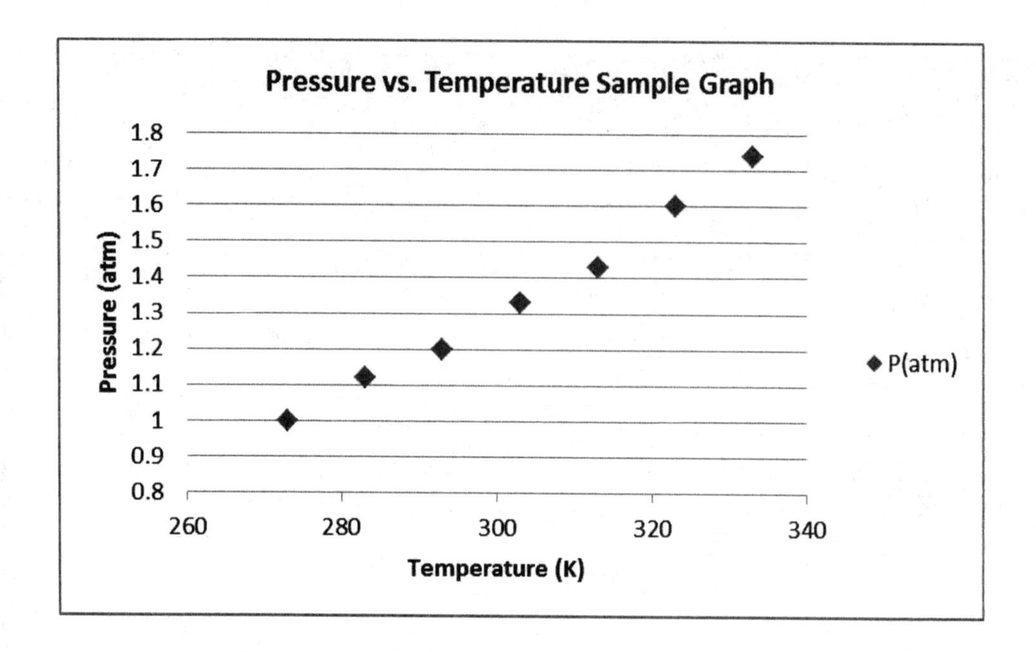

Notice that by adjusting the scale of both the X and Y axis the graph is now much larger and centered, making it easier to analyze.

The first piece of analysis you will want to complete with respect to your graph is the addition of a trend line, a "best fit" line that follows the tendency of the data and can express that tendency in an equation. For a relatively straight line like the one in our example graph, a linear trend line is the best choice.

Excel 2010

The format of the Excel program released in 2010 changed the interface for the graphing protocols. The controls for "Design," "Layout" and "Format" are all found on the main toolbar.

Adding a Trend Line

In order to add a trend line, right-click on one of the data points in the line. A box will appear. Select "Add Trendline," Then select the type of trendline you wish to add.

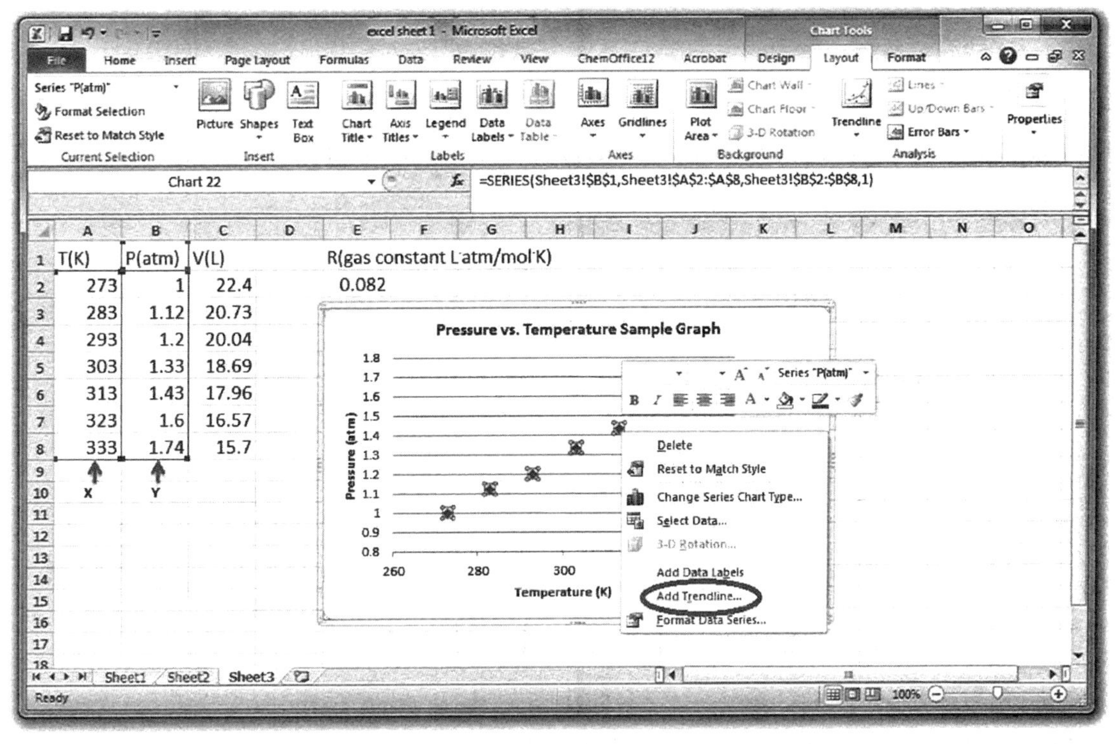

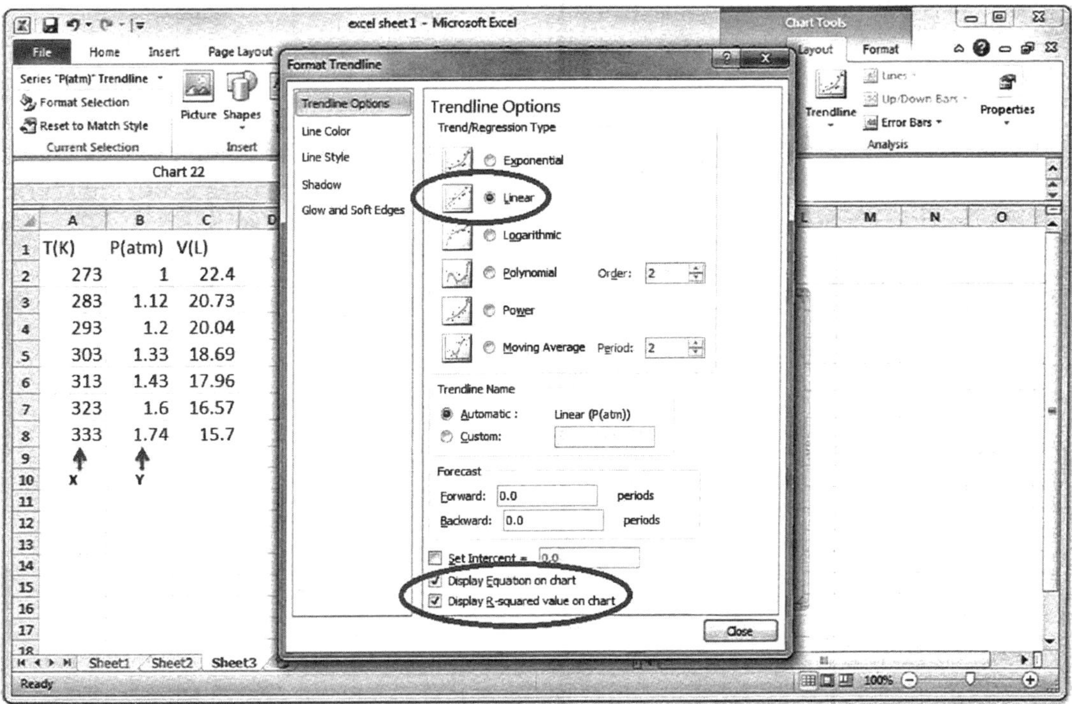

At the bottom of the box you can choose to display the slope and equation of the line to your graph. For some data sets, you will predict values using the forecast option. The R-squared value is a statistical measure of how well your data fits the y = mx + b equation; a value very close to 1 indicates a good fit. This value is often displayed to verify the credibility of the data and the graph.

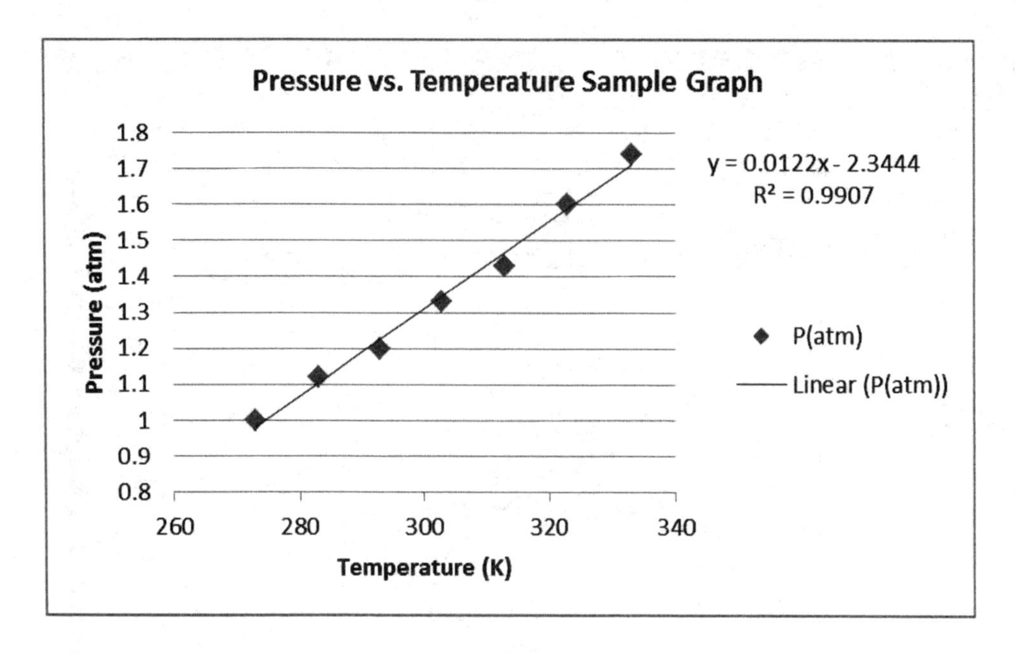

Adding Error Bars

In Excel 2010 the Error Bars functions can be found under the "Chart Tools: Layout" area. Select "Error Bars" and follow the directions below.

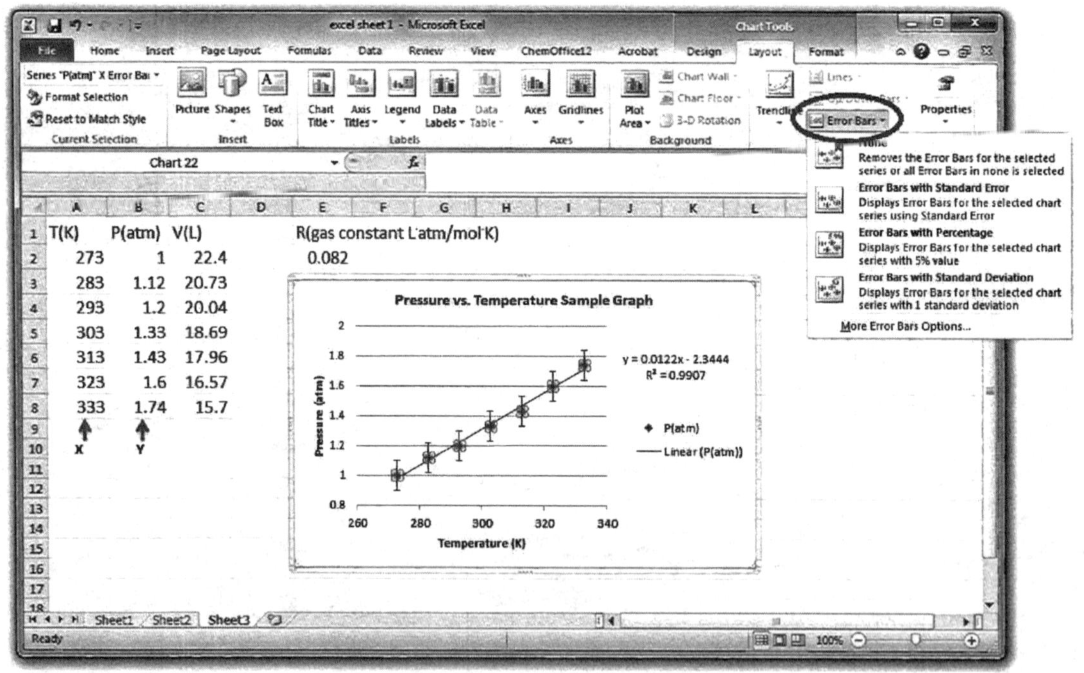

You can then select the type of error bars desired. If you have specific values you have previously determined for the error bars i.e., standard error at 90%, you will need to select "More Error Bars Options…"

Using the spreadsheet shown we select the data for 90% Confidence Interval (cells I6:I11) to display this confidence interval as the error range. Selecting the same range for both the + and – bars will give equal error above and below the data point. (This means there is a 90% probability that the "true" value will lie within the range of these bars.)

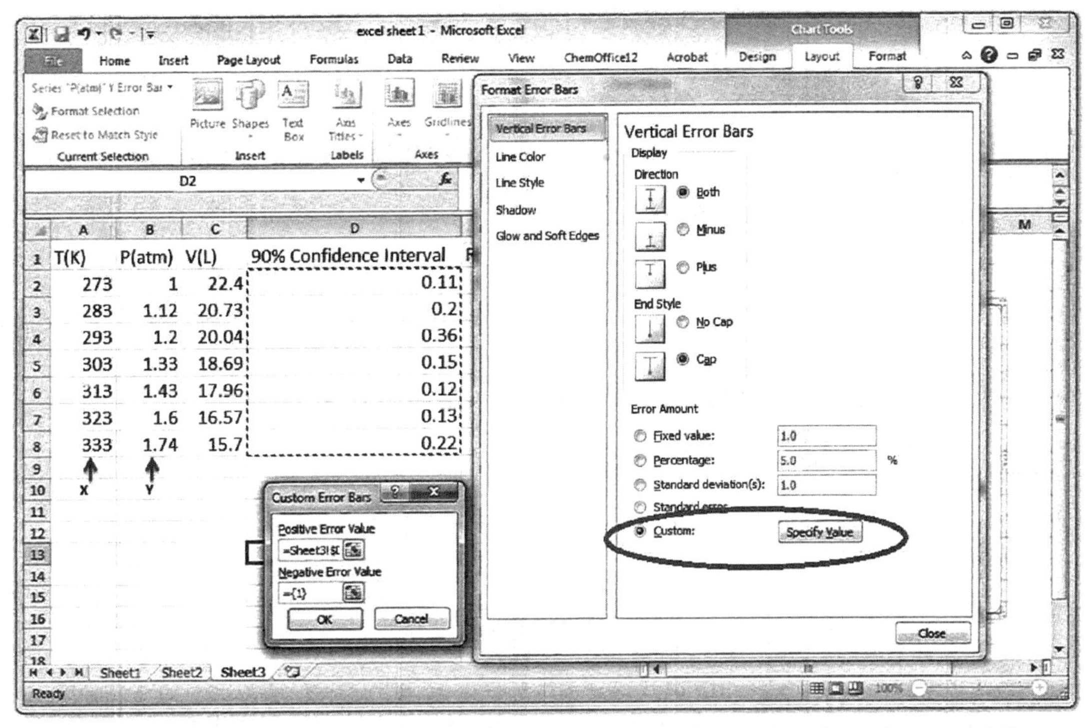

While the box above is open you can also select the style of error bar, its color, etc. If you need to go back to further format the error bars you simply double click on one of the error bars and select "Format Error Bars" to reopen the box shown above.

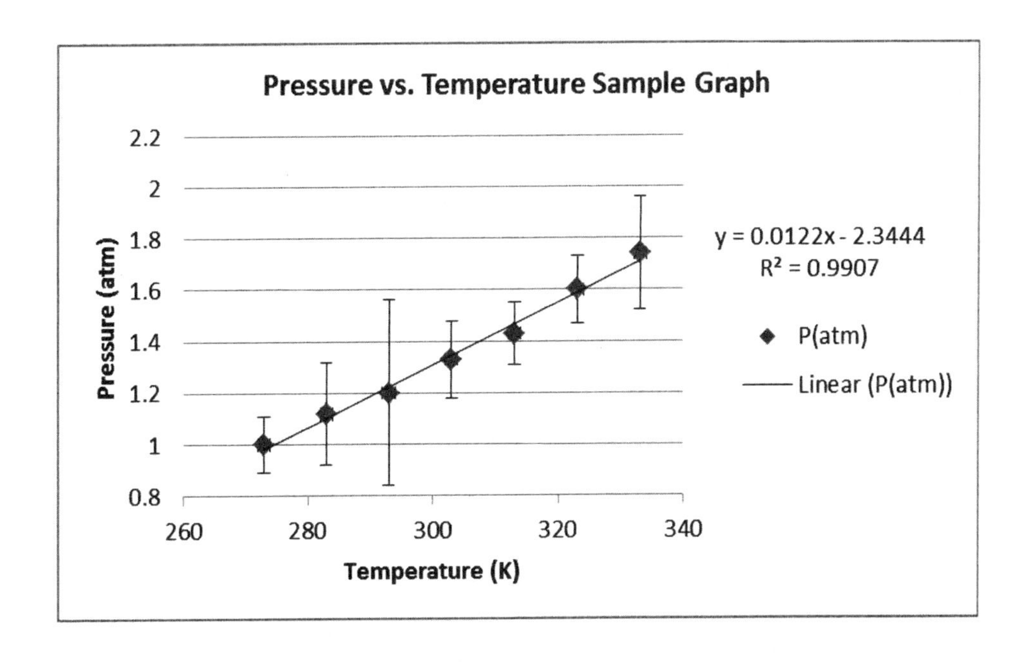

Other Functions

Spreadsheets can be used for many other calculations; a few common ones are listed in the table.

FUNCTION	DESCRIPTION
AVERAGE	Returns the average of its arguments
CONFIDENCE.T	Returns the confidence interval for a population mean, using a Student's t distribution
LINEST	Returns the parameters of a linear trend
SLOPE	Returns the slope of the linear regression line
STDEV	Estimates standard deviation based on a sample
LOG	Returns the logarithm of a number to a specified base
LOG10	Returns the base-10 logarithm of a number
PI	Returns the value of pi
POWER	Returns the result of a number raised to a power
SUM	Adds its arguments

This should get you started. Another thing to keep in mind is that Excel has a good help section. If you need an operation that is not covered in the tutorial above or you need more details than provided here, just search for the topic in the Excel help menu.

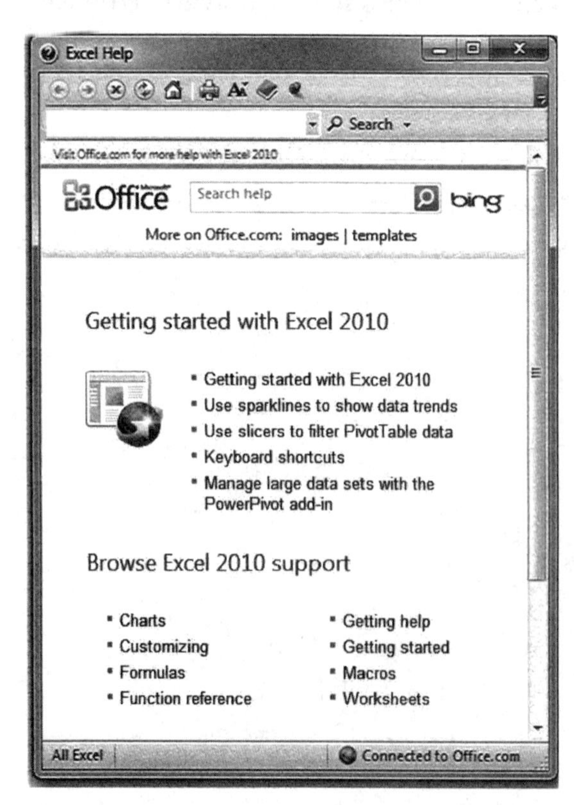

Appendix C

Volumetric Glassware

There are three principal types of volumetric glassware: *pipets*, *burets*, and *volumetric flasks*.

Markings on Volumetric Glassware

Volumetric glassware can be imagined either to contain a certain volume or to deliver a certain volume. Volumetric equipment you encounter may be marked either TC (to contain) or TD (to deliver). Temperature at which the equipment is to be used is also often printed on the equipment. When choosing a piece of equipment to use, note how it is marked.

Cleaning Volumetric Glassware

All glassware should be washed with soap solution and rinsed thrice with tap water. If your local water supply is quite hard, also rinse glassware three times with deionized water from your wash bottle. Always fill your wash bottle with deionized water from the special taps in the lab, NEVER with regular tap water. Also note: DO NOT USE THE EYE WASH AS A SINK.

Ordinary standards of cleanliness are not sufficiently stringent enough for pipets, burets, and volumetric flasks. Solutions should drain in an even film on the inside surfaces of these vessels. Slight traces of an oil coating cause aqueous solutions to break up into drops, rendering volume measurements inaccurate. If solutions do not drain evenly on the inside surfaces of the calibrated glassware, consult the lab instructor. Once clean, immediate and thorough rinsing according to the directions above will generally preserve the cleanliness of a piece of volumetric glassware.

After using NaOH solution in a buret, fill with tap water, run some of the rinse through the tip, then rinse thoroughly with tap, and then deionized water, three times each.

Reading the Meniscus

A liquid confined in a narrow tube exhibits a marked curvature of the surface called the *meniscus*. Common practice is to read the bottom of the meniscus when using volumetric glassware; with opaque solutions such as $KMnO_4$, it is necessary to read the top. It is often advantageous to highlight the meniscus by placing a white card with a heavy dark line upon it behind the container with the top of the dark band a few millimeters below the meniscus. Light reflected from the liquid surface makes the meniscus stand out effectively. The eye of the observer should be horizontal to the meniscus, noting the nearest calibration line which goes completely around the container, the observer can ascertain if the eye is at the correct level.

The Pipet

Solutions are drawn into the pipet with rubber bulbs or dial-up pipetters. (It's worth the effort to keep the inside of the bulb dry; that is, it must not become contaminated.) Standardized technique to fill and drain a pipet is necessary to minimize error due to irreproducible drainage.

First, if not dry, the pipet is rinsed with two or three small portions of sample so the entire inner wall is wetted with solution before pipetting begins. The outside of the tip of the pipet should be kept free of drops of solution, as there is the danger that drops might be transferred to the receiver.

The bulb is then put on the pipet top as loosely as possible. The solution is drawn to a point at least an inch above the calibration line, the rubber bulb or pipetter is withdrawn and the *index* finger (not the thumb) is placed over the opening. The outside of the long delivery tube is wiped off with a Kimwipe® and the solution allowed to drain down to the mark by varying the angle of the index finger. Both the finger and the pipet opening should be dry for best results.

The solution is allowed to flow freely but without spattering into the receiver. Wait five seconds after flow has ceased for the pipet to drain, then touch the tip of the pipet to the moist side of the receiver. A small residue will remain in the pipet. Do not blow this out. Follow this procedure rigorously and measured volumes with the precision of *one part per thousand* (1 ppt) will be delivered.

The Mohr Pipet

The Mohr pipet is a hybrid between a buret and a pipet. It is a pipet that has the calibrations of a buret, but does not have a stopcock. The same operating and cleaning procedures described for the pipet and buret apply.

The Buret

Burets provide a means for accurately delivering a volume of liquid; they consist of a calibrated tube and stopcock arrangement which allows a flow from the tip to be controlled. Scrupulous cleanliness is mandatory. Recall the buret is to be rinsed with the solution it is to contain and filled all the way to the tip. Also note that the stopcock should be liquid-tight, so that it does not leak! During the titration demonstration by the lab instructor, pay particular attention to the way the instructor checks the buret tip just below the stopcock for air bubbles. After solution has been delivered, note that the buret must be given about 30 seconds to drain before a reading is taken.

Teflon stopcocks are made of very chemically inert materials, but a few rules for care are in order:

1) NEVER USE ABRASIVE MATERIALS to clean either stopcock or barrel. This includes

brushes.

2) A Teflon washer is generally placed adjacent to the end of the barrel so minimal friction is created on turning the stopcock. These washers are small so be careful not to lose it when cleaning the stopcock.

3) Teflon stopcocks can be easily scored around the bore if rotated when solid particles are lodged between plug and barrel. Once scored, the stopcock may leak.

4) If Teflon stopcocks are used with liquids corrosive to glass, such as alkalis, *rinse stopcock thoroughly with water after use*. Do not allow the liquid to evaporate. The concentrated solution remaining will attack the glass surface, and the eventual solids may also mar the Teflon surface if the stopcock is then rotated.

5) When not in use, store in a dust-free space with the stopcock loose in the barrel. Although Teflon is tough and unbreakable, it is softer than glass and has a tendency to conform to the glass surface, including eventual expansion into the ports of the barrel.

The Volumetric Flask

The principal use of the volumetric flask is to prepare solutions of a specified concentration, either in preparation of standard solutions or in dilution of samples of known volumes prior to taking aliquot portions with a pipet. The calibration of the flask is usually to contain a specified volume at a temperature marked on the flask.

After transferring the solute, fill the flask about half full and swirl to mix. Add more solvent and mix again. Bring the liquid level almost to the mark, allow time for drainage down the neck of the flask, then *using a medicine dropper*, make the necessary additions of solvent. Firmly stopper the flask and invert repeatedly to assure uniform mixing. Loosen the stopper to allow solution to drain back into flask before removing stopper completely. The contents *must* be homogeneous to achieve any valid results.

NOTE: Solutions added should be at room temperature for accurate volume calibration. A 60 °C solution suffers a considerable volume loss on cooling to room temperature. Since the flask is often made from soft glass, alkali solutions should not be stored in them for indefinite periods of time.

Preparing a standard solution often requires that a known weight of solid solute be introduced into a volumetric flask. This can be accomplished in two ways:

1) Insert a powder funnel into the neck of the flask and add the solid sample. After transfer, the solid is washed off the funnel into the flask with deionized water and the resulting solution is diluted to the correct volume.

2) Dissolve the sample in a *clean* beaker or flask with a small portion of water and then transfer the solution into the volumetric flask. The beaker is then rinsed three times with small amounts of water and each rinse is added to the volumetric flask. The advantage of using this procedure is that solids that do not easily dissolve in room-temperature water can be heated in the beaker and then transferred to the volumetric flask. Be sure the solution is at room temperature before diluting to the final volume.

Appendix D:
Standard Reduction Potentials (in Volts), 25 °C

Reaction	E^o
$F_2 + 2e^- \rightarrow 2F^-$	+2.87
$Co^{3+} + e^- \rightarrow Co^{2+}$	+1.80
$PbO_2 + 4H^+ + SO_4^{2-} + 2e^- \rightarrow PbSO_4(s) + 2H_2O$	+1.69
$MnO_4^- + 8H^+ + 5e^- \rightarrow Mn^{2+} + 4H_2O$	+1.49
$PbO_2 + 4H^+ + 2e^- \rightarrow Pb^{2+} + 2H_2O$	+1.46
$Cl_2 + 2e^- \rightarrow 2Cl^-$	+1.36
$Cr_2O_7^{2-} + 14H^+ + 6e^- \rightarrow 2Cr^{3+} + 7H_2O$	+1.33
$O_2 + 4H^+ + 4e^- \rightarrow 2H_2O$	+1.23
$Br_2 + 2e^- \rightarrow 2Br^-$	+1.07
$NO_3^- + 4H^+ + 3e^- \rightarrow NO + 2H_2O$	+0.96
$Hg^{2+} + 2e^- \rightarrow Hg$	+0.85
$Ag^+ + e^- \rightarrow Ag$	+0.80
$Fe^{3+} + e^- \rightarrow Fe^{2+}$	+0.77
$I_2 + 2e^- \rightarrow 2I^-$	+0.54
$Cu^+ + e^- \rightarrow Cu$	+0.52
$Fe(CN)_6^{3-} + e^- \rightarrow Fe(CN)_6^{4-}$	+0.36
$Cu^{2+} + 2e^- \rightarrow Cu$	+0.34
$Cu^{2+} + e^- \rightarrow Cu^+$	+0.15
$Sn^{4+} + 2e^- \rightarrow Sn^{2+}$	+0.15
$2H^+ + 2e^- \rightarrow H_2$	0.00
$Fe^{3+} + 3e^- \rightarrow Fe$	-0.04
$Pb^{2+} + 2e^- \rightarrow Pb$	-0.13
$Sn^{2+} + 2e^- \rightarrow Sn$	-0.14
$Ni^{2+} + 2e^- \rightarrow Ni$	-0.25
$Co^{2+} + 2e^- \rightarrow Co$	-0.29
$PbSO_4 + 2e^- \rightarrow Pb + SO_4^{2-}$	-0.359
$PbI_2 + 2e^- \rightarrow Pb + 2I^-$	-0.365
$Cr^{3+} + e^- \rightarrow Cr^{2+}$	-0.40
$Cd^{2+} + 2e^- \rightarrow Cd$	-0.40
$Fe^{2+} + 2e^- \rightarrow Fe$	-0.41
$Cr^{3+} + 3e^- \rightarrow Cr$	-0.74
$Zn^{2+} + 2e^- \rightarrow Zn$	-0.76
$2H_2O + 2e^- \rightarrow H_2(g) + 2OH^-$	-0.83
$V^{2+} + 2e^- \rightarrow V$	-1.18
$Mn^{2+} + 2e^- \rightarrow Mn$	-1.18
$Al^{3+} + 3e^- \rightarrow Al$	-1.66
$Mg^{2+} + 2e^- \rightarrow Mg$	-2.37

Appendix E
Glossary of Terms

Absorbance (A): a logarithmic function of the percent transmission of a wavelength of light through a liquid.

Absorbent: anything which absorbs.

Aliquot: a sample that is a definite fraction of the whole.

Barometer: an instrument that measures the pressure of the atmosphere.

Buret: an apparatus for delivering measured quantities of liquid, or for measuring the quantity of liquid or gas received or discharged. It consists essentially of a graduated glass tube, usually furnished with a small aperture and stopcock.

Capillary action: the movement of liquid within a material against gravity as a result of surface tension.

Chromatography: techniques for separating molecules based on differential absorption and elution. Also a term for separation methods involving flow of a fluid carrier (mobile phase) over a stationary absorbing phase.

Cuvette: a transparent or translucent box-shaped container with precisely-measured dimensions for holding liquid samples to be put into a spectrophotometer.

Dry ice: solidified carbon dioxide.

End point of a titration: that point at which the indicator shows a change.

Error of the titration: the difference in volume of the titrant used between the equivalence point and end point.

Extinction coefficient: the molar absorptivity or molar extinction coefficient ε of a chemical species at a given wavelength is a measure of how strongly the species absorbs light at that wavelength. It is an intrinsic property of the species; the actual absorbance of a sample is dependent on its thickness L and the concentration c of the species.

Hard water: water containing soluble salts of calcium and magnesium, and sometimes iron.

Indicator: a chemical compound which exhibits an observable change near the equivalence point.

Meniscus: the curved surface of the liquid at the open end of a capillary column.

Mobile phase: the phase that moves along the stationary phase. It is the solvent in paper chromatography and thin-layer chromatography.

Molar Absorptivity (ε): see extinction coefficient.

pH range for an acid-base indicator: the pH interval over which the indicator shows a change in color.

Primary standard: a highly purified, stable chemical compound that can be weighed in a form having a known composition.

Qualitative: observations that do not involve measurements and numbers.

Quantitative: observations that involve measurements and numbers.

Retention factor (R_F): in chromatography, the distance traveled by the compound divided by the distance traveled by the solvent.

Separatory funnel: a piece of laboratory glassware used in liquid–liquid extractions to separate (partition) the components of a mixture between two immiscible solvent phases of different densities.

Serial dilution: a series of dilutions which amplifies the dilution factor quickly, beginning with a small initial quantity of material. The source of dilution material for each step comes from the diluted material of the previous step.

Spectroscope: an optical instrument for producing spectral lines and measuring their wavelengths and intensities, used in spectral analysis.

Standard cell: an electrochemical cell whose half-cells are both in the standard state of 25 °C, 1.0 atm and immersed in electrolytes at a concentration of 1.0 M.

Standardization of a solution: the process of finding the exact concentration of a solution.

Stationary phase: in chromatography, the non-mobile phase contained in the chromatographic bed.

Titrant: a titration reagent added from the buret to the sample solution. When referring to a specific titration, the titrant is placed second.

Transmittance (%T): the fraction of radiant energy that passes through a substance.

Vernier scale: a small, movable scale that slides along a main scale; the small scale is calibrated to indicate fractional divisions of the main scale.

Volumetric method: a method in which the analysis is completed by measuring the volume of a solution of established composition necessary to react completely with the unknown substance.